New Crops for Food and Industry

Edited by

G.E. Wickens

Royal Botanic Gardens,
Kew, Richmond, Surrey

N. Haq

Department of Biology,
University of Southampton

P. Day

Center of Agricultural Molecular Biology,
Rutgers University, New Brunswick, USA

London   New York
CHAPMAN AND HALL

International Symposium on New Crops for Food and Industry

ORGANISING COMMITTEE

Professor P. R. Day (Chairman), Plant Breeding Institute

Dr N. Haq (Secretary), University of Southampton

Mr V. E. Gale, Industry Council for Development

Professor D. O. Hall, King's College, London

Mr T. S. Jones, Booker Agriculture International

Mr J. Meadley, Rural Investment Overseas

Mr R. W. Smith, Overseas Development Administration

Dr G. E. Wickens, Royal Botanic Gardens

Mr C. Wicks, Bioresources/Earthlife Foundation

INTERNATIONAL COORDINATING COMMITTEE

Dr A. Grobman, Peru

Dr K. F. S. King, United Nations Development Programme

Professor J. O. Kokwaro, Kenya

Professor Sanga Sabhasri, Thailand

Dr N. Vietmeyer, USA

Professor J. T. Williams, International Board for Plant Genetic Resources

SPONSORS

The Organising Committee expresses its gratitude to the following sponsors, without whose help the Symposium would not have been possible.

Agricultural Genetics Co. Ltd.

Booker Seeds Ltd

British American Tobacco Co.

British Council

Commonwealth Development Corporation

Commonwealth Foundation

Deutsche Gesellschaft fur Technische Zusammenarbeit (GTZ) GmbH

Dulverton Trust

Heineken N. V.

I.C.I. Ltd

International Board of Plant Genetic Resources

International Foundation of Science (IFS), Sweden

Shell International Petroleum Co. Ltd.

Tate & Lyle Ltd.

Technical Centre for Agricultural and Rural Co-operation (CTA)

Unilever plc

United Nations Environmental Programme

University of Southampton

First published in 1989 by
Chapman and Hall Ltd
11 New Fetter Lane, London EC4P 4EE

Published in the USA by
Chapman and Hall
29 West 35th Street, New York NY 10001

Printed in Great Britain at
the University Press, Cambridge

ISBN 0 412 31500 9

British Library Cataloguing in Publication Data

New crops for food and industry
  1. Crops
  I. Wickens, G. E.   II. Haq, N.   III. Day, P.
  631

  ISBN 0-412-31500-9

Contents

List of Participants   5

Preface
    P. Day   10

MARKETING AND DEVELOPMENT

1.   Food Industry and Agriculture
    T. Thomas   13

2.   The Commercial Implications of New Crops
    J. Meadley   23

3.   Management Problems in the Introduction of New Crops: The
   Zambian Experience
    Jameson H. Muchaili   29

4.   New Crops; a Suggested Framework for their Selection, Evaluation
   and Commercial Development
    E. S. Wallis, I. M. Wood and D. E. Byth   36

5.   Assessment of New Crops for Plantations
    R. H. V. Corley   53

6.   Criteria for the Selection of Food Producing Trees and Shrubs in
   Semi-arid Regions.
    H. J. von Maydell   66

7.   Opportunities and Requirements for the Development of New
   Essential Oil, Spice and Plant Extractive Industries
    C. L. Green   76

8.   Herbal Drugs: Potential for Industry and Cash
    J. V. Anjaria   84

9.   The Potential Value of Under-exploited Plants in Soil Conservation
   and Land Reclamation
    J. Smartt   93

10. Green Biomass of Native Plants and New, Cultivated Crops for
   Multiple Use: Food, Fodder, Fuel, Fibre for Industry, Phytochemical
   Products and Medicine
    R. Carlsson   101

Contents

11. Strategy for Development of a New Crop
    L. Lazaroff   108

CROPS – REGIONAL ASSESSMENTS

12. New Crops for Food and Industry: the Roots and Tubers in Tropical Africa
    B. N. Okigbo   123

13. New Plant Sources for Food and Industry in India
    R. S. Paroda and Bhag Mal   135

14. Potential New Food Crops from the Amazon
    D. B. Arkcoll and C. R. Clement   150

15. Potential Multipurpose Agroforestry Crops Identified for the Mexican Tropics
    J. L. Delgardo Montoya and E. Parado Tejeda   166

16. Food, Fuel and Jobs from Sugar Cane and Tree Legumes
    T. R. Preston   174

17. Alternative Crops for Europe
    R. S. Tayler   185

SPECIFIC CROPS

18. *Vernonia galamensis*: a Promising New Industrial Crop for the Semi-arid Tropics and Subtropics
    R. E. Perdue, Jr., E. Jones and C. T. Nyati   197

19. *Santalum acuminatum* Fruit: a Prospect for Horticultural Development
    D. E. Rivett, G. P. Jones and D. J. Tucker   208

20. Commercial Exploitation of Alternative Crops, with Special Reference to Evening Primrose
    P. Lapinskas   216

21. Chenopodium Grains of the Andes: a Crop for Temperate Latitudes
    J. Risi C. and N. W. Galwey   222

22. Ethiopian T'ef: a Cereal Confined to its Centre of Variability
    M. R. Cheverton and G. P. Chapman   235

23. New Small-grained Cereals which may have Value in Agriculture
    C. N. Law   239

24. Crop Plants: Potential for Food and Industry
    N. Haq   246

Contents

25. Cocona (*Solanum sessiliflorum*) Production and Breeding Potentials of the Peach-tomato
    J. Salick   257

26. Review of the African Plum Tree (*Dacroydes edulis*)
    J. N. Ngatchou and J. Kengu   265

27. Study on the Processing of *Balanites aegyptiaca* Fruits for Drug, Food and Feed
    I. M. Abu-Al-Futuh   272

28. Variation of Fruit Production and Quality of Different Ecotypes of Chilean Algarrobo *Prosopis chilensis* (Mol.) Stunz
    M. Pinto and E. Riveros   280

29. Development of *Prosopis* Species Leguminous Trees as an Agricultural Crop
    R. M. Saunders and R. Becker   288

30. Rattan as a Crop for Smallholders in the Humid Tropics
    J. Dransfield   303

31. The Pejibaye Palm: Economic Potential and Research Priorities
    C. R. Clement and D. B. Arkcoll   304

32. Native Neotropical Palms: a Resource of Global Interest
    M. J. Balick   323

33. Food Production by Selective Algal Biomass in the Desert
    K. Shinohara, Y. Zhao and G. H. Sato   333

SCIENCE AND MISCELLANEOUS

34. Plant Introduction and International Responsibilities
    J. T. Williams   345

35. Kairomones – Chemical Signals Related to Plant Resistance Against Insect Attack
    H. Rembold   352

36. Wild Plants a Source of Novel Anti-insect Compounds: Alkaloidal Glycosidase Inhibitors
    M. S. J. Simmonds, L. E. Fellows and W. M. Blaney   365

37. Bioactive Phytochemicals – the Search for New Sources
    P. G. Waterman   378

38. Water Relations of Guayule and their Effect on Rubber Production
    D. Mills, A. Benzioni and M. Forti   391

3

Contents

39. The Possible Contribution of Ethnobotany to the Search for New
    Crops for Food and Industry
    J. F. Barrau   402

40. Economic Botany and Kew in the Search for New Plants
    G. E. Wickens   411

General Index   424

Taxonomic Index   436

4

List of Participants

Al-Futuh, A. I. M.	Faculty of Pharmacy, University of Khartoum, PO Box 1996, Khartoum, Sudan
Amara, D. S.	Soil Science Dept., Njala University College, University of Sierra Leone, PMB, Freetown, Sierra
Anjaria, J. V. & N. J. C.	5, Sonarika Apartment, Atira Road, Ahmedabad 380 015, India
Anthony, K.	Hillandale, 18 East Hill Road, Oxted, Surrey, RH8 9H2
Arkcoll, D.	Min. de Agricultura – MA, EMBRAPA, Centro Nacional de Pesquisa de Technologia, Av. das Americas, 29.501 Guaratiba, RJ, CEP 23020, Brazil
Armstrong, K. B.	Commonwealth Development Corporation, 1, Bessborough Gardens, London SW4 2JQ
Aronson, J. A.	Centre Emberger, CNRS, B.P. 5051, 34033 Montpellier, France
Balick, M. J.	New York Botanical Garden, Bronx, New York 10458, USA
Bandara, M. S. P.	Dept of Biology, Southampton University, Building 44, The University, Southampton, Hants SO9 5NH
Bandara, Mrs S. I.	Postgraduate Institute of Agriculture (Sri Lanka); Flat No. 10, Hartley Court, 40 Glen Eyre Road, Southampton, Hants SO9 5NH
Barrau, J. F.	Laboratoire d'Ethnobiologie-Biogeographie, Museum National d'Histoire Naturelle, 57 rue Cuvier, 75231 Paris
Bell, E. A.	Royal Botanic Gardens, Kew, Richmond, Surrey TW9 3AB
Bisby, F.A.	Dept. of Biology, Building 44, University of Southampton, Southampton, Hants SO9 5NH
Boulter, D.	Dept of Botany, Durham University, Durham DH1 3LE
Breyer, F.	Rathausgasse 20, Postfach 172, A-6700 Blundenz, Austria
Bryant, D	International Seed Producers, Tayfen Road, Bury St. Edmunds, Suffolk IP32 6BH
Burbage, M. B.	Overseas Development Natural Resources Institute, 56–65, Grays Inn Road, London WC1X 8LU
Buruga, J. H.	Botany Department, Makerere University, P.O. Box 7062, Kampala, Uganda
Carlsson, R.	Dept. of Plant Physiology, Box 7007, S-220 07 Lund, Sweden
Castillo, R. O.	Plant Biology Dept., University of Birmingham, Birmingham, Warwick B15 2TT
Chapman, G.P.	Wye College, University of London, Wye, Ashford, Kent, TN25 5AH
Cherry, M.	c/o Facilities Unit, Room 641 SEW, Bush House, Strand, London WC2B 4PH
Chewe, C. M.	Biology Dept., Building 44, University of Southampton, Southampton, Hants SO9 5NH
Clement, C. R.	INPA, CP 478, 69000 Manaus AM, Brazil
Corley, H.	Unilever Plantations Group, Unilever House, Blackfriars, London EC4P 4BQ
Crowley, J. G.	Agricultural Institute, Oakpark Research Centre, Oakpark, Carlow, Ireland
Day, P. R.	Center for Agricultural Molecular Biology, Cook College, Rutgers University, P.O.Box 231, New Brunswick, New Jersey, 08903 USA

De Groot, P.	Department of Biology, King's College London, Campden Hill Road, London W8 7AH
Delgardo, J. L.	Bioticos/INIREB, Apdo. Postal No. 63, 81000 Xalapa, Ver, Mexico
Dixon, D.	c/o Facilities Unit, Room 641 SEW, Bush House, Strand, London WC2B 4PH
Dransfield, J.	Herbarium, Royal Botanic Gardens, Kew, Richmond, Surrey TW9 3AB
Edwards, S.	Ethiopian Flora Project, The National Herbarium, Biology Dept., Addis Ababa University, Box 3434, Addis Ababa, Ethiopia
Farrow, B.	New Projects Development, Holmewood Hall Field Station, Holme, Peterborough, Hants PE7 3PG
Fenner, B. J.	Department of Research and Specialist Services, P.O. Box 8108, Causeway, Harare, Zimbabwe
Field, D. V.	Economics & Conservation Section, Herbarium, Royal Botanic Gardens, Kew, Richmond, Surrey TW9 3AB
Fordham, R.	Wye College, Ashford, Kent TN25 3AB
Fuchs, A.	Laboratory of Phytopathology, Agricultural University, 6709 RZ Wageningen, The Netherlands
Fuller, M. P.	Seale-Hayne College, Newton Abbot, Devon TQ12 6NQ
Gale, V.E.	Industry Council for Development, Pope's Close, Greenfield, Watlington, Oxfordshire
Galwey, N.	Dept. of Applied Biology, University of Cambridge, Pembroke Street, Cambridge, Cambs CB2 3DX
Gibbs, J.	Dept of Biology, Building 44, University of Southampton, Southampton, Hants SO9 5NH
Gnecco D., S.	Depto. de Química U. de Concepcion, Casilla 3-C, Concepción, Chile
Gohl, B.	SAREC, S-105 15 Stockholm, Sweden
Green, G. L.	Spices and Plant Extraction Section, Overseas Development Natural Resources Institute, 56–62 Grays Inn Road, London WC1X 8LU
Guyer, D. W.	Wye College, Ashford, Kent TN25 5AH
Hall, J. L.	Department of Biology, Building 44, University of Southampton, Southampton, Hants SO9 5NH
Haq, N.	Centre for Underutilized Crops, Dept. of Biology, Building 44, University of Southampton, Southampton, Hants SO9 5NH
Harris, B. J. & Mrs.	Dept. of Biological Sciences, Ahmadu Bello University, Zaria, Kaduna State, Nigeria
Harris, P. J. C.	Dept. of Biological Sciences, Coventry Polytechnic, Priory Street, Coventry, Warwick CV1 5FB
Harvey, B. L.	Dept. of Crop Science and Plant Ecology, University of Saskatchewan, Saskatoon, Sask. Canada S7N OWO
Hastings, R.	Economic & Conservation Section, Herbarium, Royal Botanic Gardens, Kew, Richmond, Surrey TW9 3AB
Haynes, B. H.	W. S. Atkins London and Water Management, Girton Road, Cambridge, Cambs CB3 0LN
Heemskerk, P. A. M.	Consultatn Agro-Industrial Projects, P.O. Box 500, 2380 BA Zoeterwoude, The Netherlands
Hide, G. P.	38, Elmfield Road, Newcastle-upon-Tyne, Northumberland NE3 4BB

Hjorth, M.	Dept. of Plant Physiology, Box 7007, S-220 07 Lund, Sweden
Jellings, A. J.	Seale-Hayne College, Newton Abbot, Devon TQ12 6NQ
Jenkins, G.	Cereals Dept., Maris Lane, Trumpington, Cambridge, Cambs CB2 5HW
Jideani,	School of Sciences, Abu Ballor Jafawa Belwa University, P.M.B. 0248 Bauchi, Nigeria
Johnson, D.V.	USDA Forest Service, P.O. Box 96090, Washington DC 20090, USA
Jokinen, K.	Plant Tissue Culture Dept., P.O. Box 44, SF-02271 Espoo, Finland
Jones, T. S.	c/o Bookers Agriculture International Ltd., Masters Court, Church Road, Thame, Oxon OX9 3FA
Kapoor, P.	Commonwealth Science Council, Marlborough House, Pall Mall, London SW1Y 5HX
Kelly, L.	Dept. of Biology, Building 44, University of Southampton, Southampton, Hants SO9 5NH
Kengue, J.	IRA, POB 2123, Yaounde, Cameroon
Kenwright, P. A.	Dept. of Applied Biology, University of Cambridge, Pembroke Street, Cambridge, Cambs CB2 3DX
Kessler, C. D. J.	School of Plant Biology, University College of North Wales, Deiniol Road, Bangor, Gwynedd, LL57 2UW
Khan, M. F.	Dept. of Physiology and Environmental Science, Faculty of Agriculture, University of Nottingham, Sutton Bonington, Notts.
Kimani, P. M.	Dept. of Crop Science, P.O. Box 30197, Nairobi, Kenya
Kokwaro, J.	Dept. of Biology, P.O. Box 30197, Nairobi, Kenya
Lapinskas, P.G.W.	Efamol Ltd., Efamol House, Woodbridge Meadows, Guildford, Surrey GU1 1BA
Law, C.	Plant Breeding Institute, Maris Lane, Trumpington, Cambridge, Cambs CB2 2LQ
Lazaroff, L.	ICDUP, 18 Meadow park Court, Orinda, California 94563, USA
Lee, P. G.	Shell Research Ltd, Sittingbourne Research Centre, Sittingbourne, Kent ME9 8AG
Lewis, I. U.	c/o GTZ, FB 151, Postfach 5180, 6236 Eschborn, Fed Rep of Germany
Linington, S.	Royal Botanic Gardens, Kew, Wakehurst Place, Ardingly, Nr. Haywards Heath, West Sussex RH17 4NS
Loughton, A.	Ontario Ministry of Agriculture & Food, Box 587, Simcoe, Ontario, Canada N3Y 4N5
Lungu, D. M.	University of Zambia, School of Agricultural Sciences, Box 32379, Lusaka, Zambia
Madom, M.S.B.	Dept. of Horticulture, Wye College, University of London, Wye, Ashford, Kent, TN25 5AH
Mal, Bhag	National Bureau of Plant Genetic Resources, Pusa Campus, New Delhi, 110 012 India
Maydel, H.J. von	Inst. für Weltforstwirtschaft und Ökologie, Bundesforschungsansalt für Forst-und Holzwirt-schaft, Leuschnerstr. 91, 2050 Hamburg 80, W. Germany
Meadley, J.	Rural Investment Overseas Ltd., 8a, Lower Grosvenor Place, London SW1 0EN

Mills, D.	Institute for Applied Research, Ben-Gurion University of the Negev, PO Box 1025, Beer-Sheva 84110, Israel
Mithen, R. F.	Dept. of Biological Sciences, University of Zimbabwe, PO Box MP 167, Mount Pleasant, Harare, Zimbabwe
Muchaile, M	Tate & Lyle Agribusiness, Enterprise House, 45 Homesdale Road, Bromley, Kent
Ogbe, F.	Botany Dept., University of Benin, Benin-City, Benin State, Nigeria
Okigbo, B. N.	I.I.T.A., PMB 5320, Oyo Road, Ibadan, Nigeria
Panfilo C. Tabora, Jr	PB 2067, San Pedro Sula, Honduras, C.A.
Perdue, R.E. Jr	Germplasm Introduction and Evaluation Lab., Room 334, Bldg. 001, Beltsville, MD 20705 USA
Phillips, L. D.	High Value Horticulture PLC, Colne House, Highbridge Industrial Estate, Oxford Road, Uxbridge, Middlesex UBB 1UL
Pinto, M.	Laboratorio de Bioquimica, Production Agricola, Fac. de Agronomia, Univ. de Chile, Casilla 1004, Santiago, Chile
Poulter, N.	Overseas Development Natural Resources Institute, 56–62 Grays Inn Road, London WC1X 8LU
Preston, T. R.	Fundacion para el Desarrollo Integral del Valle del Cauca, Calle 8a, No. 3–14, Piso 17, Apartado Aereo 7482, Cali, Colombia
Rabbich, P.O.	Shuttleworth College, Old Warden Park, Biggleswade, Beds SG18 9DX
Ray, D.T.	Dept. of Plant Sciences, University of Arizona, Tucson, Arizona 85721, USA
Rembold, H.	Max-Planck-Institut für Biochemie, D-8033 Martinisried bei Munchen, W. Germany
Rijkens, B. A.	Inst. for Storage & Processing Agricultural Produce (IBVL) PO Box 18, 6700 AA Wageningen, The Netherlands
Rivett, D. E.	CSIRO, Div. of Wool Technology, 343 Royal Parade, Parkville, Victoria 3052, Australia
Roy, R. C.	Agriculture Canada, Research Station, Box 186, Delhi, Ontario, Canada N4B 2W9
Ruhee, K. C.	University of Mauritius, Reduit, Mauritius
Saunders, R. M.	800 Buchanan Street, Albany, California 94710, USA
Schauer, A	Rodale Research Center, R. D. 1, Box 323, Kutztown, Pa 19520, USA
Schechter, J.	Applied Research Institute, Ben-Gurion University of the Negev, PO Box 1025 Beer-Sheva 4110, Israel
Scheld, H. W.	707, Texas Avenue, Suite 101-E, College Station, Texas 77840, USA
Seaton, O. M.	Booker Agriculture International Ltd., Masters Court, Church Road, Thame, Oxon OX9 3FA
Sergeant, A. T. H.	Columbus House, Trossacks Drive, Bath, Somerset, BA2 6RR
Shields, R.	Unilever Research, Colworth House, Sharnbrook, Beds MK4 4LQ
Shinohara, K.	Jodrell Lab., Royal Botanic Gardens, Kew, Richmond, Surrey TW9 3AB
Smartt, J.	Dept. of Biology, Building 44, University of Southampton, Southampton, Hants SO9 5NH
Smith, B. G.	49, Roding Lane, Buckhurst Place, Essex 1G9 6BJ

Smith, R. W.	ODA, Eland House, Stag Place, London SW1E 5DH
Soest, van L. J. M.	Rijks-kwaliteitsinst, v-land-en Tuinbouwprod, Bornsesteeg 45, 6708 pd Wageningen, The Netherlands
Southwell, K. H.	Overseas Development Natural Resources Institute, 56–62 Grays Inn Road, London, WC1X 8LU
Sullivan, F.	World Wildlife Fund, Panda House, 11–13 Ockford Road, Godalming, Surrey GU7 1QU
Tayler, R. S.	Dept. of Agriculture, The University, PO Box 236, Earley Gate, Reading, Berks RG6 2AT
Taylor, R. D.	Booker Agriculture International, Masters Court, Church Road, Thame, Oxon OX8 3FA
Thackray, D. J.	Dept. of Biology, Building 44, University of Southampton, Southampton, Hants SO9 5NH
Thomas, T.	Unilever Ltd., Unilever House, Blackfriars, London EC4P 4BQ
Von Franz, J. C.	TAD, Kotak Pos 140, Samarinda 75001, Indonesia
Waidyanatha, U. P. de S.	British Council, 49 Alfred House Gardens, Colombo 3, Sri Lanka
Wallis, E. S.	Dept. of Agriculture, University of Queensland, St. Lucia, Queensland 4067, Australia
Waterman, P. G.	Phytochemistry Research Labs., Dept of Pharmacy, University of Strathclyde, Glasgow G1 1XW
Watson, G. A.	Highmeads, Upperfield, Midhurst, Sussex GU28 8AE
Watson, J. S.	Minster Agriculture Ltd., 13 Upper High Street, Thame, Oxon OX9 3HL
Wholey, D. W.	Minster Agriculture Ltd., 13 Upper High Street, Thame, Oxon OX9 3HL
Wickens, G. E.	Economic and Conservation Section, Royal Botanic Gardens, Kew, Richmond, Surrey TW9 3AB
Wiles, T. L.	T. L. Wiles and Associates Ltd., PO Box 43, Chichester, West Sussex PO19 1UJ
Williams, J. T.	International Board for Plant Genetic Resources. c/o FAO, Via delle Terme de Caracalla, 00100 Rome, Italy
Williams, L. R.	Macquarie University, Sydney, New South Wales 2109, Australia
Winter, D.	100, Avenue Road, London NW3 3TP
Wood, B. J.	Sime Darby Plantations, Ebor Research, PO Box 202, 40000 Shah Alam, Selangor, Malaysia

Preface

P. Day

In early 1986 a small committee was formed to explore the possibility of organizing an International Symposium on New Crops for Food and Industry. The objective was to review progress in the identification and development of new materials with potential from all over the world. At the same time, the Symposium would provide an opportunity for the growing band of widely dispersed individuals dedicated to new crops to discuss results, compare notes and consider the opportunities for future work. In spite of the need for alternative crops in the developed world, and the need to utilize more fully their potential in the developing world, new crops continue to be neglected and have to struggle hard in the competition for support. The established crops are better known, have assured markets, and the risks in growing them are better understood. In comparison, new crops and their markets are mostly under-developed. There are often many competing for attention, there is often little information on large scale cultivation, and the general perception of their potential is one of risk.

The Symposium was planned to provide an opportunity for industry and development agencies to participate as well as those directly working with new crop species. 139 participants from 38 countries participated in the Symposium. The dialogue was intended to assist both groups in the important task of identifying priorities.

The meeting, and this record of the Symposium proceedings, would not have been possible without the generous help of the sponsoring organizations. A special note of appreciation is due to the University of Southampton for their assistance in hosting the meeting, to Sandra Wilkins for retyping the edited text and to Dr. David Newman for preparing the index and desktop publishing the camera-ready copy.

MARKETING AND DEVELOPMENT

1

Food Industry and Agriculture

T. Thomas

INTRODUCTION

When I was asked to participate in this symposium I was a bit apprehensive because 'agriculture' has become an activity in acute crisis in most parts of the world and there are no short-term solutions that anyone other than politicians can offer. In the 1970s and the earlier part of the 1980s, manufacturing industry (as distinct from agriculture) went through an economic crisis. Because of its structure of ownership and management 'industry' was able to take several steps to resolve the crisis, sometimes in conjunction with politicians and trade unionists, but very often by taking the initiatives appropriate to the economic trends and, in many cases, coordinated on an international scale. But now that agriculture is in crisis, the initiative is not with the farmers who are the owners and producers. It seems to be diffused among politicians and bureaucrats. It is far from being coordinated on an international scale and continues to arouse greater emotions along narrow nationalist lines than industry does. A major difference between agriculture and industry is that in agriculture a farmer cannot move his major means of production, viz land, internationally, whereas in industry you can move your activity internationally to the most optimal location. Therefore farmers think nationally. In years to come even farming may have to move and think internationally!

I do not intend to discuss possible solutions to the problem, but only to highlight the context in which we are meeting. But it is possible that until 'agriculture' acquires some of the features of 'industry' with regard to ownership, management and internationality it is likely to remain vulnerable to intervention based on political expediency. One area where 'agriculture' would gradually move closer to the characteristics of 'industry' is where it has a closer interphase with the food industry. As the produce of agriculture becomes more closely and precisely linked to the needs of the food industry with production being related to demand in the market place as assessed by marketeers and not speculatively, it will be possible to bestow agriculture with more economic stability, better technology, more internationality and less dependence on political considerations. Therefore I thought it would be interesting and necessary for you who are associated with agriculture to get some idea about what the food industry expects from agriculture.

LINKAGE BETWEEN AGRICULTURE AND FOOD INDUSTRY

The link between agriculture and food is a well established one. But it is not often recognized that this linkage is an evolving one. In primitive societies, the farmer and consumer were either the same family or were neighbours who bartered their products and services. As society evolved it added other links. Today the chain starts with the farmer, followed by a trader who trades in the farm produce, a manufacturer/marketeer who converts it into an article of food, a retailer who sells it in retail shops and finally the consumer who buys the food.

A very important new link that has been introduced into the chain in this century and more intensely in the last 50 years is the role of the scientist. Scientists as breeders, plant biologists, nutritionists and chemists have made a very powerful input contribution into farming and food manufacture. They along with marketeers have also assessed more clearly and linked the needs of the consumer to the produce from agriculture.

With this evolution of the linkage one can almost see agriculture and food as a continuum in a business sense. Although no single large business group is involved in all links in this chain it is interesting to note that large retailers in the UK, like Marks & Spencer, are now establishing direct links between retailing and farming of vegetables, fruits, chickens, eggs, fish, etc. Could this be a new trend that could make the whole system more dynamic and more efficient? It does make a lot of sense for those who are closest to the consumer to assess her needs and interpret them directly to the primary producer.

But the evolution and strength of this chain from farmer to the consumer differs among countries and regions depending on the trends in agriculture and in food. We shall now examine these trends.

TRENDS IN WORLD AGRICULTURE

World agriculture has reached a stage of global surplus in several key commodities. This has not happened because of some fortuitous events. It has been in the making for the last several decades. One strong impetus was the technological one – improved varieties, better nutrients, more irrigation, better protection from disease, improved storage, handling and processing, distribution, etc. The other strong impetus has come from government policies which are essentially protectionist and domestically promotional in several leading economies like the USA, EEC, Japan, as well as China and India, which have chosen to subsidize farmers even while knowing that they are producing commodities in global surplus. So the surplus has been created through deliberate policy although not with the intention of creating the present embarrassment.

An interesting consequence of this surplus situation is that the worst sufferers from the surplus are the farmers of the world. Their income and assets have declined in the developed countries and they seem to be on a permanent dole of subsidies. The consumer, the retailer and even the food manufacturer have on the whole benefited from the lower cost of agricultural produce.

The question is whether the fortunes of agriculture are going to change for the better, significantly and in the short term. The answer has to be found by analysing the underlying trends for the future.

In the case of potential technological inputs we are not anywhere near the limit. The tremendous gains that have been achieved through conventional breeding could well be dwarfed by what will be achieved through application of cytology and molecular genetics. It may well take another 25 years but we are definitely on the way. The improvements in yield could be at least 20 per cent or perhaps even more dramatic. More importantly it could create new varieties and crops that could influence the choice of the consumer which in turn could render older commodities less acceptable.

With regard to government policies that encouraged greater production, the trends are somewhat mixed. In the developed regions like the USA and EEC, despite the relatively small percentage of farmers in the population (under 5 per cent in most cases) society in general seems to be willing to pay for preserving agriculture. The reasons are partly sentimental, and partly rational. The rational elements are to sustain the several other industries that are suppliers to agriculture, to preserve the countryside and rural communities as a natural heritage, to avoid further urbanization and to ensure continued national self-sufficiency in a vital segment of the economy. Society is willing to pay a price in terms of subsidy for this. The question is only how much it can afford to pay in the longer term. This is yet to be determined. Therefore the politicians will continue to be ambivalent with regard to agricultural subsidies.

In the less developed countries there are a variety of approaches. In the bigger ones like China and India pricing mechanisms and subsidies have been used to increase output. With a very large percentage (over two thirds) of the population engaged in agriculture neither of these large countries can afford to neglect agriculture as the single largest source of employment. The social consequences of any sharp decline in agricultural employment can be catastrophic. So they can be expected to continue their policies and contribute more to the overall global surplus. They have ceased to be markets for the developed world's surplus and could well become significant competitors in export trade.

At the other extreme there are some less developed countries which try to use the price mechanism to favour the more vocal urbanized segment of the population, which result in uneconomic prices to the farmers with consequent decline of agriculture, depopulation of the countryside and growth of potentially explosive urban slums. The folly of this situation is bound to be recognized over a period of time. But whether adequate corrective action can be taken in good time is still an open question. In fact there is a possibility that some of these poorer countries could fall into what has been called the demographic trap. The earlier theory of demographic transition postulated three stages, viz. the first stage when birth and death rates are high and population grows very slowly; the second stage when health measures and food production combine to produce fewer deaths and more births leading to a rapid growth in population; and the third stage when economic and social gains and awareness lead to a drop in birth rate and population growth slows down. Several of the poorer countries seem to be trapped at the second stage with population growing at over three per cent per annum with distinct prospects of their running out of resources and reverting to the first stage of poverty. However, even in such countries, governments will have to become supportive of policies that stimulate agriculture. The number and significance of such recalcitrant countries in the global context are so small that it will not detract from the general scenario of continuing surplus and relatively low prices.

This scenario of global surplus has some important consequences:

(a) Firstly it could stimulate consumption. This will be not only through lowering of prices but also by the granting of subsidies to the consumers by governments which will have to dispose of their mountains and lakes of surpluses.

(b) Secondly it will stimulate greater efficiency in agriculture to produce at lower costs and better quality levels. This in turn will look for innovative inputs from science and technology. The period 1920–1950 onwards was the age of machines in agriculture. From 1950 onwards we saw the age of chemicals in agriculture. Now we have entered the age of biotechnology in agriculture. This is a new watershed in agricultural productivity. Biotechnology will be more research and know-how intensive and probably not as capital intensive as the machine or chemical phases of agriculture. Thus its benefits can flow faster even into poorer countries which do not have the capital. Therefore its impact could be faster, wider and more significant.

(c) A third possible consequence of continued agricultural surplus is that there will be greater innovation in crops to satisfy new needs which the consumer will begin to recognize and the mission of agriculture will become more of providing choices for the consumer. This in turn could give better added value to the farmer. It is not a coincidence that this symposium with 'New Crops' as its theme is being held at a time when there is a global surplus and economic pressure on farming.

Thus in a paradoxical way the continuing surplus of agricultural production in the world will stimulate a further significant modernization of farming both in its technology and its production.

TRENDS IN THE FOOD INDUSTRY

Food has evolved over the centuries to satisfy an increasingly sophisticated consumer need.

16

(a) At the most primitive level, food is required to satisfy hunger.

(b) At the next level of sophistication, food is eaten more for pleasure than to satisfy hunger. It is fun to have a coke or an ice cream. This is characteristic of the consumer who has enough money to spare after satisfying his basic hunger.

(c) The third level of food consumption is that of discrimination according to nutritional value rather than pleasure alone. This requires a more sophisticated information system and creation of awareness which is provided by scientists and food manufacturers.

(d) Arising mainly from the pleasures of overeating, we arrive at the next level of discrimination in food consumption, i.e. the one based on healthy eating. At this level people are not looking for nutrition. They are looking for well being which could mean looking slim and trim or avoiding cardiovascular diseases, cancer, etc. Thus there is the new market for products that constitute healthy eating. Awareness about these products is created by the medical profession and by the food industry.

(e) Superimposed on these trends in the more prosperous economies, there is also the trend towards exotic ethnic foods. It may be Chinese or Indian or Lebanese or Mexican.

We can place countries or classes of population at different stages of sophistication in food consumption ranging from hunger to healthy eating, and relate this to the stage of economic development that the country or the group has reached. The poorer countries will have no market for the more sophisticated products. These markets will have food as a commodity to satisfy hunger or minimal nutritional need. Conversely, in the richer economies the market and added value is not for commodities but for the more sophisticated products.

FACTORS AFFECTING TRENDS IN THE FOOD INDUSTRY

Now let us analyse some of the factors that have influenced the trends in the food industry in the more recent past.

Need for Convenience and Creativity

Increased prosperity, the disappearance of domestic help and the increasing trend for women to work outside the home created the need for convenience foods. This has enabled people to spend more of their time on several leisure activities. Interestingly, as sophistication levels have advanced further in choice of foods, cooking itself has become more of a creative hobby for women and even some men in the developed countries instead of being a chore as in the past. One can therefore expect that the food industry will have to provide not only for convenience but also for creativity and fun in cooking which has become a hobby.

Internationalization of Food

In the last few decades there has been a tremendous internationalization of food habits. Coca Cola, hamburgers, pizzas, tandoori chicken, chop suey, etc., are examples of this. It has happened mainly because of much better international communication and the flow of people across boundaries through travel and through emigration. Another contributory factor has been the international food companies which transferred products across boundaries, adapted them to local tastes where necessary and created the market.

Technology of Food Processing

From being a commodity, food items are increasingly sold either as prepared or ready to eat. This has been made possible by the development of technology to process, preserve, package and transport food in more sophisticated forms. The evolution has been from cold stores to frozen products and now chilled products which are even nearer to the fresh and natural. In packaging there has been continuous evolution in terms of keepability, cost and convenience with new packaging materials, gas packing, etc. In the actual cooking of food again there have been improvements with the microwave oven as the most recent innovation. All these changes in technology have influenced the food industry in all parts of its operation from raw material procurement and processing, through to packaging and distribution and marketing. And we have not by any means seen the end to these changes.

Increasing Sophistication with Income

As economies develop, food claims a lower proportion of the total income of the household. Other items like housing, education, entertainment, travel, medical expenses, retirement provisions, etc., become more significant. In a curious way because food is less significant as an item of expenditure people become more willing to spend on better and more sophisticated foods. Therefore the food industry has tremendous scope to create such products in the more advanced economies.

Restaurant and Hotel Supplies

Another significant trend is the increase in the habit of eating out in hotels and restaurants. This is stimulated by travel, business and the availability of disposable income. As the hotel and restaurant industry developed, food manufacturers have developed products specially designed to cater to this segment of the market. This again will be a growing feature in the marketing and product development activities of the food industry.

Food Industry's Expectations from Agriculture

Against this background of the trends in agriculture and food industry it will now be worthwhile for us to examine some of the key expectations of the food manufacturer from the agriculturist. Most of them are well known to you but it is interesting to see them through the eyes of the manufacturing industry.

The key players in the chain of activities that connect food and agriculture are (a) the farmer, (b) the manufacturer who converts the farmed commodity into food and (c) the retailer who sells it, and (d) the consumer. There are several other agencies that play supportive roles, e.g. the scientists, the suppliers of key inputs into farming and manufacturing and the distributors. But let us concentrate on the four main players.

While all of them have a very strong common interest in the development of the whole span of activities, in practice there is also a genuine conflict of interest if each of them looked at it from a compartmentalized viewpoint.

The farmer is interested in getting the best return from his produce, which usually equates to maximum price for unlimited quantities. The manufacturer wants the least cost, best quality produce from the farmer so that he can sell it at the best possible price and yet remain competitively viable. The retailer wants from the manufacturer high quality reliable supplies at the most competitive prices. The consumer needs the best product at least cost. As you will realize all of them have conflicting interests.

In an ideal world there must be a kind of strategic partnership among the four interest groups. Each group would even realize intellectually that in the long run one of the four groups will find it difficult to survive profitably without the other groups also being able to do the same. But in real life, attitudes are not those of the ideal world or of the longer term. It is focused more on the shorter term and in preserving the interest of each group. It is only by allowing each group to take care of its interests, that a balanced longer term partnership can evolve. We must bear this in mind when we consider what the food industry expects from agriculture.

The primary objective of a food manufacturer is to ensure that his products are preferred by the consumer so that he can build a profitable business. To ensure this, manufacturers strive to create innovative specialized products which they then market as branded products in order to build consumer loyalty to their brands. Therefore the branded product is the corner stone for any successful food business. In order to create a successful brand it is necessary to find out and define the needs of the consumer. Then the food manufacturer has to interpret the needs of the consumer not only to his own product innovators but also increasingly to the farmers directly. Batchelors in peas, Birds Eye in peas and beans, Heinz in tomatoes are all examples of the direct link that the food industry finds necessary to establish with farmers.

In establishing this link with agriculturists, the food industry has the following types of expectation:

Quality

Like in all other consumer products, the food industry can build up brand loyalty only on the basis of delivering high quality consistently. As the market develops more towards healthy eating and natural products, the quality of the raw material from agriculture becomes even more critical. Agriculture can therefore expect to face increasing emphasis on quality. Equally well agriculture can expect to share in the better return for innovative improvements in quality.

Cost

Next to quality will come cost. With increased facility to scour the world for raw materials, the food industry is in a position to find the lowest cost source for any given level of quality. Methods of analysing quality have become faster and more reliable. International trade and transportation by air have removed another source of uncertainty. Therefore international cost competitiveness is a key expectation. For the food manufacturer, the country in which he manufactures or markets has no longer to be the source of agricultural produce. The world is becoming his source of supply. This is a big change in the competitive environment for agriculture which the farming community has to realize because they have so far been largely cocooned in their respective domestic markets.

Non-seasonality

Agricultural products were traditionally seasonal in their production and hence availability to the consumer. With modern methods of farming and with much wider international reach the food manufacturer does not have to operate seasonally. Nor can he afford to do so as the working capital in seasonal operations can be a crushing burden. One can now get a wide variety of fruits, vegetables, or Atlantic salmon throughout the year. This will of course mean that farming of such crops will have to compete in terms of reducing seasonality or fitting into a pattern of seasonal competitiveness.

Reliability

A manufacturer who has invested a lot in building up his brand will be very keen to get reliable supplies in terms of quality, timing and cost. Agriculturists will have to conform increasingly to this key parameter of reliability in order to strengthen the link with the food industry.

Health Connotation

Eating and health (well being) are going to be increasingly connected in the food manufacturer's and consumer's mind. Therefore farmers will also have to turn their attention to the health connotation of what they choose to grow. There are two aspects of health to be taken into account. Firstly, there is the understanding of what is good for the consumer's health, e.g. yoghurt and brown bread are 'healthy' as compared to cream and white bread. Industry will move more towards the healthy eating inputs and will expect agriculture to respond. Secondly, there is the whole area of avoiding chemicals like pesticides and herbicides which the consumer is increasingly concerned about. Farmers may be able to produce without using such chemicals but at an extra cost; but then manufacturers have to justify the premium to the consumer if the farming without chemicals is going to cost more. The consumer and the food industry will expect farmers to produce without chemicals but also without any additional costs. This again will be a challenge to agriculture.

Processing

Ease of processing will become an increasingly important expectation of the food industry. Like all industries reduction in cost of capital equipment, wages and inventories is a key expectation of the food industry. For instance farmers who can deliver on the 'just in time' principle will be a great advantage in reducing working capital and space requirements. Farmers who can do part of the processing while harvesting or transporting will be adding another advantage. Crops that are specially bred and designed to facilitate processing (e.g. fruits without seeds, chicken without feathers) will be another type of advantage that the food industry could expect from agriculture.

Product Differentiation

In competitive brand marketing, the food industry has to innovate continuously to create new products that are different from and superior to existing ones of their own or of competitors. It is such product differentiation that underpins the creation and competitive position of branded products. Although product life cycles have been long in the food industry in the past, one must expect increasing pace of innovative changes even in the food business as technology and competition intensify. The scope for innovation has been traditionally in the conversion of agricultural produce through factory processing. While this will continue to be an important area for innovation, manufacturers will increasingly tend to look for innovative changes in the agricultural produce itself to provide the basis for product differentiation. It may be in terms of novel tastes, improved texture, more attractive shapes, etc.

CONCLUSION

The trends in agriculture and in the food industry have been surveyed and some of the things that the food industry is looking for from agriculture have been outlined. As you would realize the possible response from agriculture to the demands made by the food industry is going to depend increasingly on agricultural scientists. The link of the agricultural scientist has been traditionally stronger with the farmer and the agricultural inputs trade and almost non-existent with the food industry. Agriculture itself had been insulated from the food industry through layers of traders and brokers. This situation is changing with the food industry finding it more advantageous to interact more directly with the farmers. This is bound to inject some of the consumer need-oriented culture of the food industry into the farming sector. That will ultimately benefit the consumer, the farmer and the food manufacturer. And in all this the agricultural scientist with his increasingly versatile array of tools will play a significant role. It is hoped that the deliberations of this symposium will give us all a better appreciation of the opportunities that lie ahead of us, despite all the problems that exist today in agriculture.

2

The Commercial Implications of New Crops

J. Meadley

INTRODUCTION

I would now like to take you from the scientific into the commercial world. How will the risktakers – the growers, traders, processors and consumers react to the ideas of the scientists; those who put their money on the line, who are concerned with profit and loss, gross margins and satisfying the accountants? Accountants, like football referees, are necessary if sometimes unpopular individuals.

THE IMPLICATIONS OF CHANGE

New crops imply change – and I want to briefly consider the implications of change – at the farm level, in industry and in the market place. I also want to distinguish between those crops with international implications, such as soya bean, and those crops which are developed essentially for local use and for helping farmers to achieve a wider spread of income or security when faced with unfavourable ecological conditions. I will be concerned mainly with the former.

Few people like fundamental change because it creates uncertainty and is generally expensive. A cosmetic change – yes. A fundamental change – no. British Airways wants to stay with Boeings because it has the parts, the servicing facilities and the staff experienced with these aircraft. To change to another make implies much more than just the capital costs involved.

There is a similar fundamental resistance to change in the agricultural and agro-industrial sectors. Farmers like the crops they know and which are adapted to their soils and machinery; which have established markets and grades – unless there is a sugar daddy like the EEC offering incentives to make the change worthwhile.

There are, nonetheless, strong intellectual arguments for change – the narrow genetic base of cultivated agriculture and its vulnerability; the scope for developing new products; establishing or re-establishing natural products; better ecological adaptation. And these arguments form part of the justification for saving the tropical forests – in addition to the arguments based on watershed management. But many of the biological arguments fall down in the cold light of commerce and competition.

It is not long ago that many people were promoting the production of ethanol from sugar or molasses. But the cost of production is such that it is only truly competitive when crude oil sells at between $30–40/barrel; a price made even less attractive by the need to modify engines to use this fuel. Thus, the production and use of this fuel becomes essentially a political rather than a commercial decision.

COMMERCIAL IMPLICATIONS AT THE FARM LEVEL

Does it pay?

The farmer is a free agent and raises crops and livestock to feed himself and his family – directly or indirectly. Either way, revenue must exceed expenditure; the returns must exceed the effort. If a cash crop then the farmer will want to know if he can sell it. With rice, palm oil, corn, etc. there is basically no problem, although he might not like the price at the time of sale. There are local, national and international markets where all the norms are known. But with new crops there are no traders; no future markets; no accepted grades.

Does it fit into the cropping pattern?

In our own work on the problem of aflatoxin in maize in Thailand, where maize is planted at the beginning of the rainy season and harvested in the middle of the rainy season to be followed by a second crop of mung beans or peanuts, we have shown that aflatoxin can be controlled by drying the maize within five days of harvest. Alternative suggestions have been put forward including:

(a) Leaving the maize in the field to dry for up to a month. Whilst this does reduce the incidence of aflatoxin, the farmers are not interested because they want to plant a second crop immediately.

(b) Planting later in the season so that the maize matures in the dry season. Again farmers are not interested because they do not have a crop to grow in the first half of the season.

Thus apparently sensible biological arguments can be invalidated by commercial realities.

Does it fit in with the available machinery?

Farmers who invest in machinery must be able to maximize its utilization, from field production through handling, drying and storage. If you have the machinery for cereals you will probably be able to grow and handle oilseed rape, lupins and sunflowers – but not peas, apples or potatoes. Does the new crop fit in mechanically? Or is it even suited to mechanization, the right height or the right shape, or even maturing at the right time or at the same time? I worked some time ago with Nazmul Haq on the development of the winged bean. It is a vine crop which has to be staked on bamboos which are costly to buy and need a lot of labour to install and harvest. Further, there was a wide variation in the period of maturity. The cost of bamboo and labour proved too high for the commercial yield and there was, at that stage, no hope of mechanization. A substantial investment in research and development is needed to convert that plant into a commercial crop.

Is there the volume and uniformity of produce to justify a market?

To trade, especially internationally, you need a large quantity of consistent produce with the grades clearly defined. Many new crops suffer from the very diversity which is their biological attraction. Commodity markets are now fully internationalized and linked with the financial markets. Futures in maize and soya bean are traded by people who drive a Porsche rather than a Massey Ferguson, and they can do that because the grades are clearly defined and consistent.

On the question of future markets, the newspapers have recently been full of the high salaries paid to youthful financial dealers. If you are asked the difference between a Eurobond and a Eurobond dealer, then the answer is that a Eurobond eventually matures. Thus if you introduce small amounts of an unknown commodity of inconsistent quality into the market you will find little interest. The development of new crops must be market led and developed to meet specific needs more cheaply and effectively, fitting into the market and the industry without requiring any major changes in the market grades or industrial equipment.

THE INDUSTRIAL PROCESS

Most agricultural commodities are processed in some way. Thus, maize is shelled or ground, rice and wheat are milled, sugar and oil palm are crushed and refined, cocoa is dried and fermented, grass is digested by livestock, sunflower is crushed for oil, strawberries are made into jam, and geranium is distilled into essential oil. In every case there is a market and an industrial process. Therefore any new crop or commodity is likely to be meeting an existing need or, possibly, replacing an existing source of supply. No matter how biologically or agriculturally interesting, the commodity must be able to enter into the existing marketing/industrial infrastructure without modification – unless the product is startlingly cheap or effective or meets a very specific need in order to justify that change.

As an example, consider the introduction of North Sea Gas into the UK with its massive investment in piping and in the modification of every gas appliance in the country. This was economically justified because it was very cheap; because there were huge quantities available; and because it was safe. Similar advantages are necessary if new crops are to compete with the old faithfuls. Again on the winged bean, in 1977 we were able to interest a US bank in financing the development of the crop, until soya bean farmers threatened to pull out their deposits because they feared the winged bean would be future competition. At this point the bank lost interest in funding our work. Similarly you may be aware that at present the soya bean lobby is pushing for the labelling of vegetable oils in the US. Ostensibly to inform the consumer about the nature of the oils, it will have a serious impact on the market for coconut oils. This, combined with the measures being imposed by the EEC against aflatoxin in copra, will have a devastating effect on the rural economy of the Philippines. Like the US Farm Bill, this measure is designed to protect US producers but it will have a massive impact on the rural economies in those very countries which the US wishes to befriend politically. The impact of these market decisions will far outweigh any benefits from financial aid.

THE NON INDUSTRIAL CONSUMER

When I was a student at Durham we learnt about potato blight and that there were then (20 years ago) some resistant varieties. Despite that, most farmers grew King Edward potatoes because that was the variety which the 'housewife' wanted and bought. We are all aware of the massive breeding programme for sorghums and millets which have ended up with unacceptable grain colour and texture. 'Coca Cola' found the strength of consumer resistance when they introduced 'new' coke and then had to reintroduce 'traditional' coke. A new commodity has no value to the farmers unless he can sell it to someone to eat or process.

THE PROCESS OF DEVELOPING NEW CROPS

Many crop species which we might like to develop are undomesticated, have a modest yield, mature irregularly and are an inappropriate shape for harvesting. To be converted into a functional plant they need to be submitted to plant breeding. It was Lord Mancroft who observed that "Cricket is a game which the British, not being a spiritual people, had to invent in order to have some concept of eternity." The word 'cricket' might be replaced by the words 'plant breeding'. It needs a long time frame to develop new crops and the question arises 'Who pays? and if you pay – How do you get any benefit? What is to stop others from taking the benefits?'

Which brings us back to the basic question of 'Why develop new crops?' Is it an intellectual or a humanitarian interest, or is it strictly a commercial one? If it is strictly commercial then the commercial users should be involved in planning, managing and funding the programme, and they will only do this if they are confident that they can control the product – be it hybrid seed, patented micropropagation process or confidential manufacturing process. They will, understandably, not be interested in developing new crops for small farmers. This does not invalidate the case for developing new crops for the benefit of rural communities and because they have the potential to meet market need competitively. But it is naive to expect individual companies, or even developing countries, to fund this alone.

I recently visited the JET (Joint Energy Torus) project at Culham, which is researching nuclear fusion technology. It is a massive three phase programme with the first phase completed in 1992. Fourteen nations are only to be co-financing this programme with its universal benefits. No single country could afford it alone.

You may also be aware of the programme of funding for pre-competitive research being underwritten by the Department of Trade and Industry, in which DTI will put up 50 per cent of the funding for research, which itself is of no immediate commercial value (i.e. is pre-competitive), on the understanding that at least three UK companies will together put up the balance of funding. Individually no company would fund this work, but together they all get the fundamental information from which they are free to develop commercial applications. Thus the funding of the development of new crops cannot be considered as the onus or responsibility of individual countries, companies or institutions, except where a particular company sees commercial benefit. In general the benefits are more widespread and the funding must be equally widespread. If you are involved in developing a new crop you must ask the following ten questions:

–   Why?
–   Who will grow it and who will buy?
–   Who will process?
–   Who will consume?
–   What demand is being met?
–   What commodities are being replaced?
–   On whose toes will you be stepping?
–   Who should pay for the development costs?
–   How will they recoup those development costs?
–   Can they control the intellectual property?

Which means that the development work:

(a)   must involve growers but above all the end users, i.e. the new crops should be pulled by potential demand rather than pushed by enthusiastic biologists, and

27

(b)   can only be considered by industry if it can control the crop, the genetic
      material, the process or the market. Otherwise, how does it get its
      money back?

There are sound reasons for developing new crops on a speculative basis to
meet the needs of small farmers, but industry will not pay for that, and hence
my strong support for Nazmul Haq's initiative with the new Centre for the
Development of New Crops at Southampton, including pre-competitive
research. Those interested in developing new crops must be single minded if
they are to succeed, but be careful; Max Beerbohm described those participating in the University Boat Race as 'Eight people with a single thought – if
that!'. Being single minded does not mean being concerned only with
biochemistry or biology, but rather with the totality of producing a new crop
which is growable, saleable, processable and consumable profitably and
competitively in the market place.

3

Management Problems in the Introduction of New Crops: The Zambian Experience

Jameson H. Muchaili

INTRODUCTION

In my paper I would like to outline the problems arising from the introduction of new crops such as tea for many Zambian farmers. These are as follows:

(a) Reluctance of the farmer to accept new crops about whose potential he is unsure.

(b) Complexities of the husbandry practices and availability of specific inputs such as fertilizers.

(c) Marketing of the crop.

(d) Farmers' knowledge and experience.

(e) Financial resources at the farmer's disposal for him to meet the costs involved.

(f) Lack of proper extension services and necessary back-up in terms of applied research.

Farming can be classified as the original indigenous industry. The characteristics being production- oriented is not, therefore, unique. Because of their isolation from the market place, it is understandable that farmers adopt a selling attitude, i.e. getting rid of what they have got rather than a marketing approach, i.e. having what the customer wants. We are all too well aware that a transformation in attitude or approach is not an overnight process.

PURPOSE OF GROWING A CROP

There may not be the appropriate degree of compatibility between the national Zambian objective of generating opportunities to earn profitable foreign currency in export and that of Zambia's many farmers to earn an income to support a family who may possibly be located some hundreds of miles from our capital, Lusaka. How, therefore, can we have a greater degree of convergence of approach?

The seeds of any new crop like that of any new idea will fall on barren ground unless the introduction is properly planned and sustained through strategic marketing and the implementation of a marketing strategy. I would like to explain what I mean. The strategic marketing approach endeavours to identify market segments or product market niches while a marketing strategy is the application of the mix of the four P's, i.e. product, place, price and promotion. We must remember that amateurs hope for results but professionals plan for them.

One of the major characteristics associated with the introduction of a new crop is the degree of confidence and commitment that the originators can generate. No team can win unless it has the confidence in a professional commitment to the product's success. By professional commitment, I do not mean initially an over-concentration on the productions aspects, i.e. yield per hectare, etc., but in identifying the markets and evaluating if we can compete in export markets in relation to cost, quality, packaging and question whether we have a unique product to sell.

How far down the marketing road do we have to bring our producers/farmers? This ultimately depends on the nature of their investment. If it is a once-off type of investment, so what? However, to achieve a sustainable competitive advantage in the market place that investment commitment has to be for the long term.

FARMERS' ATTITUDES

We must recognize that not just Zambian farmers can be classified as conservative but that it is an almost universal characteristic of farmers. Their crop production educational input has been passed on from generation to generation. Their behavioural pattern is therefore very much associated with a specific crop rotation procedure which has yielded true and tried results over the years. We must also be conscious of the traditional cultural patterns of the region. We could, perhaps, classify farmers into three types which in diagram terms could be regarded as a cone. At the base, we have a considerable number that could be classified as accepting farming as a way of life. The next band we would perceive as those who regard farming as a commercial operation and, subsequently, conform to commercial criteria.

In Zambia a similar classification would be as follows:

Commercial farmers	—	200 ha and over
Emergent farmers	—	10 – 20 ha
Subsistence farmers	—	less than 10 ha

The emergent and subsistence farmers produce between them 60 per cent of the staple food, maize. Since they are the least mechanized they opt to grow more maize than other cash crops. Maize as most of you know is very easy to grow. This underscores the point that small scale farmers, because of their numbers, can have an impact in any economy if they adopt a new crop with high potential. Hence the need to plan carefully the introduction of a new crop.

However, the trauma for many farmers to move to new crop production must not be underestimated. Conservatives view change as a risk and I am not talking within the British political scene. But we must remember what the management scholar Peter Drucker said, 'Results are gained by exploiting opportunities not by solving problems.' We, therefore, must not allow fear of the future to paralyse our innovative proposals.

Of course, there are always problems and negative people feed on them while the marketing person only perceives opportunities. However, to accommodate change we must endeavour to seek a balance between both threats and opportunities.

MARKETING PROBLEMS

It has been stated that 22 per cent of new food products were discontinued after test marketing and 17 per cent were withdrawn after they had been introduced into the market place. We, therefore, must be aware of the complexities of introducing a new product in the market place and the substantial hill we have to climb to ensure its success. The following six reasons have been suggested, together with their percentage of occurrence for new product failures:

1.	Inadequate market analysis	32%
2.	Product defects	23%
3.	Higher costs than anticipated	14%
4.	Poor timing	10%
5.	Competitive reaction	8%
6.	Inadequate marketing effort (includes weaknesses in sales force, distribution, and advertising)	13%
		100%

A breakdown of these figures indicates that problems or inadequacies are the cause of most failures (inadequate market analysis, poor timing, competition and sales force, distribution, and advertising). Bruzzell and Nourse also found that a substantial percentage (approximately 80 per cent) of the reasons given for discontinuing a product was a result of marketing misjudgments or inadequacies.

The market place

I therefore return to my earlier point, the importance of the market place. Within this context, I would suggest we review the proposal within the following project framework (Fig. 3.1).

Figure 3.1: Project Framework.

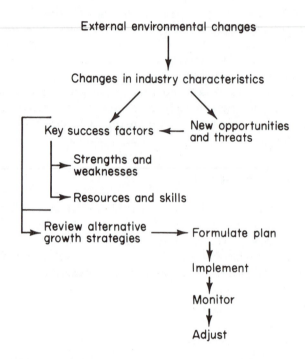

CULTURAL PRACTICES

If the new crop demands mechanization as opposed to using human labour the farmer is likely to reject it. In some cases though the crop demands human labour; if the cultural practices are too cumbersome, as in tobacco and tea, then the crop also stands the same chance of being rejected. Tea in Zambia has been difficult to introduce to small scale holdings surrounding the Kawambwa Tea Company due to some of the above reasons.

Conflicting interests and the normal way of life in an area could also pose problems when introducing a new crop. The following example, though unrelated, illustrates this point. In 1976 I was in charge of a rural milk production scheme in the west of Zambia, along the fertile Zambezi River plains. This was an ideal area in that grazing was abundant and farmers already had local cattle which we crossbred with high milk yielding breeds. However, we overlooked one point, the fact that these farmers were also fishermen. At the time when they were supposed to milk the cows, at five in the morning, they were miles away checking their fish nets. This affected their income from milk and therefore the scheme was not successful. This could happen not only in Zambia but world-wide, especially in developing countries.

Farmers' experience

The fact that the farmer or his manager has the knowledge and experience of a new crop makes it easier for him to adopt the crop. If the opposite is the case, some resistance could be encountered. In relation to this is the availability of a properly run extension service to help the farmers; this is very vital indeed. The farmer or producer needs it, and in most cases urgently, otherwise it may not be as useful. When unsuspected problems do occur applied research is required to assist the farmer solve them. This too is vital, because the buildup of pests and diseases takes time after the new crop is established.

For instance, though the Kawambwa Tea Company has been producing tea from 1969, there have been no problems with pests and diseases. However, this does not mean that the Kawambwa Tea Company will not eventually have these problems. It is sad to note that tea research in Zambia is not well organized and developed. This could be disastrous in the event of a problem arising.

Inputs

Another problem that new crops face when introduced into an area is the lack of specific inputs such as fertilizers, herbicides, fungicides, pesticides and other disease-controlling chemicals. This is true of tea in Zambia. This is because the crop is not widely grown and therefore the country cannot afford specifically to formulate and produce inputs for this crop.

The tea grower therefore has to make do with the nearest input available. The end result is that however high yielding the clones, their genetic potential will not be fully exploited. This may lead to the rejection of a high potential crop such as tea.

Resources may not be the only limiting factor in availability of these inputs. There is also the problem of lack of expertise to formulate and produce such inputs. Fortunately for countries such as Zambia, Malawi has a properly developed research institute, the Tea Research Foundation of Central Africa. This could be of help in such matters.

Costs

We must not ignore the substantial cost involved in the introduction of a new crop. This usually involves the injection of new capital and the provision of initial cash incentives to the farmers. We already have a good example in

Zambia of an Irish company, Masstock Ltd., proposing to invest US$10 million in a wheat-growing project. It is hoped this will make Zambia self-sufficient in this relatively new crop within 5 to 10 years.

Clearly, therefore, the cost of introducing a new crop is high. In order to meet this expense the farmer has to borrow from lending institutions. This is not only costly in Zambia or elsewhere but the farmer has also to convince his bank manager of the potential of the crop in question. Bank managers are not the easiest people to convince, especially where money is concerned. Perhaps joint ventures within the overall national policy could be one way of raising the necessary capital and expertise required in the introduction of new crops.

New approach

I would not be far from the truth if I said that most of the food crops available in the world were developed in the research fields and greenhouses of Europe. It is not surprising therefore that the introduction of these crops, such as potatoes, has not been very successful in the developing world. What scientists have been doing is manipulating the genetic pool of a foreign crop to suit a foreign environment. This, combined with the problems enumerated above, has resulted in a negative food situation within developing countries.

It must also be appreciated that there is a limit as to what extent it is possible to manipulate the genes of a developed crop without sacrificing some important characteristics, such as taste. The developing countries have abundant natural resources for crop production; they also have several high potential under-developed crops such as cassava. Until now, these have been regarded as items for the poor man's diet. I therefore challenge scientists the world over to address their research to such crops for the good of mankind. In this way they will be developing the right crops in the right environment for appropriate management.

CONCLUSIONS

In the final analysis, the success of the product will depend on its management. The quality of the ultimate profits generated will reflect on the management input along every link in the chain, i.e. marketing, processing and production.

It has been stated that 'growth and progress are related for there is no resting place for an enterprise in a competitive economy'. We do not regard Zambia as a resting place, except perhaps for some European tourists who would wish to avail themselves of our excellent game parks and reserves. I can assure you we will earnestly strive for that growth and progress in the world's export market. This, we recognize, cannot be instantly achieved but with that Zambian characteristic of persistence and determination we will achieve our objectives.

ACKNOWLEDGEMENTS

I wish to thank most sincerely Tate & Lyle Technical Services (TLTS) of the United Kingdom for sponsoring me to present this paper. In particular I am indebted to Messrs. Peter Cheshire, Commercial Director (TLTS), Simon Wynn and Bob Davey both of TLTS for their encouragement and making all the necessary travel arrangements for me.

The views in this paper are my own and do not necessarily represent those of Tate & Lyle Technical Services nor the Export Board of Zambia.

4

New Crops; a Suggested Framework for their Selection, Evaluation and Commercial Development

E. S. Wallis, I. M. Wood and D. E. Byth

INTRODUCTION

Australia is in the unique position that every major crop that is grown commercially had, at some stage, the status of being a new crop. There was no established agriculture at the time of European settlement in 1788 and none of the plants used by the aboriginals has been developed for commercial production. Consequently all the current commercial crops have been introduced at some time during the past 199 years. This of course does not mean that these introduced crops were intrinsically new. In all cases they were established crops at some other location and were only 'new' to Australia. In many cases such new crops were poorly adapted to the Australian environment and agricultural research during the past century has involved improvement programmes to adapt these crops to the local environment.

In recent years a number of factors has provided a stimulus to work on new crops within Australia. The marketing problems with wheat and other temperate cereals during the 1960s highlighted Australia's dependence on these crops and clearly demonstrated the need to diversify production. A high priority has been given since then to the development of alternative crops and several, such as soybean, lupin and sunflower, are now well established components of Australia's rural production.

A further stimulus to new crops research in recent years has been the increasing demands by industry for animal feedstuffs. As intensive production of livestock (poultry, pigs, beef cattle and fish) has expanded throughout the world the need to produce balanced rations at the lowest cost has greatly increased. In some instances this has merely meant the diversion of some currently produced crops to meet these needs. However, it has also prompted attention to the possibilities of developing new end uses for other crops. An example here is the increased used of food legumes, such as soybean, mungbean and pigeonpea, as sources of energy and protein in a compound feed ration to replace cereal grains. We return to this possibility later.

Experience has shown that research and development of new crops is both a time consuming and expensive exercise. Therefore it is important to develop procedures that are cost effective, and to ensure that research and development are concentrated on those potential new crops that offer the best prospects for commercial development. In this paper we propose a logical framework for assessing potential new crops and for conducting research and development aimed at their commercial development. We begin by defining new crops and offer a simple classification of new crop research. We discuss some of the difficulties involved in establishing and conducting programmes of research and development of new crops and offer suggestions for overcoming or avoiding these. We conclude by discussing the progress of work on three crops new to Australian agriculture which are at different stages of development and commercial acceptance.

WHAT IS A 'NEW CROP'?

The term 'new crop' is used in two different senses. The first is in the sense of a crop that it is 'new' to a country, a state or a region. Most of what is generally termed new crop research is of this type and involves research to assess the adaptation in a particular environment of what is an already established crop at some other location. In some instances the establishment of a new crop in an area will prompt its utilization for the production of new products.

The second refers to a crop which has been, or is being, developed *de novo* from a previously unexploited plant. The interest in such new crops generally arises because of the presence in the plant of a chemical of pharmaceutical or industrial value. During the past two decades there has been some very interesting work in the USA (Anon 1984) to develop such 'new crops' from a number of native species which have been shown to produce compounds that have industrial applications. These include: crambe (*Crambe hispanica* (syn. *C. abyssinica*)) for oil and erucic acid; guayule (*Parthenium argentatum*) for natural rubber; jojoba (*Simmondsia chinensis*) for industrial oil, pharmaceuticals, cosmetics, etc.; Stokes aster (*Stokesia* sp.) for epoxy oil; meadowfoam (*Limnanthes alba*) for oil that is used in cosmetics and as specialty lubricant. Undoubtedly other examples will be discussed at this symposium.

A SUGGESTED CLASSIFICATION OF NEW CROP RESEARCH

New crop research can take a number of different forms and it is useful to consider a simple classification of these. The research can involve work which is primarily directed at either the crop or the product produced from that crop. The following classification is based on whether the crop and/or the product is established or new.

(a) *Established crop/established product:* This category covers work to adapt an already established crop in a new area. While some transfer of technology is often possible, this generally requires a selection or breeding programme to identify genetic material adapted to particular climatic and soil conditions. An example here is the work on selection and development of wheat cultivars for the different wheat growing regions in Australia. In other cases, the research may involve selection for particular social conditions or particular biotic factors. Chickpea (*Cicer arietinum*) is a food legume that has long been grown in India under a peasant farming system in which the crop is harvested and threshed by hand. In the adaptation program that lead to the establishment of a commercial chickpea industry in Australia, one of the most important requirements was for a genotype that was amenable to mechanical harvesting. This requirement was met by the cv. Tyson, first released in 1980 and now sown on over 50000 ha (Hamblin 1987) in southern Queensland.

(b) *Established crop/new product:* This category covers the research seeking the development of a new product from an already established crop. The work to utilize the bast fibre crop kenaf (*Hibiscus cannabinus*) for paper pulp production (Wood 1984) and the forage crop sweet sorghum for alcohol production (Ferraris 1986) are examples in this category.

(c) *New crop/established product:* Examples of this category of new crop research are guayule for rubber production, *Sesbania* for galactomannan production and jojoba for the production of 'sperm whale oil'.

(d) *New crops/new product:* The work conducted to develop crops for the production of special industrial oils and special pharmaceutical compounds is an example of new crop work in this category. A specific example is the development of velvet bean as a source of the compound L-dopa, which is used in the treatment of Parkinson's disease.

Generally, the complexity and difficulties of the research and development programme increase, and the prospects of commercial success decrease, from (a) to (d).

THE PROBLEMS OF NEW CROP RESEARCH AND DEVELOPMENT

Only a small proportion of the programmes of research and development of new crops can be expected to be commercially successful, and this parallels the experience with research and development of new innovations in industry. However, there are a number of special difficulties associated with crops that put them at a greater disadvantage compared with new industrial innovations. These include the seasonal nature of crop production which acts to retard the rate of progress, and the effect of biotic factors (insects and diseases) which can be a major limiting factor with new crops (e.g. *Phytophthora* root rot disease of guayule).

Several other important factors disadvantage research and development work on new crops. These include:

(a) There is no active lobby for new crop research. Farmer organizations tend to have a somewhat parochial outlook. While not opposing research on new crops, they do favour research to improve the profitability of their existing crops.

(b) In most developed countries growers of particular crops are prepared to provide funding for research on that crop. Such funding is unavailable for work on new crops, so that the problem of funding is centripetal.

(c) Research on new crops generally needs to be multidisciplinary and may require agronomists, plant breeders, entomologists and pathologists for the field programme and chemists, engineers and technologists for the industrial programme.

(d) Because of the uncertainty of successful commercial adoption of new crops and the often long lag time between research and commercial development for those crops that are successful, researchers may work for many years without any apparent success. This can be very discouraging for those involved and can also adversely affect their promotion prospects.

The difficulties in establishing a new crop rapidly escalate as the degree of industrial processing increases. Where industrial processing is a major part of the production system it is often necessary to develop a whole new system of field production, transport and storage, processing marketing and consumption and there are often problems in reconciling the various facets of the system. Growers commonly are prepared to produce a product only if they have a reasonably assured market, especially if additional or specialized equipment needs to be purchased. On the other hand, those involved in the marketing will be prepared to undertake the necessary sales promotion programme only if they can be assured of a continuity of supply. The processor requires both assured markets and supplies of raw materials. The reconciliation of all these requirements requires close coordination and cooperation between all the different groups involved, and failure to do so can delay or prevent the development of new crops and industries.

A SUGGESTED FRAMEWORK FOR NEW CROP RESEARCH AND DEVELOPMENT

Our observations of new crop research in a number of countries suggests that the genesis of the work generally falls into one of the following categories.

(a)   A market shortfall of a product leading to increased prices and representations from an entrepreneur anxious to establish a new and profitable industry.

(b)   An entrepreneur becomes interested in an agricultural product which he believes could be produced locally, either for local use or export.

(c)   A scientist becomes aware of a new product that could be produced from a particular crop either through his research or his reading of the scientific literature.

The first of these can be taken as a response to market demand (i.e. market pull) while the third is an example of technology push; the second may fall in either of these categories. Where market demand is the primary incentive, it is possible that the shortfall may have been overcome in the lag time between initiation of the research and development programme and the attainment of commercial production. Consequently, prices will have declined and the new commercial venture may either fail to materialize or experience financial difficulties. This *ad hoc* approach to new crop development is far from satisfactory and the need for, and potential value of, new crops warrants a more ordered approach. In this section we suggest a logical framework for the conduct of a research and development programme on new crops. We propose a four stage process in which potential new crops are subjected to increasingly intense scrutiny and additional selection criteria, with a progressive reduction in the number of crops entering the more advanced stages of the programme. The following are our suggested stages:

(a)   Initial identification of potential new crops
(b)   Preliminary testing
(c)   Research and development
(d)   Commercialization.

Initial Identification of Potential New Crops

The cost of undertaking a field evaluation programme on any plant species is high and inevitably involves considerable resources in land and labour. It is important, therefore, that there be a preliminary screening process to identify crops with reasonable prospects of being commercially viable. Wood *et al.* (1974) proposed that an initial desk study be conducted in which each potential crop is considered on the basis of three criteria which must be satisfied before it is advanced to the next stage. The criteria are:

(a)   That it is (or could be) adapted to the environment at the proposed site.

(b)   That production would be profitable for the farmer and, if it is an industrial crop, for the processor.

(c) That markets are available for the product. These markets need to be medium or long term to ensure an adequate return on the costs of the research and development.

These assessments require the use of any information from the literature on agronomic requirements of the crop and the likely yields, as well as estimates of probable costs of production and profitability and of market prospects. It is also valuable at this stage to determine the extent of interest by industry in the crop or crop product.

We consider that all potential new crops should be assessed on the basis of these criteria and only those that meet these criteria should be considered in the next phase. While these assessments must be somewhat subjective and subject to error, it is our experience that many crops can be clearly eliminated on the basis of these criteria.

Publication of the findings is an essential part of this initial screening phase. A published record provides a firm basis for any subsequent review if circumstances suggest that a previously rejected crop may warrant further consideration. It also avoids unnecessary duplication of research in the future.

Preliminary Testing

The objectives in this phase are:

(a) To assemble an appropriate germplasm collection.

(b) If necessary, test the collection in quarantine and screen it for the incidence of disease.

(c) Make a preliminary assessment of the collection in respect of growth, yield, resistance to disease and insect pests.

(d) Bulk up seed supplies of promising genotypes for more detailed testing in phase 3.

(e) If appropriate, conduct preliminary quality tests on the crop product.

(f) Reassess the commercial viability of the crop.

The assembly of an adequate germplasm collection is a key element in this phase. The genetic variation in the collection should be representative of the species, so that the field study can provide an adequate indication of the range of adaptation of the crop to the target environments.

The initiation of the larger research and development programme in phase 3 should be dependent on a satisfactory outcome from this assessment. If the decision is made not to continue the work on to the next phase the results should again be documented and the germplasm collection should be catalogued and stored under long-term storage conditions.

Research and Development

In this third phase we propose a research strategy that is focused on the resolution of primary limit(s) to production, productivity or utilization. This does not necessarily infer establishment of a programme concentrated narrowly and in a consequentive manner on individual limitations in order of priority. Research into other, less fundamental problems may be included in the strategy but should be predicated on the absence (through resolution) of the primary limits.

It is clear that the limitations to increased production and use of a crop can be quite diverse and numerous, ranging from the physical/climatic/biological constraints to productivity and adaptation, to socio-economic constraints, to inclusion of the crop in the farming system or to utilization. With respect to crop improvement, this approach is designed to focus on the potential productivity of a crop in a particular range of farming systems. Limitations to achieving this potential can then be addressed in a logical framework. The adoption of a philosophical approach to research centred on crop improvement should ensure that real needs are addressed, and that the potential dangers of a narrowly-based disciplinary approach are avoided, while exploiting its strengths and contributions as appropriate. This concept of plant improvement inevitably leads to consideration of the farming system as a whole and we contend that all research should be conducted in the context of the farming system of interest. However the distinction is drawn between 'farming systems research' and integrated research which addresses the needs of the crop within the numerous and complex farming systems used by farmers.

As an example of the approach we advocate as a framework for research and development we present a synopsis of the findings of an International Workshop entitled 'Food Legume Improvement for Asian Farming Systems' held in Thailand in 1986 (Wallis & Byth 1987). This Workshop considered the problem of undertaking a programme of research and development for a group of crops which is important in the farming systems of Asia, both from the ecological point of view and in terms of human and animal nutrition. At least 18 species of the food legumes are considered to be important within the region and these differ markedly in their agronomic characteristics, cultural needs, uses and roles in the farming systems. There is only limited scientific knowledge on the production, improvement and use of many of these crops. As a result there are special difficulties in planning and organizing structured improvement programmes. Many of these crops are new in terms of production and end use in particular parts of Asia and elsewhere.

An important aspect of new crop research, as demonstrated clearly in the legume crops, is that once a genetic advance is made, e.g. earliness, then a whole new range of options become available to fit the crop into new or existing farming systems. This may require considerable research effort.

The basis of the programme for research and development discussed at the Workshop is the concept of an integrated multidisciplinary crop improvement programme. This approach is essentially an application of systems analysis which has found wide acceptance in industry and commerce but has, until recently, been little used in agriculture. This concept of crop improvement implicitly includes a consideration of the social, political, economic, physical and biological factors that limit production. The limits that exist may vary in relative importance in different crops and different farming systems and the research strategy that is developed needs to take this into account.

The adoption of this approach requires the formation of multidisciplinary teams which require careful and flexible management and strong leadership. Just as productivity is a function of the direct effects and interactions of many factors, the effectiveness of multidisciplinary research depends on the establishment of a coordinated framework relevant and responsive to the needs of the target production environment.

This approach is described in detail by Byth *et al.* (1987), and will not be repeated here. However, we believe that the approach has many attributes that make it a useful framework in which to plan the research and development of new crops.

COMMERCIALIZATION

Although this is considered here as a separate phase, it in fact overlaps with the earlier stages in that close collaboration should be maintained with industry at all stages of any new crops programme. Industry must have the opportunity for input into questions of markets, production systems and economic assessments, and any special requirements of industry in respect of product quality must be considered at an early stage.

The commercialization phase can be considered to begin with the first commercial production. An important question is the duration of research investigations prior to the commencement of commercial production. Ideally the research programme should anticipate problems and develop the optimum system of production before commercial production begins. Unfortunately, this commonly is not feasible with agriculture. Wood & Hearn (1985) considered this question in respect of cotton production in the Ord River Valley in north-western Australia. They highlighted the dynamic nature of the cropping ecosystem and the important interactions between cultural practices and the biotic factors of insect pests and disease. Because the incidence of biotic factors is so dependent on scale of operations it is virtually impossible to develop an appropriate management system in the absence of the commercial operations.

Clearly commercial-scale farming is essential for relevant research on the production system. Commercial development should therefore not be seen as the logical end of the research and development programme, but rather as the start of a new phase of the programme which concentrates on the development of the cropping system as opposed to research on the crop *per se*. Research and development in the commercialization phase must be seen as an ongoing programme in which the objectives are determined by those problems and by technological possibilities which are identified.

PARTICULAR PROBLEMS OF THE DEVELOPING WORLD

The introduction of new crops into the developing world is particularly difficult. Resources, both human and financial, certainly are limited, but in addition the farmers are often the poorest section of the population. At this level the adoption of a new crop, or even of a new technology, must be carried out very carefully. The resource-poor farmer generally cannot accept additional risk as his livelihood and that of his family are at stake.

However, there have been a number of successful introductions of new crops in the developing world which have substantially improved the living standards of farmers. The adoption of cassava cultivation by the farmers of north-east Thailand is commonly cited as an example. Recent price reductions for cassava have not drastically affected production because it is now established as an integral part of the local farming system. New markets and new end uses need to be developed.

The *ad hoc* trial-and-error approach which has characterized most past work on new crops must be avoided. It is clear that the recent advances in crop growth modelling will greatly assist in selecting areas of greatest potential, and in selecting appropriate genotypes and production systems. However, a series of 'best best' trials in the areas of interest may provide a cheaper and quite effective way of assessing the likely viability of a potential new crop. Such trials can quickly identify the major potential limitations and so focus subsequent research efforts.

In this context, the International Agricultural Research Centres can play an important role as they develop their data bases of agroclimatic information. In many cases the introduction of the mandate crops of these Institutes to countries unfamiliar with their culture can be viewed in the new crop context. New uses may be required for existing crops and a wide range of socio-economic aspects must be considered. An example here is the expansion of pigeonpea (*Cajanus cajan*) from the Indian subcontinent to other areas of the world where it is currently unknown but has considerable potential (Whiteman *et al*. 1985).

SELECTED CASE STUDIES

Three case studies of new crops for Australia will be considered briefly. They are chosen to illustrate three different stages of commercial adoption:

Soybean	commercially adopted and now the most important summer growing legume crop; 69 000 ha in 1986 (Anon 1987).
Pigeonpea	three cultivars released and commercial plantings beginning but not yet widely adopted; 8000 ha in 1986 (Hamblin 1987).
Kenaf	the background research and development completed but not as yet adopted (no commercial crop; pilot plant sowings of 500 ha planned for 1987–88).

In both soybean and pigeonpea the agronomic and ecophysiological factors utilized to adapt the crop to Australian agricultural environments are similar – basically because both crops are quantitative short- day plants.

The introduction of both crops depended on the availability of markets. In the case of soybean, oil extraction technology had to be available prior to the dramatic expansion of the late 1960s and early 1970s.

For pigeonpea the impetus for development is the need for a summer growing crop legume suitable for broadscale agriculture in rain-fed conditions in Australia. Soybean is not considered sufficiently hardy for these conditions and early maturing cultivars of pigeonpea have the potential to fit this niche.

In commencing the research programmes on both soybean and pigeonpea, the multidisciplinary approach, in a simpler form, was adopted. It is the experience in the development of these crops which has led to the refinement of our ideas on utilizing the holistic crop improvement approach described previously.

A short summary of the problems and foci of the crop improvement programmes for soybean, pigeonpea and kenaf follow as examples of the approach.

Soybean

BACKGROUND

Limited soybean production occurred in the subtropics of Australia prior to the late 1960s. Since then the crop has expanded rapidly to 69 000 ha in 1986. Most of the production occurs in the subtropics and warm temperate regions of eastern Australia (26–35 °S). In recent years, there has been some limited production in the tropics of Queensland (Emerald, 22·5 °S; Burdekin, 20 °S) and in north-west Australia (Ord and Douglas/Daly Rivers 14–15 °S) (Lawn & Byth 1979). Most production is conducted with full or supplementary irrigation or in the higher rainfall areas. Very substantial potential exists for the expansion of soybean production in Australia, mainly in the tropics and subtropics and under non-irrigated culture.

ADAPTATION

The primary climatic variables of interest are daylength, temperature and water availability. It was early realized that the phenology, or phasic development of the plant, was influenced heavily by photoperiod and temperature or both, and that these are therefore major factors influencing varietal adaptation to change in latitude, sowing date and moisture availability. Most variation among varieties in phenology is associated with photoperiod, and this provides a basis for prediction of phenology in different environments. Most soybean genotypes are quantitative short day plants and become earlier in flowering and maturity as day length is shortened by later sowing or culture in lower latitudes. In equatorial regions, most genotypes are insensitive, and relatively early in flowering and maturity, to the prevailing short and relatively constant photoperiodic regime. As a result, in increasingly wide range of genotypes from the higher latitudes become potentially 'adapted' to the lower latitudes.

Such reduction in crop duration is accompanied by reduced plant size, so that the optimum row spacing is narrower and the optimum plant density greater, than for longer season crops. As a result of this cultivar × daylength (sowing date, latitude) × plant density/arrangement interaction, a wide range of agronomic recommendations and production systems can be developed within and among regions which allow flexibility of sowing time to complement particular crop rotations.

Soybean improvement in the tropics is a more complex challenge. For wet season tropical culture, similar basic principles apply to those in the subtropics. Although most accessions from the subtropics are unsuitable agronomically for mechanized culture in low latitudes, it has been possible to select effectively in the subtropics of Australia for acceptable agronomic merit and phenology appropriate to wet season sowings in the tropics. Selection within the region of influence would be even more effective. However, dry season sowings in the tropics present new and major challenges to soybean improvement – short and decreasing daylengths during vegetative growth, relatively long and increasing daylengths during reproduction, low night temperatures during vegetative growth and the early phases of reproduction, and very high temperatures during late reproduction of late sown crops. These form a combination of challenges not experienced in any other region of soybean culture. Under these conditions, most accessions are too short-statured for mechanical culture. Some tropical accessions have acceptable phenology but tend to exhibit unacceptable agronomic habit, susceptibility to shattering and high sensitivity to low night temperature. Currently, no lines have been identified which are phenologically and agronomically acceptable for mechanized production under these conditions (Lawn *et al.* 1986).

Soybean improvement for short-day culture in the low latitude tropics is a fundamental challenge, which has major international significance. Current research is designed to elucidate the ecophysiology of such adaptation, to identify characters of adaptive significance, and thus to define appropriate parentage, breeding strategies and screening methods. It appears inevitable that soybean improvement for such usage will require incorporation of genes for acceptable habit, shattering, disease and pest resistance, seed size and quality, and tolerance of low temperature into genetic backgrounds with appropriate phenology under the prevailing photoperiod and temperature regimes. This is a major scientific exercise, aspects of which have recently been initiated in Australia, which potentially could contribute substantially to soybean adaptation throughout the tropics.

Pigeonpea

BACKGROUND

The pigeonpea is a short-lived perennial crop legume grown widely in the tropics and subtropics, particularly in the Indian subcontinent, parts of Africa and South East Asia, and in the Caribbean. It is consumed mainly as the dry split pea, or as a green vegetable, and rarely enters world trade. The protein content of the dry seed is 20–25%.

Pigeonpea has several advantages over other leguminous crops for broadscale production in the semi-arid tropics. These include drought tolerance, lodging and shattering resistance, greater resistance to pre-harvest weathering, and perenniality which allows the possibility of ratoon cropping.

Despite these advantages, the crop has not, until recently, been considered seriously as an Australian crop. There are several reasons for this, including susceptibility to waterlogging and frost damage, a possible requirement for seed drying after mechanical harvesting, and susceptibility to attack by insects (particularly pod borers). In addition most cultivars are of long duration (7–11 months) which restricts adaptation and incorporation into crop rotations and negates mechanical harvesting.

PRODUCTION SYSTEMS

Most pigeonpea genotypes are quantitative short-day plants, and phenology is influenced by photoperiod, temperature and their interaction. An extremely wide range of phenology exists (<60 – >200 days to flowering), including insensitivity to photoperiod. As a result, pigeonpea is used in a wide diversity of production systems internationally. These range from the use of photoperiod-sensitive genotypes grown as long-season (9–11 month) or full-season (6–8 month) crops to short-season (3·5–4 month) crops of photoperiod insensitive genotypes (Byth *et al.* 1981).

47

Such large differences in phenology obviously have major implications for vegetative development and canopy structure, and thus for agronomic use. Only some of these production systems have relevance to Australian agricultural environments.

Change of sowing date (or latitude) has substantial influence on the phenology and vegetative development of photoperiod-sensitive genetic material. As a result, large sowing date × plant density/arrangement interactions exist, and this makes management an important determinant of seed yield. For example, photoperiod-sensitive cultivars such as cv. Royes can be sown at or after the longest day in order to reduce the pre-flowering period and thus avoid excessive vegetative growth. Even then, at 28° S plant population has to be varied from 50 000 plants/ha for sowing at the longest day to 250 000 plants/ha for sowings 2–3 months after the longest day, in order to obtain optimum canopy development and maximum yield.

Photoperiod insensitive cultivars that flower in approximately 60 days have recently received considerable research interest. These cultivars will flower and mature in approximately the same time regardless of sowing date, providing temperature is not limiting. Ratoon cropping is feasible. Little plant improvement has been attempted in this production system so far. Despite this, extremely high seed yields have been obtained experimentally, with line mean yields from a plant crop in excess of 8 t/ha under favourable conditions (Wallis *et al.* 1983). Plant populations of 400–500 000 plants/ha are necessary and such canopies are suitable for mechanical harvesting.

IMPROVEMENT

Nomination of clear objectives for improvement is particularly difficult in pigeonpea because of the wide diversity of phenology and habit and use in quite contrasting production systems. In fact greater differences exist in growth and development among genotypes adopted to the various production systems than there are between many other crops. As a result, it is probable that quite different physiological limitations will exist between systems, and that genetic variability and parameters will also differ. Thus scientific knowledge must be considered system-specific until proven otherwise, and improvement within different systems is likely to pose quite different problems.

Recent research has demonstrated that pigeonpea can exhibit both remarkable tolerance of limiting environments (moisture stress, low soil fertility, insect damage) and extremely high seed yield potential in favourable environments. Both forms of adaptation need to be exploited and, if possible, combined.

Since 1977, pigeonpea research at the University of Queensland, initially in collaboration with ICRISAT and currently with support from the Australian Centre for International Agricultural Research (ACIAR), has been directed at the improvement of short-season, photoperiod-insensitive (or nearly so) genotypes and the development of appropriate agronomic systems. Such material allows entirely new production systems, considerably broadens the adaptation of the crop, enables mechanical harvesting, and can result in very high seed yields in favourable environments with effective crop management (Wallis *et al.* 1983).

Despite these advances, scientific knowledge of the physiological, agronomic and genetic limits to productivity in these systems is limited, as it is for pigeonpea in general. Simply, pigeonpea is under-researched and relatively primitive in its development for domesticated agriculture. This is related, in part, to its perennial habit. Current research at the University of Queensland is designed to provide a better understanding of the scientific basis of yield accumulation and improvement within the short-season production systems, and of their extrapolation to other environments. The research is multidisciplinary and includes studies of growth and development in different production environments, regional evaluation of genotype × environment interactions and environmental adaptation, genetic analysis of yield and its components, definition of appropriate strategies and methods of improvement, as well as studies on the utilization of the crop and its products, including human and animal uses for Australia and in Thailand, Indonesia and Fiji.

Kenaf for Pulp Production

Kenaf is a traditional fibre crop that has been extensively grown in Asia for its bark fibre which is used for the production of sacks, cordage and burlap. Work conducted in the USA between 1957 and 1980 showed that the stems of kenaf could be used to produce a wide range of pulp and paper products. Kenaf is well adapted to the tropics and following a detailed appraisal was selected as having potential as an agro-industrial crop for the Ord Irrigation Area (OIA) in northern Western Australia (Wood & Angus 1974). Subsequently a field programme was conducted by the CSIRO Division of Tropical Crops and Pastures at Kimberley Research Station in the OIA between 1972 and 1980. This was supported by research on the pulping and papermaking properties of kenaf conducted by the CSIRO Division of Chemical and Wood Technology. Harvesting studies with sugarcane harvesters were conducted in Queensland in collaboration with Toft Bros and Massey Fergusson Pty Ltd, the manufacturers of sugarcane harvesters. Economic analyses of the costs of production and processing and market appraisals of the pulp were also undertaken.

The field and industrial studies, and the harvesting trials were promising, as were the economic and market assessments. However, there was little commercial interest in establishing a pulping industry in northern Australia and in 1980 the decision was made to terminate the research programme.

Since then circumstances have changed and in recent years there has been an upsurge of interest. The changes that have occurred are:

(a) A sharp drop in the value of the Australian dollar which has substantially increased the cost of imported pulp and improved the competitiveness of Australian- produced pulp on world markets.

(b) Strong falls in the world price of sugar which has encouraged the sugar industry and governments to examine alternative crops that might be grown instead of sugarcane.

(c) Establishment in 1982 of a pulp mill in north-east Thailand to pulp kenaf. This mill has proved the commercial feasibility of pulping kenaf and marketing the pulp.

For the past three years an Australian company has been evaluating the possibilities of growing kenaf for pulp production in the Burdekin. A pilot pulp plant is planned for installation early in 1988 to process the production from 500 ha planned to be sown in late 1987. If the pilot plant trials prove successful the company plans to establish a pulp mill in the area.

This research with kenaf in Australia well illustrates some of the difficulties with new crop research and development that we have mentioned previously. These include:

(a) The uncertainty of commercial success.

(b) The necessity for a multidisciplinary research and development programme.

(c) The long lead time that is often involved between the initial studies and commercial production.

THE DOs AND DON'Ts OF A NEW CROP PROGRAMME

We draw together and summarize our main recommendations with a list of do's and don'ts for a new crop programme:

DO : Carefully review past work on the crop and build on that experience.

DO : Plan a multidisciplinary integrated programme from the outset.

DO : Involve industry (farmers, processors, etc.) from the outset.

DO : Start at an appropriate spot in the crop improvement framework.

DO : Develop a staged strategic plan of R & D.

DO : Stop the programme at a propitious time but make sure that staff are given time to complete their research.

DO : Write up a collated report on the work including comment on experience, difficulties, conclusions, etc.

DO : Retain the germplasm collection.

DO : Remember the long lead time before commercial development.

DON'T :   Underestimate the time and difficulties.

DON'T :   Underestimate the importance of having a close involvement with industry.

DON'T :   Rush into the release of a new cultivar until there has been an adequate appraisal of quality, disease, resistance, etc.

DON'T :   Forget that many agricultural problems such as insects and disease are a function of scale of operations.

DON'T :   Forget that many plant chemicals can be synthesized.

DON'T :   Forget that a new crop and new technology can be utilized by other countries – a comparative advantage is desirable.

New crops offer exciting prospects for increasing agricultural productivity and meeting the needs for industry and the community. However new crops research is expensive and there is no certainty that a particular crop will prove to be a commercial success.

ACKNOWLEDGEMENTS

The financial assistance provided by the Australian Centre for International Agricultural Research (ACIAR) to enable one of us (ESW) to attend this symposium is gratefully acknowledged.

REFERENCES

Anon (1984) *Growing Industrial Materials: Renewable Resources from Agriculture and Forestry. An Initial Report of the Task Force on the Role of American Agriculture and Forestry in Maintaining Supplies of Critical Materials.* USDA Critical Materials Taskforce, USDA, Washington, DC.

Anon (1987) Oilseed Industry. In: *Australian Agriculture*. Morescope Pty Ltd., Camberwell, Victoria.

Byth, D. E., E. S. Wallis and K. B. Saxena (1981) Adaptation and Breeding Strategies for Pigeonpea. In: *Proceedings of International Pigeonpea Workshop, ICRISAT Patancheru, India 15–19 December 1980*, pp. 450–465. ICRISAT, Patancheru.

Byth, D. E., E. S. Wallis and I. M. Wood (1987) The Workshop – An Overview. In: *Proceedings of Food Legume Improvement for Asian Farming Systems*, E. S. Wallis and D. E. Byth (Eds.), ACIAR Proceedings Review No. 18, Australian Centre for International Agricultural Research, Canberra.

Ferraris, R. (1986) Review: Some aspects of utilization of sorghum forage in Australia. In: *Proceedings of 1st Australian Sorghum Conference, Gatton, February, 1986*, M. A. Foale and R. G. Henzell (Eds), pp. 2.14–2.20. Queensland Branch, Australian Institute for Agricultural Science, Gatton.

Hamblin, J. (1987) Grain Legumes in Australia. In: *Proceedings of the 4th Australian Agronomy Conference, La Trobe University, Melbourne, Victoria, August, 1987*, pp. 65–82. Australian Society of Agronomy, Melbourne.

Lawn, R. J. and D. E. Byth (1979) Soybean. In: *Australian Field Crops 2*. J. V. Lovett and A. Lazenby (Eds). Angus and Robertson, North Ryde, NSW,

Lawn, R. J., J. D. Mayers, D. F. Beech, D. E. Garside and D. E. Byth (1986) Adaptation of soybean subtropical and tropical environments in Australia. In: *Soybean in Tropical and Subtropical Cropping Systems,* S. Shanmugasundaram and E. W. Sulzberger (Eds). AVRDC Shanhua, Taiwan.

Wallis, E. S. and D. E. Byth (Eds) (1987) *Proceedings of Food Legume Improvement for Asian Farming Systems.* ACIAR Proceedings Review No. 18. Australian Centre for International Agricultural Research, Canberra.

Wallis, E. S., D. E. Byth, P. C. Whiteman and K. B. Saxena (1983) Adaptation of Pigeonpea (*Cajanus cajan*) to mechanized culture. In: *Australian Plant Breeding Conference, Adelaide, 14–18 February, 1983,* pp. 142–145. University of Adelaide, Adelaide.

Whiteman, P. C., D. E. Byth and E. S. Wallis (1985) In: *Grain Legume Crops,* R. J. Summerfield, and E. H. Roberts (Eds), pp. 658–698. Collins, London.

Wood, I. M. (1984) Kenaf and guar. In: *Proceedings of the Symposium Jojoba, guayule or what? New crops – factors for survival,* J. R. Cook (Ed.), pp. 33–45. AIAS Occasional Publication No. 16.

Wood, I. M. and J. F. Angus (1974) *A review of prospective crops for the Ord Irrigation Area. II. Fibre Crops.* CSIRO Australian Division of Land Use Research of Technology, Paper No. 36.

Wood, I. M. and A. B. Hearn (1985) Fibre crops. In: *Agro-research for the Semi-arid Tropics,* R. C. Muchow (Ed.) University of Queensland Press, St. Lucia.

Wood, I. M., J. F. Angus and A. G. L. Wilson (1974) *A Review of Prospective Crops for the Ord Irrigation Area. I. Regional Data and Method of Review.* CSIRO Australian Division of Land Use Research and Technology, Paper No. 35.

5

Assessment of New Crops for Plantations

R. H. V. Corley

INTRODUCTION

The humid tropics are climatically well suited to agriculture, with abundant rainfall, reasonable radiation levels and favourable temperatures throughout the year. However, the majority of tropical soils are poor in nutrients and fragile in structure. Many tropical soils are described as unsuited to continuous intensive agriculture (e.g. Jordan & Herrera 1981), and it is in the wet tropics that some of the most ecologically disastrous agricultural developments have taken place. Although the climax vegetation on such soils is usually forest, and while the soils may be unsuited to cultivation of annual crops, trees can be grown successfully. The ecological arguments in favour of perennial cropping in the tropics are several.

First, trees, because they do not have to be replanted and established every year, take full advantage of the favourable climatic conditions in the tropics by maintaining a complete leaf canopy throughout the year, maximizing interception of solar radiation, upon which plant growth depends. Mature oil palms intercept up to 95 per cent of incident radiation, while average interception over a 25 year planting cycle is between 75 and 88 per cent (Squire & Corley 1987). In contrast, tropical annuals may intercept as little as 20 per cent (Monteith 1972). Thus, the yields obtainable from perennials are appreciably higher than those from annual crops (Corley 1985).

Another advantage lies in the conservation of soil nutrients. Jordan & Herrera (1981) showed that most of the calcium in a tropical rain forest ecosystem was bound up in the vegetation and the organic matter of the leaf litter layer. Roots within the leaf litter form a highly efficient recycling system, so that loss of nutrients by leaching is negligible. Once the forest is destroyed, the organic matter decays quickly (or it may be burnt at the time of clearing) and nutrients are then rapidly lost. Within a few years after clearing, the soil may become too infertile to support unfertilized annual crops (Sanchez 1976). Recently, Teoh & Chew (1987) have shown that up to 75 per cent of the potassium in an oil palm plantation is held within the palm biomass, the ground vegetation and decaying leaf litter, just as for calcium in the rain forest (Table 5.1). As in the forest, loss of nutrients by leaching may be negligible (Chang & Chow 1985).

Table 5.1: *Distribution of Nutrients in Two Tropical Ecosystems*

	Amazonian rain forest[1] % of total Ca	Malaysian oil palm plantation[2] % of total K
Trunks, branches	35	41
Leaves	8	11
Roots	18	15
Fruit	–	8
Leaf litter and humus	4	4
Total in organic form	65	79
Soil exchangeable	35	22

1. Data from forest on tropical humus podsol in Venezuela (Jordan & Herrera 1981).
2. Data from 15 year old plantation, without added fertilizer, on Tropeptic Haplorthox (Munchong series) soil (Teoh & Chew 1987).

Another problem is soil erosion; with very intense rainfall in the tropics, erosion of bare soil is rapid. The relatively fertile top soil is washed away, rivers are silted up, and flooding of low-lying areas follows. A complete ground cover will prevent erosion, but with annual crops the soil is bare or partially bare for much of the year. A tree crop may not stop erosion; under a high leaf canopy raindrops still fall with sufficient force to loosen soil particles, which are then lost in runoff. Good management is the answer. A leguminous cover crop grown between young trees will both control erosion and improve soil nitrogen status. Under mature trees, sensible distribution of crop residues and prunings as a mulch to protect the soil can eliminate erosion completely.

Finally, concern is often voiced about the increase in atmospheric carbon dioxide level which follows forest clearing, and the consequent greenhouse effect on world climate. Establishment of a tree crop, which will build up a large standing biomass, will help to minimize this effect.

For these reasons, therefore, I believe we should not consider monocropping of annual species in the humid tropics (with the possible exception of rice in wet lowlands). Annual crops do have a place in agricultural systems based on perennial crops; they can be intercropped between immature trees, or under-planted in mature plantations or 'agroforestry' systems.

There are several well established tropical plantation crops: oil palm, rubber, cocoa, coffee, tea, coconuts, bananas. These crops are widely grown, both on large plantations and by small farmers. Why, therefore, should we look for new tropical perennial crops? Answers to this question should help us decide which crops are likely to be worth developing.

First, the products of the major tree crops are not an adequate basis for a healthy diet. There is nothing wrong in growing cocoa, selling the product, and using the cash to buy surplus grain from Europe, but if we are to take full advantage of the favourable agricultural environment in the tropics, a wider range of perennial food crops needs to be developed.

Next, although all the above crops have been profitable in the long term, commodity prices vary widely, and in individual years prices are sometimes below production cost. There are usually some commodities whose prices move against the trend (Fig. 5.1), but introduction of new crops should help to increase protection against price fluctuations.

Figure 5.1: Prices of Three Commodities on an Arbitrary Scale to Illustrate Relative Price Movements

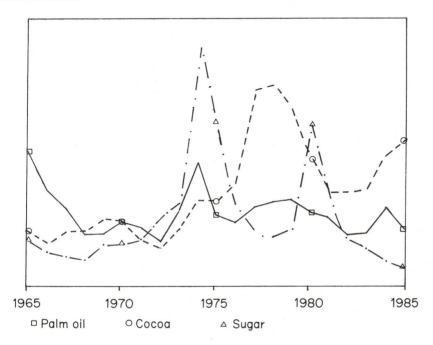

□ Palm oil   ○ Cocoa   △ Sugar

The third major reason for crop diversification is to extend the range of environmental conditions which can be cultivated. For example, there are large areas of swamp in the tropics where some tree crops cannot be grown at all, while others can only be grown after extensive and expensive drainage. A swamp-dwelling species such as sago could be grown in many such areas, while for brackish, tidal swamps *Nypa fruticans* is a possibility. The inflorescences of this palm can be 'tapped' to yield a sugar-rich sap (Paivoke 1985).

A new crop as an alternative source of an existing commodity must have some special attribute if it is to succeed. The established crop probably has a long history of breeding and agronomic research behind it, which will place any alternative at an insuperable disadvantage. For example, many different species of oil bearing palm have been proposed as potential crops, but plant breeders have increased the yield of *Elaeis guineensis* by at least 140 per cent over the past 60 years (Hardon *et al.* 1982). A new crop would start with unselected material, comparable to the oil palms of 60 years ago.

55

One situation in which a new crop producing an existing commodity might be worthwhile is to extend the range of cultivable environments. For example, *Mauritia flexuosa* and *Manicaria saccifera* are both swamp-dwelling palms which produce a vegetable oil. If these can be grown in areas unsuitable for oil palm, they could be useful.

Given three good reasons for crop diversification, I will now discuss the criteria which a new crop must meet if it is to be adopted. Although most of my examples are from tropical tree crops, much of what I have to say should apply to new crops.

The main criterion will be profitability. The new crop must have a good chance of giving a return comparable to that which can be obtained from well known crops; if it does not, those crops will be preferred. Depending on circumstances, the grower may look for maximum return per hectare, per dollar invested, or per man-day of labour.

THE MARKET

Many papers on potential new crops do no more than describe the product. A grower will need to know what yield of product he can obtain, and what price he will get for it.

Sometimes, there already exists a trade in the product, perhaps harvested from wild stands. Sago starch, illipe nuts (fruits of dipterocarp species, containing a cocoa butter substitute) and rattan cane for furniture are examples. In such cases, historic price data will be available, and may be used to estimate future trends, but there are several complicating factors.

The quality of a plantation-produced product may be very different from that of the same material harvested from wild stands. For example, sago starch is often not extracted until days or weeks after the tree is felled, and is usually dirty and partly degraded. As a result, it is one of the cheapest forms of starch. According to Jones (1983), the CIF price for Malaysian sago starch in the EEC between 1975 and 1980 was 22 per cent less than that for Thai tapioca starch, and in Japan 30 per cent less. In a pilot plant in Papua New Guinea, trunks are processed on the day of felling (Power 1986). This gives a much better quality starch, very clean, and with quite different properties to those expected for sago starch. Quoted prices for the standard product would be no guide to the value of this new material.

The next point to be considered, where a market already exists, is the volume of trade in the product. Commodities such as spices and perfumes often command high prices, but in very small volume. Sandalwood oil sells for up to £180000 per ton, but world production is probably less than £200 ton/year. In contrast, palm oil at present sells for only £220/ton, but with world production of about 8 millions tons, the total value produced is over 50 times that of sandalwood oil.

56

The investor must therefore consider the extent to which the planned plantation will increase the total volume of the commodity available. Will this increase be sufficient to trigger a price collapse? Unless there is some flexibility in demand for the product, the main effect of a large scale project may be to put small and inefficient producers out of business, certainly not a desirable result of any investment. Conversely, the concept of 'the market' is a passive one. If the grower is prepared to play a constructive role in marketing the product, he may well be able to increase total demand.

Where a crop produces a substitute for an existing commodity, such as a vegetable oil, it may be important to ensure that the new product is truly interchangeable with the old. If it differs from the standard commodity, is the difference disadvantageous, or will it enhance the value of the new product?

If no market in the product exists, then the investor must either be committed to creating a market himself, or be convinced that some other agency will do this for him. A good example here is the kiwi fruit, where the little known Chinese gooseberry was renamed and, by skillful marketing, turned into a luxury product.

YIELDS

An investor must have estimates of yields per hectare, in order to calculate the gross return on his investment. As Carlowitz (1986) has pointed out, such data are rarely available and where they do exist may be difficult to interpret. It is often not clear whether yield data refer to dry or fresh weight of product, at what age and stage of development the trees were recorded, how large was the sample, nor how frequent the harvesting. Yields of tree crops vary greatly from year to year, so data from one or a few trees in a single year are of little use. Yields also vary enormously with climate and soil type, with significant genotype × environment interactions in many crops. Planting density may have large effects within a single environment. Yields from single trees are almost meaningless, without some indication of tree density per unit area. Yields per hectare estimated by multiplying the yield per tree by an assumed, and often apparently arbitrary, number of trees per hectare are equally meaningless.

In indeterminate species, such as most perennial crops, yields of fruit are strongly dependent on planting density and the level of interplant competition (Corley 1983). The highest yield per hectare will usually be obtained with close planting, where yield per tree is much less than the maximum possible (Table 5.2).

Given these uncertainties, it is perhaps not surprising that the yields quoted for a single species may range over up to two orders of magnitude (Carlowitz 1986). The only yield data of practical value to the potential investor are likely to come from properly designed field experiments, including a range of plant densities, and preferably of sufficiently flexible design to allow other aspects of

Table 5.2: *Yield of Oil Palms at Two Planting Densities in Papua New Guinea*

	Trees/ha	Yield of Fruit	
		kg/tree	t/ha
Isolated trees	56	285	15·9
Optimal stand[1]	127	181	23·0

1. Optimum estimated by fitting curve to a range of densities.
Source: Breure (personal communication).

field management to be studied. A recent series of papers on roselle (*Hibiscus sabdariffa*), an annual fibre crop grown in Thailand, shows the sort of simple experimentation which is needed initially (Sermsri *et al.* 1987a,b&c).

THE IMMATURE PERIOD

All crops pass through an immature period before harvest of the commercial product starts. For tree crops, this period ranges from less than two years, for cocoa under good conditions, to 15 years or more for some tropical fruit trees. The length of this period is critical in determining the payback period for the investment. At present, the immature period appears discouragingly long for many potential crops. Flach (1977) estimates that sago can be first harvested eight years after planting, while Purseglove (1972) quotes the peach palm (*Bactris gasipaes*) as starting to fruit five to eight years after planting.

With established crops, research has usually brought about large decreases in immature period. Dwarf × tall hybrid coconuts come into production about three years sooner than the old tall varieties, while application of fertilizers can increase early yields significantly (Table 5.3). Over seven years changes in planting techniques have brought the immature period for rubber down to between four and five years. Improved management of young palms in the field has reduced the immature period for oil palm from four years to less than 2·5 years, and recent developments in nursery techniques could bring this down even further (Nazeeb *et al.* 1987). Bud-grafted fruit trees often have much shorter immature periods than seedlings.

Table 5.3: *Effect of Nitrogen and Potassium Fertilizers Applied at Different Levels during the Immature Period, on Early Yield of Hybrid Coconuts, 4·5 to 5·5 years after Planting in the Solomon Islands*

		Yield (tons dry copra/ha)			
Nitrogen level		0	1	2	3
Potassium level	0	0·82	1·26	2·17	2·01
	1	1·40	2·23	2·56	2·51
	2	1·20	2·40	2·01	2·81
	3	1·08	2·44	3·02	3·03

Source: Lever Solomons Limited.

PRODUCTION COSTS

Given estimates of yield, of the value of the product, and of the period before production starts, the investor then needs some idea of establishment and production costs, to give the other side of the cash flow equation.

Harvesting

Harvesting of most tropical tree crops is a labour intensive operation, but labour has been cheap in the tropics in the past. This situation will not persist, so with any potential new crop, it would be sensible to consider the possibilities of mechanizing harvesting. Looking at existing crops, tea plucking can already be done mechanically, while mechanical harvesting of coconuts is easily envisaged, at least for those varieties where ripe nuts fall to the ground. Mechanization of oil palm harvesting is receiving a lot of attention at present, but with little success as yet. As for rubber tapping, it is difficult to see how this could be completely mechanized, though various gadgets have been developed to make the tapper's task easier.

Frequency of harvesting is an important variable, ranging from every other day with some rubber tapping systems, to less than once a month for coconuts. The cost of harvesting varies from less than 30 per cent of total production costs with oil palm and cocoa, to over 50 per cent with rubber.

Transport

Transport of the product between field and factory may present problems. Flach (1977) estimates that sago could yield up to 25 tons of starch per hectare per year. The data of Sim & Ahmed (1978) indicate that the fresh weight of trunks to give such a starch yield could be over 200 tons. If the crop is grown in swamps, construction of suitable roads for transport of such a tonnage would be prohibitively expensive. The obvious alternative, currently used by sago collectors, is to float the logs out, but construction of canals in a plantation might cost nearly as much as draining the area to plant a dry land crop.

The cost of transport from factory to consumer should not be forgotten. Many tropical plantations have poor communications, and transport costs may price the product out of the market.

Processing

The cost of post-harvest processing of the product, and of the investment in processing equipment, may be considerable. First, the investor must identify, or be prepared to develop, suitable equipment for large- scale processing. There may be a choice of products: with coconuts, fresh nuts can be sold if there is a local market. If not, the copra must be extracted and dried, and the grower then has the option of further processing to produce coconut oil and copra cake, or other products such as desiccated coconut and coconut cream. Future peach palm planters will have to choose whether to produce starchy or oily fruit (Arkcoll & Aguiar 1984), and may also wish to harvest palm hearts. Potential by-products should not be forgotten: the coconut grower can produce coir from the husk, which at times may have a greater value than the copra itself.

Where a complicated process is involved, there is likely to be a minimum scale of operation below which the process cannot be done economically. For example, it is generally considered that the minimum size of oil palm plantation to supply a processing mill is about 2000 ha, though clearly this will depend on the yield per hectare. On the other hand, good quality cocoa can be produced by crude fermentation methods from a smallholding of a few hectares. The total cost of the processing plant will depend on the complexity of the process. An oil palm mill for 2000 hectares would cost over £2 million, while a cocoa fementary for the same area of plantation might be built for less than £200 000.

Fertilizers

Some indication of fertilizer requirements must be established in early trials with any new crop. Prediction, without experiment, may prove difficult. In most areas of the world, oil palms require phosphate, and often show no response to other fertilizers unless phosphate is also supplied (Foster & Goh 1977). In the Llanos orientales of Colombia, on soils previously planted with rice, oil palm yield is actually reduced by phosphate fertilizer (Vallejo, personal communication).

The cost of fertilizers, applied at the most profitable rate, may range from zero, for oil palms planted after forest on young volcanic soils, to over £100/ha/year for oil palms and cocoa on many poorer soils.

Pollination

Natural pollination does not always ensure adequate fruit-set, and some crops, such as vanilla, are normally hand pollinated. Hand pollination comprised over 10 per cent of the production cost of palm oil in Sabah and Papua New Guinea, until the discovery and introduction of the pollinating weevil (Syed 1979; Syed *et al.* 1982).

When an entomophilous crop is introduced to a new area, therefore, it will be important to ensure that suitable pollinators are present. However, it is worth noting that the oil palm was cultivated successfully in Asia for 60 years before its pollination was understood.

Pruning

Regular pruning is essential with some crops. Old leaves must be pruned from oil palms to allow ripe fruit branches to be seen and harvested. If cocoa is not pruned, vegetative growth predominates, and yield may deciine drastically. Costs of pruning range from nothing with coconuts, where old leaves fall naturally, to over 10 per cent of production costs with cocoa.

Pest and disease control

Costs here are difficult to predict. In general, tropical plantation crops remain relatively disease free, and pesticide use is negligible. Even with well- known crops, though, unexpected pest or disease outbreaks may cause disastrous losses. The complete disappearance of Britain's elm trees and the damage caused to maize by southern corn blight in the 1970s are examples.

Jennings & Cock (1977) have pointed out that pest disease problems tend to be greatest in the area of origin of the species. Thus it is impossible to grow rubber commercially in some parts of South America, its area of origin, because of South American leaf blight. In general, therefore, new crops might have relatively few problems provided they are planted outside their area of origin. However, this is not always the case. The African oil palm suffers several severe diseases in South America, and more than one plantation has been completely wiped out by disease (Turner 1981).

Resistance breeding is likely to be the most effective approach to disease problems, as with Fusarium wilt of oil palm (Meunier *et al.* 1979; Franqueville & Greef 1987), but this will not usually start until the second phase of research, when the value of the crop has been established.

Planting

If estimates of price, yield and production costs indicate that an investment would be worthwhile, then a source of planting material must be found, and methods of planting investigated.

Choice of varieties

In the early stages of development of a new crop, the grower may have to take whatever planting material is available, but any crop already exploited is likely to have been subject to some degree of selection. The sago farmers in Southern Malaysia select high yielding, spineless palms and propagate them by suckers (Flach 1977). As already noted, some peach palm varieties produce an oily fruit, whereas others are starchy (Arkcoll & Aguiar 1984).

In a discussion of tropical fruits, Proctor (1985) advises that in choosing varieties one should consider ease of harvest, susceptibility to post-harvest disease losses, ease of shipping and storage, shelf life, size, colour, flavour and uniformity. Most of these criteria will be applicable to any crop, though data may often be inadequate to allow assessment.

Propagation

There are several possible methods of propagation. The simplest, and usually the cheapest, method is to grow plants from seed. This can be done with almost all crops, but, depending on the breeding system of the species, may have some disadvantages. Naturally cross-pollinated species, such as the oil palm, are highly heterozygous; in such crops there will be much variation in yield between trees within seedling progenies, and lack of uniformity of the product may cause problems in harvesting, processing or marketing. With some annual crops, such as maize, inbred lines have been developed, and are crossed to produce uniform F1 hybrids, but this approach has not been taken with tree crops, because of the long generation time in breeding programmes.

With crops where the product is part of the vegetative plant, such as tea, sago and timber species, seed production from commercial trees is usually undesirable, and often impossible. Further, because vegetative and reproductive growth are often competitive processes within the plant, the best yielding trees may be the poorest seed producers.

The alternative to seed is vegetative propagation. Standard horticultural methods (cuttings, bud-grafting) are used with rubber, tea, cocoa, bananas, and other crops. The advantage of vegetative propagation is the uniformity obtained, but skill is needed in successful bud-grafting, and may not be available. With trees such as cocoa and coffee, where most buds are on plagiotropic branches, shaping of the tree may present difficulties.

With some trees, such as the oil palm, vegetative propagation cannot be achieved by the traditional methods. With such crops, micropropagation by tissue culture may be possible. A technique has been developed for the oil palm (Corley *et al.* 1977), and extensive trials with clones have been planted (Corley *et al.* 1987). Some work has also been done on peach palm (Arias 1985), sago (Z. C. Alang, personal communication) and rattan (Umali-Garcia 1985).

The advantage of micropropagation is that it can be applied to almost any species, but development of a technique can require a major research effort, and the culture facilities needed are expensive. With some species somaclonal variation may occur, so that the plants produced are not genetically uniform. While this is a disadvantage, it should be remembered that seedling populations usually include plants which yield poorly or not at all. Provided that the proportion of off-types from micropropagation is low, the method will still be useful. With oil palm, somaclonal variation is sufficiently rare that it has not been a problem (Wooi *et al.* 1982), but some palms have flowered abnormally as a result of an epigenetic (non-heritable) change induced by the plant hormones in the culture medium (Corley *et al.* 1986).

Nursery requirements

For most crops, nursery techniques are quite straight- forward. The objective is to raise the plants under more or less protected conditions, to a size where survival in the field will be assured, and to minimize the immature period in the field. Disease outbreaks are common in nurseries where management is sub-optimal; it is important to ensure that water and fertilizer supplies are adequate.

Field planting methods

In many environments the planting season must be carefully chosen. The plant will be transferred from a regularly irrigated nursery to dependence on rainfall alone, so must be given time to establish a good root system before the next dry season. The method of field planting, and the treatment the palm receives after planting, can have a large effect on the immature period, as noted above.

CONCLUSIONS

Much useful research can be done by universities and government institutes, but such organizations may not be able to answer all the questions which a potential investor will ask. I believe, therefore, it is essential that the private sector becomes involved in the development of new crops at a fairly early stage. Many plantation companies already undertake extensive research, and should be prepared to adapt their programmes to include trials with new crops. Such trials would be aimed at filling the gaps in the general picture. For example, it might be necessary to define yields or production costs more precisely, to identify suitable varieties or to produce samples for quality assessment and test marketing.

Historically, the private sector has made major contributions to the development of the major plantation crops, and experience with these crops shows that innovations from the private sector are rapidly adopted by small farmers. A new crop taken up by the plantation industry will have a much greater chance of success than one promoted only by aid donors and academic research workers.

ACKNOWLEDGEMENTS

Unilever PLC gave permission for publication.

REFERENCES

Arkcoll, D. B. and J. P. L. Aguiar (1984) Peach palm (*Bactris gasipaes* H. B. K.), as a new source of vegetable oil from the wet tropics. *J. Sci. Food Agric.* 35: 520–526.

Arias, O. (1985) Organogenesis del pejibaye (*Bactris gasipaes* H. B. K.) *in vitro*. Paper presented at Simposio de interciencia biotechnologia en las Americas II. Aplicaciones en agricultura tropical, San Jose, Costa Rica, July 15–17.

Carlowitz, P. G. von (1986) Multipurpose tree yield data – their relevance to agroforestry research and the current state of knowledge. *Agroforestry Systems* 4: 291–314.

Chang, K. C. and C. S. Chow (1985) Some questions of the leaching losses of soil nutrients from mature oil palm fields in Malaysia. *Oleagineaux* 40: 233–243.

Corley, R. H. V. (1983) Potential productivity of tropical perennial crops. *Exp. Agric.* 19: 217–237.

Corley, R. H. V. (1985) Yield potentials of plantation crops. In: *Potassium in the agricultural system of the humid tropics. Proceedings of the 19th Colloquium International Potash Institution*, pp. 61–80. International Potash Institution, Bern.

Corley, R. H. V., J. N. Barrett and L. H. Jones (1977) Vegetative propagation of oil palm via tissue culture. In: *International Developments in Oil Palm*, D. A. Earp and W. Newall (Eds), pp. 1–8. Incorporated Society of Planters, Kuala Lumpur.

Corley, R. H. V., C. H. Lee, I. H. Law and C. Y. Wong (1986) Abnormal flower development in oil palm clones. *Planter, Kuala Lumpur* 62: 233–240.

Corley, R. H. V., C. H. Lee, I. H. Law and E. Cundall (1987) Field testing of oil palm clones. Paper presented at International Oil Palm Conference, Kuala Lumpur, June 23–26.

Flach, M. (1977) Yield potential of the sago palm and its realisation. In: *Sago-76. Proceedings of the 3rd International Sago Symposium*, K. L. Tan (Ed.), pp. 157–177. Kemajuan Kanji, Kuala Lumpur.

Foster, H. L. and H. S. Goh (1977) Fertiliser requirements of oil palm, in West Malaysia. In: *International Developments in Oil Palm*. D. A. Earp and W. Newall (Eds), pp. 234–261. Incorporated Society of Planters, Kuala Lumpur.

Franqueville, H. de and W. de Greef (1987) Hereditary transmission of resistance to vascular wilt of the oil palm: facts and hypotheses. Paper presented at International Oil Palm Conference, Kuala Lumpur, June 23–26.

Hardon, J. J., R. H. V. Corley and C. H. Lee (1982) Breeding and selection for vegetative propagation in the oil palm. Paper presented at Symposium on Breeding of Vegetatively Propagated Crops, Long Ashton, Sept. 13–15.

Jennings, P. R. and J. H. Cock (1977) Centres of origin of crops and their productivity. *Econ. Bot.* 31: 51–54.

Jordan, C. F. and R. Herrera (1981) Tropical rain forests: are nutrients really critical? *Amer. Naturalist* 117: 167–180.

Jones, S. F. (1983) *The world market for starch and starch products with particular reference to cassava (tapioca) starch*. Report G173, Tropical Development Research Institute, London.

Meunier, J., J. L. Renard and G. Quillec (1979) Heredity of the resistance to Fusarium wilt in the oil palm, *Elaeis guineensis* Jacq. *Oleagineaux* 34: 555–561.

Monteith, J. L. (1972) Solar radiation and productivity in tropical ecosystems. *J. Appl. Ecol.* 9: 747–766.

Nazeeb, M., S. G. Loong and B. J. Wood (1987) Trials on reducing the non-productive period at oil-palm replanting. Paper presented at International Oil Palm Conference, Kuala Lumpur, June 23–26.

Paivoke, A. E. A. (1985) Tapping practices and sap yields of the Nipa palm *Nipa fruticans* in Papua New Guinea. *Agriculture, Ecosystems and Environment* 13: 59–72.

Proctor, F. J. (1985) Post-harvest handling of tropical fruit for export. *The Courier (EEC)* 92: 83–86.

Power, A. P. (1986) Strategy for sago development in the East Sepik province, Papua New Guinea. In: *Sago-85. Proceedings of the 3rd International Sago Symposium*, N. Yamada and K. Kainuma (Eds), pp. 105–108. Sago Palm Research Fund, Tokyo.

Purseglove, J. W. (1972) *Tropical Crops – Monocotyledons*. Longman, London.

Sanchez, P. A. (1976) *Properties and Management of Soils in the Tropics*. Wiley, New York.

Sermsri, N., C. Duyapat and Y. Murata (1987a) Studies on Roselle (*Hibiscus sabdariffa* var. *altissima* L.) cultivation in Thailand. II. Effect of planting time, harvesting time and climatic factors on fibre yield. *Jap. J. Crop Sci.* 56: 64–69.

Sermsri, N., S. Jatuporupongse and Y. Murata (1987b) Studies on Roselle (*Hibiscus sabdariffa* var. *altissima* L.) cultivation in Thailand. III. Effect of row spacing on fibre yield at a definite population density. *Jap. J. Crop Sci.* 56: 70–72.

Sermsri, N., S. Tipayarek, and Y. Murata (1987c) Studies on Roselle (*Hibiscus sabdariffa* var. *altissima* L.) cultivation in Thailand I. Growth analysis. *Jap. J. Crop Sci.* 56: 59–63.

Sim, E. S. and M. I. Ahmed (1978) Variation in flour yields in the sago palm. *Malay. Agric. J.* 51: 351–358.

Squire, G. R. and R. H. V. Corley (1987) Oil palm. In: *Tree Crop Physiology*, M. R. Sethuraj and A. S. Raghavendra (Eds), pp. 141–147. Elsevier Science Publishers, Amsterdam.

Syed, R. A. (1979) Studies on oil palm pollination by insects. *Bull. Ent. Res.* 69: 213–224.

Syed, R. A., I. H. Law and R. H. V. Corley (1982) Insect pollination of oil palm: introduction, establishment and pollinating efficiency of *Elaeidobius kamerunicus* in Malaysia. *Planter, Kuala Lumpur* 58: 547–561.

Teoh, K. C. and P. S. Chew (1987) Potassium in the oil palm eco-system and some implications to manuring practices. Paper presented at International Oil Palm Conference, Kuala Lumpur, June 23–26.

Turner, P. D. (1981) *Oil Palm Diseases and Disorders*. Oxford University Press, Kuala Lumpur.

Umali-Garcia, M. (1985) Tissue culture of some rattan species. In: *Proceedings of the Rattan Seminar*, K. M. Wong and N. Manokaran (Eds), pp. 23–31. Forest Research Institute, Kuala Lumpur.

Wooi, K. G., C. Y. Wong and R. H. V. Corley (1982) Genetic stability of oil palm callus cultures. In: *Proceedings of the 5th International Congress on Plant Tissue and Cell Culture*, pp. 749–750. Tokyo.

6

Criteria for the Selection of Food Producing Trees and Shrubs in Semi-arid Regions.

H. J. von Maydell

INTRODUCTION

Trees, shrubs and palms are expected to play an increasing role in future food supply, especially on marginal lands and in semi-arid regions. Although species diversity at a first glance appears to be low under unfavourable site conditions and destructive to land use practices, there are surprisingly many woody species available, already utilized and/or having a potential for future food production.

However, people tend to be amazingly selective even under stress situations of drought, famine and political pressure when they use food from trees. Moreover, their expectations may be quite different, including higher quantitative availability, improved quality, diversification of diets, bridging seasonal or occasional temporary gaps of supply, and low cost or even free access.

Most woody perennials may be classified as 'multipurpose trees', i.e. various benefits may be obtained from them. Not all these benefits are compatible with food production, and some may even be detrimental. Thus many different criteria will have to be observed prior to establishing new plantations or preserving and managing natural resources in a selective way. These criteria may be grouped in different ways, e.g. those that mainly refer to the ecology or site-orientation and those based on demand-orientation, biological, technical, economic, cultural and other characteristics, or, last but not least, the individual evaluation by the people concerned. In recommending food producing species for semi-arid Africa, a simple scheme has been developed over the past 15 years (Fig. 6.1).

For proper assessment of the qualification of a specific species following the scheme shown in Fig. 6.1, a corresponding level of information would be crucial; however, there are still big gaps in our knowledge. It is for this reason that the International Council for Research in Agroforestry (ICRAF) is working on a comprehensive programme of computerized data collection, processing and dissemination (Carlowitz 1985, 1987 & in press; see also Webb *et al.* 1984).

Figure 6.1: *Relevant Criteria in the Selection of Food Producing Trees and Shrubs.*

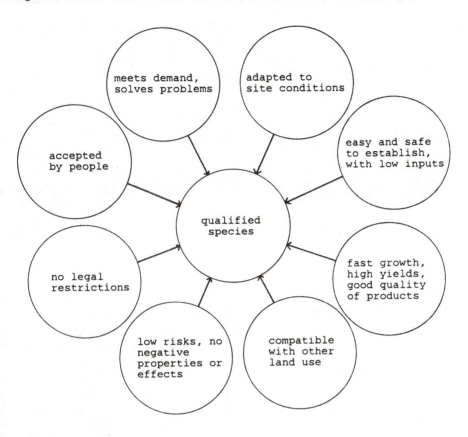

Source: Maydell (1986).

CHARACTERISTICS OF SPECIES TO BE CONSIDERED FOR QUALIFICATION

These important features are derived from various data collections, literature, and empirical observation. Such characteristics include physiology and morphology, phenology, specific potentials and limitations within the biophysical range of a species. All these may be further subdivided. For fruit or leaf yielding trees it may be important to know whether the plant is a single or multi-stemmed tree or shrub, the average height at maturity, flowering and fruiting months, rooting habit, crown form, leaf retention, etc. Thorns and spines, toxic parts of the species, allopathy and incompatibility with other crops may limit its uses.

Species should meet demand, solve problems

In a very unconventional way, 'demand' can be expected where no (or insufficient) alternatives exist to cover prevailing or foreseeable needs of the people concerned, 'problems' arise when the demand reaches levels which become critical ('supply bottlenecks') to the carrying capacity of a region and its further development.

People will only start selecting trees if there is a demand which can be rightfully expected to be met by benefits from such trees. Therefore demand orientation has to be the first criterion for selection. This has been overlooked from time to time, especially by foresters, consequently leading to conflicts and disappointment. In fact, the first question to be asked and answered should be: 'Why do we wish to have trees?' This is especially important in semi-arid regions with marginal site conditions and overall limited resources, including labour and funds. The theme of the 8th World Forestry Congress *'Forests for People'* should be given first rank in planning forestry as well as in tree cultivation outside forest lands.

However, identifying demands and/or problems, and (sometimes even more important) what is promising, may not be easy. There are various planning levels to be considered, from the individual land owner or tenant, over family/ household units, clans, communities, districts, state governments to national and even international bodies. It makes a difference whether one starts from 'the top', e.g. the government, with the identification of goals and making final decisions, or from the 'grassroots', and there is a lot of discussion about which way should be chosen.

Moreover, as a rule more than one benefit will be derived from the trees selected, especially in integrated rural land use systems like agroforestry, where multipurpose trees are considered of prime importance. If so, however, conflicts may arise because of incompatibility of these uses, e.g. fuelwood and fruit production, browse and revegetation. The main objectives for tree planting in semi-arid regions are fuelwood, poles and posts, forage, live fences/hedges, windbreaks, shade, soil protection and soil improvement and – so far *inter alia* – food production.

Harmonizing and optimizing selection criteria and essential demand factors is thus time-consuming, sometimes controversial, however indispensable within the planning process.

Should, for example, an urgent demand exist to close the food supply gap in the dry season, a decision will have to be made on whether to aim at fruit or leaf production, taking into account the nutritional (dietary) habits of the people concerned as well as potential substitutes. The above scheme will therefore have to be further specified, for instance as follows:

Demand

1. Food
2. Energy
3. Biotechnical raw materials
4. Environment
5. Socio-economic/cultural benefits

1. *Food* (to be assessed by quantity, quality, locality, time, value)

   A. Direct food production

      (i) Edible parts

         (a) Fruit, nuts seeds
         (b) Leaves, pods, flowers
         (c) Other (roots, barks, various exudates, etc.)

      (ii) Type of food (raw, processed)

         (a) Fruits, nuts
         (b) Vegetable
         (c) Beverages (teas, juices, syrups, etc.)
         (d) Spices, food additives

      (iii) Constituents (components, digestibility, palatability)

         (a) Carbohydrates
         (b) Fats
         (c) Proteins
         (d) Vitamins
         (e) Minerals
         (f) Drugs
         (g) Water

   B. Indirect contribution to food production

      (i) Fodder

         (a) For livestock
         (b) For game, fish, etc.
         (c) Other, e.g. bee forage

      (ii) Protective and ameliorative functions on agriculture, horticulture, animal husbandry

         (a) For plants
         (b) For animals

(iii)   Auxiliary functions of food supply

   (a)   Food preservation
   (b)   Food storage, transport, packing
   (c)   Food preparation.

This checklist is by no means exhaustive and may, moreover, be further subdivided.

Species adapted to prevailing site conditions

This concerns the natural site as well as the socio-economic site. Burley & Wood (1976) have suggested specific data to be assessed with regard to the natural site type. ICRAF in its computerized data collection on multipurpose trees records:

1.   region

2.   latitude, longitude, altitude

3.   annual rainfall, seasonality, longest dry period, Koeppen codes

4.   annual mean maximum and minimum temperatures, mean temperature of coldest and warmest month, absolute minimum temperature

5.   soil texture, reaction, salinity, depth, drainage, FAO/UNESCO soil type classification.

The socio-economic site is characterized by man's activities and attitudes to his environment, such as settlement, land use, tenure, infrastructure, pollution, preservation, etc. Previously, the main criterion for species selection used to be how far the known requirements of a specific tree could be met by the prevailing natural site conditions. Some species, however, proved to grow and perform surprisingly well in entirely different environments (e.g. *Pinus radiata*), others could be adapted through massive management inputs (e.g. *Phoenix dactylifera*) or by specific breeding (e.g. *Mangifera indica*). Some species show a great flexibility/elasticity in site adaptability, others exhibit narrow limitations.

There is ongoing discussion on whether endemic, indigenous or exotic species should be preferred from the ecological point of view, and a number of good reasons can be quoted for each.

Easy and safe to establish, low inputs required

With the exception of trial plots, arboreta, etc., or in other words, if large scale plantation or successful extension is envisaged, ecological and economic considerations will have to be balanced. There is an apparent gradient from high-income (urban) centres to poor rural areas. Obviously, for ornamental plantings in a villa garden or urban park much more could be spent in terms of money, resources (e.g. water, fertilizer) and labour than for fuelwood plantations in remote villages. With regard to food producing trees, shrubs and palms, questions of diversity, specialization, new introductions, etc. will have to be considered. Again there is a better chance for more flexibility within high-income private or government plantations than in home gardens or for single trees of the rural poor.

Establishment does, moreover, not only include the production of seedlings and the planting procedure but includes the subsequent management. In order to promote lesser-known food yielding trees in semi-arid regions, particularly multipurpose woody species, potential species would have to go through the rather time-consuming and resource-absorbing process of germplasm development plus *in situ* as well as *ex situ* conservation of the germplasm (Carlowitz 1985); seeds of good quality (and the best provenance) will have to be available. Vegetative propagation might be desirable or species allowing direct sowing be preferred. Costs generally increase sharply if irrigation is needed, if fencing is necessary to avoid livestock or game damages to the young plants, or if chemical treatment, fertilizers, etc., are required. Planting with bare roots is cheaper than container-planting, however the latter may reduce management costs by faster growth, due to the possibility of inoculation with mycorrhiza, fertilizer application, better root performance and improved soil moisture preservation.

Fast growth, high yields

One of the main constraints for more tree planting in most semi-arid regions is the slow growth of trees when compared to annual crop plants. First fruit yields or other benefits can only be expected about four to five years after establishment (edible leaves somewhat earlier), or even after 15 or more years. This is too long for many people, e.g. migrating groups or tenants who may only have the right of using a specific plot over one or two years. Moreover, if investments are made, early returns should keep interest rates low. High yields or optimal other benefits will of course be expected according to the objectives defined under demand. 'High' may be defined in various ways, quantitative as well as qualitative and/or in terms of net revenue.

Firewood could be taken as one example. High yields in terms of volume produced per unit area and year are only one indicator. The fuelwood weight (closely correlated with its calorific value), the way it burns (sparking or not, with or without smoke, etc.) and finally the ratio of total production cost against market price will have to be evaluated. In the Sahel zone small wood pieces up to about 7 cm in diameter proved to be the most economical. People preferred these because no further processing (sawing, splitting) was necessary and the fuel quality was superior to wood produced in rotations oriented to maximum volume yields.

Quantitative data on food production (or the production potential) of woody perennials, other than those established in horticulture, orchards and commercial plantations, are almost non-existent (Carlowitz 1986 & 1987). Moreover, a clear distinction will have to be made between crude material harvested, the edible components only, the end product, fresh or dry weight, etc.

Food, i.e. fruit, nuts, leaves, buds, flowers, various saps, gums, etc., are specifically subject to quality requirements. High quality can be achieved by selection and breeding, by grafting and by applying optimal management techniques. Another criterion may be the seasonal availability. Early fruiting/maturing species will thus be preferred to those whose yields fall into the main production, i.e. low price periods. Outstanding success stories exist on the results of genetic improvement and intensified management for some horticulturally developed fruit trees like mango, citrus, *Ceratonia siliqua*, date palm, etc. The oil palm, *Elaeis guineensis*, extending from sub-humid to semi-arid coastal areas of West Africa (Senegal) is in this regard a tropical development success story (Carlowitz 1985).

Compatible with other land use

Compatibility is not only important for systems of integrated land use like agroforestry, but as a general rule. The mutual feedback between components of an ecosystem always include competition, complementarity and dependence. All three may have positive as well as negative effects on the overall management and yields. The minimum expectation, however, is that the selected species should not negatively affect other crops, buildings, people, livestock, etc., and should not attract harmful insects, fungi, birds or rodents. This is closely linked with the next criterion, low risks.

In many tropical countries there is a rich experience available on growing fruit and other food-yielding trees in gardens, home-gardens and orchards. However, relatively little is known about their performance in multi-species mixtures, although diversity is generally regarded as normal and desirable. Competition may be intraspecific and/or interspecific and have a strong influence on quantity and quality of food yields. Distribution of male and female individuals is, finally, important if dioecious species are cultivated.

Low risks, no negative properties or effects

Much more may be involved than the above-mentioned compatibility. The tree selected should be resistant against biotic and abiotic damage, starting with seasonal or aperiodic drought, inundation, fire, pests and diseases (including the almost ubiquitous termites), and ending with management and marketing risks. Last, but not least, some newly introduced species may turn out to become noxious and aggressive weeds. One example – although rather for semi-humid to humid regions – is *Psidium guajava*.

No legal restrictions

Many, if not most, countries have restrictions with regard to seed or plant imports, generally specified in phytosanitary regulations. These will have to be strictly observed when aiming at working with promising exotics. Moreover, laws may also exist protecting endangered native plants and thus excluding them from management and/or export.

Traditional land use rights are also important, such as specific taboos and tenural regulations. Planting trees may, in some regions, be seen as an attempt at permanent land occupation by strangers where the (annual) allocation of fields and pastures is still subject to decisions of a local chief or a community council. Such problems arise even with regard to plantations carried out by the forest services or technical aid projects. On the other hand, in some countries trees are subject to rather restrictive regulations of forest law as long as they are alive, which means that even a private landowner who planted trees, or trees of specific species, may not be permitted to utilize and finally fell those trees.

Acceptance by people

Obviously, as long as restrictions exist, people will hardly be motivated to plant trees which fall under restrictive regulations. Moreover, rural people tend to be conservative in what they grow and use. The introduction of new species may, therefore, be difficult. On the other hand, exotics may be more attractive than indigenous species, and usually their propagation is technically more highly developed and better supported by the public services.

The final choice of species by individual landowners, however, remains strongly subjective and in some ways almost unpredictable. One example is neem (*Azadirachta indica*) in parts of semi-arid West Africa. The main reasons for preferring this species for single tree plantings is a combination of various benefits; shade and insect repellence, in addition to good availability of seedlings (and the ease of vegetative propagation), fast growth and low risk. Neem is thus more often planted than fruit bearing or ornamental, fuelwood or forage yielding trees which would have similar site requirements.

With fruit trees in semi-arid regions great differences can be encountered between parts of Africa and South Asia or tropical America. Whereas in Africa exotic or at least formerly introduced species dominate and only a few species are cultivated to any noteworthy extent (e.g. *Mangifera indica, Anacardium occidentale, Citrus* spp.) the two other regions are rich in local species which have been improved by age-old traditional practices of selection and breeding.

CONCLUSIONS

In discussing criteria for the selection of trees, shrubs, and palms for planting, conservation and management in semi-arid regions a holistic approach should be followed. Even the simplified examples mentioned above may indicate that rarely one or a few criteria would be enough for optimal decisions, and that there is a marked interface between various criteria. Simple and obvious as this may sound, in past and present practice, this has been more than often overlooked.

Selection of a species, moreover, is only a first step. The intraspecific variability of many species, especially those growing over vast areas, may be very important. Thus a number of varieties and provenances (Boland 1986) will have to be distinguished, and even within these there exist qualitative genotypic and phenotypic differences. This is extremely important for seed procurement. Age, site, performance, etc. of the parent trees will have to be observed to avoid negative selection at this stage by untrained seed collectors preferring, for example, easily accessible seeds, regardless of the quality of the mother plant or plant population. Another frequent mistake is made in distributing to the people (e.g. on arbor days or for village plantations), generally free of charge, those seedlings that were too poor to be used in commercial plantations.

Throughout the previous years much effort has been made to combat desertification, re-establish a vegetative cover, prevent further degradation, promote production or ameliorate the environment by tree planting in semi-arid regions. The first enthusiasm, based on a common belief that every tree was good and valuable and the need for immediate action has often led to uncritical choices of species, resulting in disappointment or even negative ecological effects. More consciousness of the importance of selection based on hard facts and a holistic systems approach, may help to improve future tree planting activities and could obviously contribute to more and better food supplies in semi-arid as well as other regions.

REFERENCES

Boland, D. J. (1986) Selection of species and provenances for tree introduction. In: *Multipurpose Australian Trees and Shrubs*, J. W. Turnbull (Ed.), pp. 45–57. Australian Centre for International Agricultural Research, Canberra.

Burley, J. and P. J. Wood (1976) A manual on species and provenance research with particular reference to the tropics. In: *Tropical Forestry*. Department of Forestry, Commonwealth Forestry Institute, Paper No. 10, Oxford.

Carlowitz, P. G. von (1985) Some considerations regarding principles and practice of information collection on multipurpose trees. *Agroforestry Systems* 3: 181–195.

Carlowitz, P. G. von (1986) Multipurpose tree yield data – their relevance to agroforestry research and development and the current state of knowledge. *Agroforestry Systems* 4: 291–314.

Carlowitz, P. G. von (1987) ICRAF's multipurpose tree and shrub information system. *Agroforestry Systems* 5: 319–338.

Carlowitz, P. G. von (in press) *A rational approach to multipurpose tree and shrub species pre-selection for specific technologies.* Paper presented at the Technical Seminar, 10th Anniversary of ICRAF, Nairobi, 1987.

Maydell, H. J. von (1986) *Trees and Shrubs of the Sahel.* Their characteristics and uses. Schriftenreihe der GTZ, No. 196, Eschborn.

Webb, D. B., P. J. Wood, J. P. Smith and G. S. Henman (1984) *A guide to species selection for tropical and subtropical plantations.* Second edition, revised. Commonwealth Forestry Institute, Tropical Forestry Papers No. 15.

7

Opportunities and Requirements for the Development of New Essential Oil, Spice and Plant Extractive Industries

C. L. Green

INTRODUCTION

The subjects of this paper are essential oils, spices and pine gum oleoresin products. While this may seem an odd combination, they have been chosen as offering a potential for the establishment of new agro-industries or for the multi-purpose exploitation of tree resources in developing countries. They all fall into the cash crop category and are characterized by a relatively high value to low volume ratio, combined with good storage properties.

The spice group, which includes culinary herbs, are those parts of plants, usually traded in the dried state, which are used either directly or in the form of their extracts for the flavouring of foodstuffs.

The pine products of interest here are gum turpentine and gum rosin, which are obtained by distillation of the gum oleoresin exudate of pine trees. Turpentine has traditionally found usage as a solvent but, today, its main application is as a source of terpenoid isolates by the chemical industries. Rosin is used directly or after chemical modifications as a size for paper and as an ingredient of adhesives, printing inks, rubber products and surface coatings.

The term 'essential oils' embraces the products obtained by steam distillation or solvent extraction of aromatic plants, including spices and culinary herbs, which are highly concentrated forms of the steam-volatile constituents of the plant. They are, therefore, the aromatic 'essence' and should not be confused with the fatty oils, which are not volatile in steam. Essential oils and their solvent-extracted counterparts, known as 'concretes', are used directly in fragrance and flavour applications and to some extent in medicine. Additionally, a number of essential oils are exploited by the chemical industry as a source of specific compounds for subsequent transformation to a range of synthetic products, such as aroma chemicals, flavours, vitamins and pesticides. In this function, the essential oils may compete with turpentine and petroleum derivatives as starting materials for chemical synthesis.

In the context of world trade, the value and volume of essential oils, spices and pine products, even when grossed, is not large when compared with foodstuffs and certain other cash crops. The estimated value of their world exports in 1983 was:

	US$ million	
	World exports	ldc exports
spices	800	720
essential oils	490	165
gum turpentine and rosin	110	60

Moreover, the scale of trade in individual commodities ranges widely. Amongst the spices, the annual average trade in saffron is around 50 tonnes while that for black and white pepper (*Piper nigrum*) is around 120000 tonnes (ITC 1982). In the case of essential oils, a recent market study by the International Trade Centre (ITC 1986) reports annual trade volume ranging from 1 to several thousand tonnes for individual products.

However, these bald statistics understate the true economic significance of these commodities. World production of many items greatly exceeds the world trade figures owing to captive domestic consumption. Moreover, production of these commodities is important to many rural communities. In certain cases, they provide the only means of cash income owing to the difficulties associated with the transportation of more bulky or perishable crops. Other benefits can range from the intensive labour demands for production to the ability to perform value-added, further processing within the cultivation area.

Considerable attention has been devoted in this symposium to the potential for the commercial development of previously unexploited species. Scope exists for new species introduction with tropical pines and in the field of essential oils, but this is not an aspect discussed here. Instead, emphasis is upon the potential for establishing new production ventures with items already well known in trade and, also, for greater exploitation of certain tree species.

Opportunities for new production of well-known commodities arise through two main factors. Firstly, many developing countries are showing a significant increase in demand or even the creation of a completely new demand for some commodities. Typical examples are turpentine and rosin for industrial use, and certain essential oils for the flavouring of 'personal products'. Development of local production could, therefore, be beneficial for import-substitution purposes and, also, it would provide a firm base for any eventual export oriented development. Secondly, the periodic geographical relocation of production is an historic characteristic of many of these commodities, for a variety of economic reasons, and the pattern is expected to continue.

EXAMPLES OF OPPORTUNITIES

Since it is not possible to survey the whole field, this paper provides a few examples of the type of opportunities which can arise.

Herbaceous and floral essential oils

The first example is the essential oil of *Mentha arvensis*, known commonly as 'Japanese mint'. Simple steam distillation of this herb provides a 'crude oil', but this does not normally enter international trade in large volumes. Instead, the crude oil is usually subjected to a simple, further processing step to yield methanol and a peppermint-like, essential oil, prior to export. Both products are extensively used in flavouring, and *Mentha arvensis* is in fact the major source of natural menthol (Greenhalgh 1979).

Following World War II, Brazil emerged as the major supply source and its production peaked at around 5000 tonnes of crude oil annually in the early 1970s. In the more recent period, Brazilian production has undergone a decline and Paraguay and China have taken on the mantle of the principal world suppliers. Additionally, there has been some contraction in sales of natural menthol following introduction of a synthetic product.

The change in supply sources in South America arose from the form of cultivation practised. This has involved use of the herb as the first crop on newly-cleared, forest land and its cultivation without fertilizer application until oil yields fall below an attractive level; a period ranging from three to five years. Thereafter, the land use has switched to intensive farming of soya or maize, which proved more remunerative than mint grown with fertilizer application. Thus, the major sites of *Mentha arvensis* cultivation have progressively moved along with the frontier of forest colonization in Brazil from the eastern area of the Parana State westwards towards the border with Paraguay. During the 1960s Brazilian immigrants introduced the crop into Paraguay. Current annual production in that country is approximately 1000 tonnes of crude oil, but the industry is expected to decline over the next ten years as the virgin, forest land is fully exploited and 'second generation' crops are developed.

The cultivation systems for *M. arvensis* in China, by contrast, involves fertilizer application and, thus, higher input costs. Production for export has been influenced, therefore, by the need to gain foreign exchange rather than simply for profit.

No major developments geared to export have occurred elsewhere, hence, a world shortfall in supply, possibly equivalent to some 1000 tonnes of crude oil (nominally valued at US$10 million), could develop, although India, which produces for domestic needs, has occasionally entered the menthol market when world prices were sufficiently high to provide a profit.

An opportunity is therefore presented for export oriented production in new areas where the Brazilian-type cultivation system for mint would serve a useful role as a short-term, first crop for land development. The total land area requirement corresponds to some 12 500 ha. Immediately identifiable, potential beneficiaries in South America include Bolivia and Peru. Elsewhere, the transmigration scheme in Indonesia would seem to be a candidate.

The *Mentha arvensis* case is exceptional in the scale of the potential opportunity foreseen in the essential oil group. However, the scope for smaller volume production of other commodities should not be disregarded. Examples of this are geranium oil and jasmine concrete. The former is produced by steam distillation of the leaves of certain *Pelargonium* species, and up to 20 ha of land are required to produce 1 tonne of oil per year. Jasmine concrete is a solvent extract of the fresh flowers and harvesting is a particularly labour intensive operation. On average 2·8 kg of concrete are obtained from 1 tonne of flowers.

Current annual world trade in geranium oil is somewhat over 200 tonnes while for jasmine it is about 12 tonnes. The annual value of trade in each case is about US$12 million (Scarpa 1984; Robbins 1985).

At the beginning of this century, both products were produced in southern Europe, notably in France. However, other countries with lower labour costs, particularly those with strong, French connections in Africa, gradually assumed prominence as suppliers. By the mid-1970s, the major producer of both products was Egypt, which annually exported around 100 tonnes of geranium oil and 10 tonnes of jasmine concrete. A decline in production is now evident in Egypt. In the case of jasmine, the reason is simply the rising cost of labour and production now depends upon harvesting by children. The decline in the Egyptian geranium oil industry has arisen from severe disease problems reducing oil yields and the competition for land with vegetables destined for the buoyant Cairo market.

These trends in Egypt could provide an opportunity for the entrance of new producers of geranium oil and jasmine, if on a modest scale. India has already grasped the jasmine opportunity but scope possibly exists for others to benefit. Geranium oil could provide an opening for, amongst others, the resuscitation of production within some Central and Southern African countries, and would also meet a need for diversification in the lime oil producing areas of Peru, which are ecologically very similar to Egypt.

The emergence of similar opportunities with other crops merits close monitoring. Saffron is a prime candidate owing to rising labour costs for harvesting in the major production area of Spain, but the largest individual opportunity of this type is presented with gum turpentine and gum rosin.

Naval stores as pine forestry by-products

Turpentine and rosin are mainly obtained by the manual tapping of living pine trees, the so-called 'gum naval stores' industry, and as a by-product of the Kraft pulping progress. A high proportion of production is consumed domestically, but a very significant volume annually enters world trade: around 350 000 tonnes of rosin and 30 000 tonnes of turpentine. Approximately 70 per cent of the world trade consists of gum turpentine and gum rosin (Greenhalgh 1982).

The United States and several southern European countries were formally dominant as suppliers of the gum products, but their industries have dramatically declined owing to rising costs of labour for tapping the pine trees. The major suppliers are now China and Portugal and it is questionable whether these countries will be capable of sustaining current export levels in the longer term. In the case of Portugal, the cost of labour and alternative employment opportunities will be the major deciding factor, while it is possible that a higher proportion of Chinese production will be consumed domestically. A number of developing countries have already made an entrance to this market, but there are many others who could benefit from the multi-purpose exploitation of their pine forests, which have been planted for timber or pulp purposes.

Subject to the availability of suitable pine species and a favourable climate, gum tapping can be undertaken in the last few years of the rotation without detriment to the economics of timber or pulp production (Coppen *et al.* 1984). This permits a higher return from the overall forestry operation, generates extra employment and income in the neighbouring rural communities, and benefits the foreign exchange position either through import substitution or by export.

Eucalyptus oils in agroforestry

Another opportunity in the 'agroforestry' area, if on a far more modest scale, is apparent with 'medicinal-type' (cineole-rich) eucalyptus oil. Annual international trade for this item is estimated as 1700 tonnes and it is valued at US$12 million (Robbins 1983). The major current suppliers are China (45 per cent) and the Iberian Peninsula (38 per cent). The main form of production is by distillation of the leaves, recovered upon felling mature trees. This system could be adopted, therefore, by countries which have established plantations for timber production purposes of suitable eucalyptus species.

It is necessary to stress the matter of selection. Around 600 species of *Eucalyptus* are known, but less than a dozen are capable of providing medicinal oil of the requisite quality or in an economically attractive yield. Of these, *E. globulus* is frequently the only species available in plantations which have been established for timber or pulp production purposes.

An alternative form of oil production is to grow eucalyptus specifically for this purpose, by cropping leaf and stem regularly on trees grown as bushes. Around six species are known as suitable for this coppicing regime, and they span a range of climatic requirements for cultivation, including semi-arid conditions. Amongst this last group, more attention should perhaps be devoted to *E. polybrachtea*, the Blue Mallee, which appears not to be exploited outside its native Australia. An Australian colleague, E. F. K. Denny (1986), has described the typical habitat of the Blue Mallee as poor and dry. 'In stockman's terms, the land would run three lizards to the acre in a good season. In a drought, two of the lizards would die of thirst.'

The 'agroforestry' of medicinal oil bearing eucalyptus should not necessarily be considered as restricted to large plantation developments. The primary distillation operations are quite simple and small-scale production could be undertaken by farmers, each growing a few rows of bushes and following the pattern of petitgrain (bitter-orange) oil industry in Paraguay. Production may also be contemplated as a component of communal, village-scale forestry of eucalyptus wood-lots. It should be noted that the fuel requirements for distillation can be largely met by use of the leaf and twig material recovered after distillation. The annual production of one tonne lots of oil on a coppicing system would require a planted area of approximately 10 ha.

REQUIREMENTS FOR SUCCESSFUL DEVELOPMENT

Having indicated the type of opportunity which can arise, attention is now turned to some factors which must be considered when contemplating a new venture.

The first requirement is necessarily sound market research and an understanding of supply side trends. Over-reliance should not be placed on the views of a very limited number of potential buyers, since they may have only a fragmented knowledge of the total market or may be responding to short-term shortages. Also, the selection of crops should not be restricted to those with the highest unit value in list prices. Returns from production are far more important, and it may be possible to secure an equal profit with a lower unit value commodity, which furnishes a high yield or requires lower inputs for production. Moreover, it is easier and less risky for a new producer to attempt penetration of a large volume market than is the case with very high priced items traded in small quantities.

Cardamom (*Elettaria cardamomum*) provides a current example of the type of problem which can be encountered. This is one of the highest priced spices; top grades fetching around US$12 000/tonne in the recent past. Total annual world trade is around 10 000 tonnes, of which the bulk is consumed in the Arab countries, and supplies are dominated by India and Guatemala. Cardamom shortages over several years and the resultant rising price trend prompted an expansion of production or the creation of new industries in a number of countries; Honduras being a principal example of the latter type. Current supply of this spice is now in excess of demand and, even with drought problems in India, further price erosion may be expected as young plantings of this semi-perennial mature.

Socio-economic factors in the prospective production areas also require detailed examination. Comparative returns with food and alternative cash crops and other aspects of 'opportunity costs' are usually given fairly rigorous analysis, but the structure of the agricultural society and the attitudes and motivations of the prospective growers can often be overlooked. Some Pacific territories, such as the Solomons and Papua New Guinea, furnish a good case example where spices have been considered for remote communities. Ginger and turmeric are rhizomes and could clearly be readily introduced into the existing agricultural system, which is based on tuber food crops. Superficially, however, they would appear less attractive on a unit-price basis than pungent chillies. Problems arise in selecting the best crop from the fact that in many rural communities all major agricultural activities are still undertaken solely by the womenfolk who, additionally, have the tasks of child-rearing and housekeeping. The question of whether or not women will have the time to devote to the heavy demands of chili picking must, therefore, be carefully assessed.

The early development of an adequate internal marketing system is also critical. However, this aspect is often ignored and there have been cases of farmers producing new crops for identified overseas buyers while there is no organized chain of sales to the export point. Over-reliance should not be placed on traditional village merchants, who are often very cautious over handling a previously unknown product.

Finally, the requirement for adequate scientific inputs in new developments must be stressed. Effort should be devoted to the best possible selection of planting stock, for both yield and product quality in order to ensure competitiveness on the international marketplace. Cultivar selection for the production of acceptable qualities of spices and essential oils is very important. In the field of tropical pines for turpentine and rosin production and with medicinal oil bearing eucalypts (Small 1981; Barton & Reilly 1986), there remains great scope for original research on selection at the species, provenance and individual tree levels and this needs encouragement. Rapid, reliable propagation methods, including tissue culture, for selected planted stock also requires greater attention and this would be particularly valuable for tree crops.

REFERENCES

Barton, A. F. M. and C. M. O. Reilly (1986) *Eucalyptus Oil Project Newsletter*. Murdoch Univ., Western Australia.

Coppen, J. J. W., P. Greenhalgh and A. E. Smith (1984) *Gum naval stores: an industrial profile of turpentine and rosin production from pine resin*. Tropical Products Institute Report G187: HMSO, London.

Denny, E. F. K. (1986) The Australian eucalyptus oils. Paper presented at the 10th International Congress of Essential Oils, Flavours and Fragrances, Washington DC., Nov. 1986.

Greenhalgh, P. (1979) *The markets for mint oils and menthol*. Tropical Products Institute Report G126: HMSO, London.

Greenhalgh, P. (1982) *The Production, Marketing and Utilisation of Naval Stores*. Tropical Products Institute Report G170: HMSO, London.

ITC (UNCTAD/GATT) (1982) *Spices: a survey of the world market*. International Trade Centre, Geneva.

ITC (UNCTAD/GATT) (1986) *Essential Oils and Oleoresins: a study of selected producers and major markets*. International Trade Centre, Geneva.

Robbins, S. R. J. (1983) *Selected markets for the essential oils of lemongrass, citronella and eucalyptus*. Tropical Products Institute Report G171: HMSO, London.
Robbins, S. R. J. (1985) Geranium oil: market trends and prospects. *Trop. Sci.* 25: 189–196.
Scarpa, R. (1984) Jasmin and geranium: a review of production. Paper presented at the Cairo Conference of the International Federation of Essential Oils and Aroma Trades (IFEAT), October, 1984.
Small, B. E. J. (1981) The Australian eucalyptus oil industry: an overview. *Austral. Forester* 44: 170–177.

8

Herbal Drugs: Potential for Industry and Cash

J. V. Anjaria

INTRODUCTION

The last decade has seen numerous changes in the use of botanical products in pharmacy, with certain products being used less and new or other products being used more. The World Drug Market Manual (1982–83) has shown the estimated value of the world market in plant pharmaceuticals to have exceeded 76·28 billion dollars in 1980, some 13·4 per cent higher than for 1979. EEC imports for 1980 of glycosides and their derivatives, alkaloids and their derivatives, and herbal medicine products were 1055, 2777 and 80738 million tonnes while exports were 598, 8071 and 16102 million tonnes respectively (NIMEXE 1980). The import value of plant pharmaceuticals used by France, FR Germany, Netherlands, Switzerland, UK, Canada, USA, Japan, Hong Kong, Singapore, Indonesia and Thailand during 1976 was $317·97 million and in 1980 $476·64 million, a rise of about 49·9 per cent (Comtrade Database UNSO/ITC, 1981, cited by UNCTAD/GATT, 1982). Similarly, Indian market exports between 1983 and 1986 of medicinal herbs was worth $34·69 million, $63·3 million and $53·03 million, and essential oils $6·38 million, $5·87 million and $12·52 million respectively for 1983–84, 1984–85 and 1985–86 (CHEMEXCIL INDIA 1986; Gujarat Export Bulletin 1987).

There is a tendency to look at the claims of our forefathers regarding folk medicine as quaint, curious and apparently obsolete. Recent studies have shown otherwise and are reminiscent of Kipling's line in 'Our Fathers of Old'

> 'Anything green that grew out of the mould
> Was an excellent herb to our fathers of old.'

Let us also remember Osper's advice 'Knowledge grows out in such a way that its possessors are never in sure possession. It is because science is sure of nothing that it is advancing'. Let us look with vigilance to the past and the present, near and far and exploit all resources and technologies for the benefit of small farmers.

The World Health Organization in its document 'Health for all by the Year 2000' (WHO 1977) has accepted the role traditional medicine has to play in primary health care. It has been recognized that 85 per cent of the people in developing countries still rely on traditional medicine as the first line of defence in health care (WHO 1976). Since animals are also needed for the well-being of mankind, the use of traditional medicines is being advocated by FAO for treating animals, similar to what is being done by WHO for humans. FAO has initiated the work on documenting and compiling reports on traditional veterinary medicine used by small Asian farmers (FAO 1984a,b,c & 1986).

Over 400 botanical products are marketed in western Europe. Hamburg, which for some time has been the centre for the European trade in botanicals, has now emerged as the world centre for trade in medicinal plant products. The main end-users in the European pharmaceutical industry have found it advantageous to develop contractual relationships with established growers in the developing countries provided the required volume of a particular plant is sufficiently large to make the relationship economically viable. The European company will often provide technical assistance in growing, harvesting and shipping to its joint venture partner. There are, however, very few botanicals that are traded in sufficient volume to warrant such arrangements. In general the end-users purchase direct from the major traders and stockists.

A general evaluation of the European markets for the products studied in 1974 (UNCTAD/GATT 1982) showed that the commercial market for these products had either declined or were static. It was also found that all the major traders had expanded their marketing areas to promote and provide plant extracts for the food, perfumery and cosmetic industries, i.e. in botanical products that did not have to satisfy the more stringent safety, quality and efficacy regulations pertaining to pharmaceutical products.

Several new products were examined in order to determine whether an expansion of their cultivation in developing countries could be justified in terms of market potential. Plants that can be grown in temperate climates are usually cultivated within these regions even though the costs of cultivation and harvesting have risen considerably. Eastern Europe and some Asian countries still remain a major source of medicinal plants. The growers in these countries can be relied upon to produce good quality products at prices that are commercially acceptable to the end-users.

Sources of raw material include natural forests, agroforestry, primary cultivation and secondary cultivation, including home gardens and ornamental horticultural plantings.

INVOLVEMENT OF FARMERS

The gross basic picture given above stresses the need for the involvement of farmers in any country to develop their plant resources to cater for the needs of the herbal drug industry. The flow chart (Fig. 8.1) provides a general picture of the trade in medicinal herbs and indicates the definite involvement of the farmer, either through the middlemen or directly with the industry. It also emphasizes the need for the involvement of the farmers at grass roots level to identify the new crops for the pharmaceutical industry and traditional medicine.

Figure 8.1: *Flow Chart Showing the Interaction of the Farmer, Middleman and Manufacturer in the Medicinal Plants Industry.*

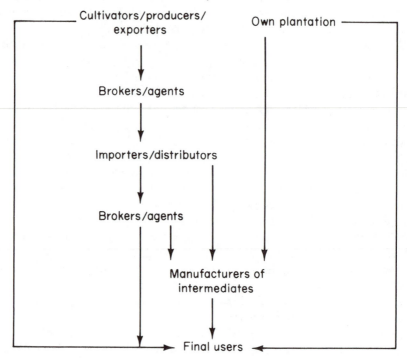

THE SOURCE AND USE OF MEDICAL PLANTS

A large quantity of plant material is used when preparing herbal remedies for both man and stock. Approximately one third of all pharmaceuticals are of plant origin and if fungi and bacteria are included, over 60 per cent of all pharmaceutical preparations are plant based. However, in most pharmacopoeias the number of medical plants and preparations has declined in proportion to that by which synthetic substances have increased. This can be explained by changes in the use pattern of medicinal plants; while many have gone out of use, the consumption of others continues to increase and yet others are finding new applications.

The veterinary profession has a long history of using medicinal plants for treating animals. They mostly started using such plants following observations of folklore claims. The ancient literature on veterinary treatment described many remedies originating from the use of medicinal plants. As new synthetic drugs develop there seems to be a trend for a decline in the use of medicinal plants. The present-day veterinarian/physician practises by prescribing proprietary indigenous drugs, then, in many cases, prescribing galenicals, which has prevented a decline in the use of at least some traditional remedies.

SOURCE OF PLANT MATERIAL FOR INDUSTRY

Plant material for the pharmaceutical industry may be obtained from the following sources:

(a) Natural resources, e.g. forests.
(b) Agroforestry.
(c) Cultivation.
(d) Horticultural by-products, e.g. ornamental gardening, etc.
(e) Herb gardens.
(f) Weeds and hedge plants, and
(g) Industrial by-products.

The use of the raw material can be categorized as follows:

(a) Raw plant material, e.g. leaves, roots, fruits, seeds, etc.
(b) Pulverized and semi-processed plant materials in the form of powders, etc.
(c) Intermediate products, such as crude and purified extracts.
(d) Active constituents such as alkaloids, glycosides, etc.
(e) Raw materials that can be used as precursors for the synthesis of compounds such as vitamins, steroids, etc.
(f) Proprietary formulations which can be marketed over the counter.

Each of the above can form a separate component and basis for the industry.

INDUSTRIAL PROJECTIONS

Folklore claims and organized indigenous drug research for animal treatment are merging towards the development of an indigenous drug manufacturing industry. The toxic hazards of the newer synthetic drugs has influenced present thinking in going back to nature. Although synthetic organic substances have achieved a substantial share in pharmaceutical applications, plant-derived substances still remain a vital tool for modern medicine.

A number of small commercial units have developed in many of the countries manufacturing indigenous human and veterinary medicines, obtaining their supplies either directly from source or through middlemen. The formulations are mostly empirical, with gunshot prescriptions where the prescribed dosages have little scientific foundation, being based on results obtained by trial and error by the practising physicians and veterinarians. There is now the beginning of a trend towards obtaining results based on properly organized research.

The small commercial components relating to the various aspects of the indigenous drug processing industry are:

(a) Collection and drying of raw material, pulverizing with quality control testing, packing and storing. This will ensure availability throughout the year.

(b) An intermediary products industry, e.g. processing of crude extracts, alkaloids, glycosides, etc.

(c) A formulations manufacturing unit capable of undertaking the manufacture of proprietary products for marketing over the counter.

(d) Agencies for collecting raw materials through forest cooperatives, etc.

(e) Agencies advising farmers on cultivation and marketing of raw material.

(f) Satellite industries for packing materials, processing machinery and spare parts, quality control laboratories, etc.

ECONOMICS

The budget for human and animal health care is usually 50–60 per cent for life-saving drugs, 25–30 per cent for supportive drugs and 10–15 per cent for indigenous traditional drugs; preventative vaccines are supplied free. There is clearly considerable scope for the traditional drug trade. Much of the trade can now be supplied through local cooperative societies which are able to supply authenticated, good quality raw material directly to the manufacturer as well as creating good rural employment opportunities.

In the past it was customary for the physician and veterinarian to prepare their own prescriptions, including herbal remedies. There is now a decline in the use of such self-compounded formulations accompanied by an increase in the use of proprietary formulations sold by indigenous drug manufacturers. Although this trend has been encouraged by the physicians and veterinarians, it has resulted in a higher price for the product and to a certain extent some deterioration in quality.

Thus, a lactogenic preparation containing *Leptadenia reticulata* with *Breynia retusa* (syn. *B. patens*) and *Asparagus racemosus* and some other drugs, would cost about 60–80 INR per treatment for bovine agalactia when given as a proprietary medicine. The same treatment would only cost 5–10 INR per animal if the crude dried powder of *Leptadenia reticulata* was given, and with similar or better results. The same is also true for antidiarrhoeals, where the cost would be lower by at least 80–90 per cent. The antiseptic preparations based solely on medicinal herbs recently prepared by the Department of Pharmacology, Gujarat Agricultural University have proved to be equivalent in efficacy to Furacin ointment (SKF) and better than Propamydin cream (M & B). The cost of 40 g of a similar preparation used as a wound healing powder in Sri Lanka and compared with Negusant powder (Hoechst) was reduced by 80–90 per cent and possessed equipotent properties. Recent work in Sri Lanka on 18 different veterinary formulations has shown that the cost of treatment could be cut by 60–75 per cent (Anjaria 1986). Thus the farmer could eventually benefit from a cheaper and equally effective product.

To summarize, the herbal drug trade would aid the economy of the country by harnessing the natural resources in the service of man and his livestock, thereby saving in foreign exchange. By establishing an export market it would offer better employment opportunities for both urban and rural labour, and provide an incentive for farmers to cultivate new crops and practise agroforestry. Cheap, effective, easily available, non-toxic herbal remedies without side-effects would be made available to the masses and would provide an incentive to develop herbal processing, manufacturing units and satellite industries as well as supporting both national health care and animal health programmes.

RESEARCH

Ongoing indigenous drug research has created an impact in the use of proved and tested herbal formulations. Herbal remedies with almost equipotent results without adverse side-effects are now available for use as antibacterials, wound healing and treatment of agalactia, fertility, cardiovascular, urinary, reproductive and respiratory disorders, etc. Some herbal remedies, such as non-hormonal galactagogues and ovulatory preparations and herbal bacterials have made spectacular advances compared with allopathic remedies.

Product Profiles

It is not feasible to provide product profiles of the ca. 400 or more species of medicinal plants that are grown and traded in Asia and Western Europe. Many of such plants traded in the developed countries are used more in applications other than allopathy, their use in pharmaceuticals being secondary or even incidental. It is estimated that approximately 50 per cent of the plants traded are used in the food industry; this includes medicinal infusions and unlicensed herbal remedies under Government Food and Drug Control Rules. A further 25 per cent are used in the cosmetic industry, 20 per cent in medicinal applications and 5 per cent for miscellaneous uses such as insecticides, etc. Some of the industrial product profiles are given below.

MEDICINAL AROMATIC PLANTS WITH ESSENTIAL OILS

There are many flourishing industries based on the production of essential oils and related secondary products. They are arranged in the following broad groups:

(a) **Forest products:** cedar wood, sandalwood, sassafras, star anise (*Illicium verum*) and turpentine oils. Other essential oils are obtained from agar-wood (*Aquilaria malaccensis*), *Artemisia* spp., asafoetida (*Ferula* spp.), bay oil (*Pimenta* spp.), benzoin (*Styrax* spp.), calamus (*Acorus calamus*), camphor (*Cinnamomum camphora*), cassie (*Acacia farnesiana*), ginger

(*Zingiber officinale*), grass (various aromatic grasses), juniper berry, licorice, myrrh, ginger grass or palma rosa (*Cymbopogon martinii*), palm, thyme, valerian, Peru and tolu balsams (*Myroxylon balsamum* var. *pereirae* and var. *balsamum*), galangal (*Alpinia* spp.), *Cyperus rotundus* and *C. scariosus*, oplibanum or frankincense (*Boswellia* spp.) and pine oils.

(b) **Cultivation products:** major essential oils in this group are from cinnamon bark, jasmine, lavender, lemon grass (*Cymbopogon citratus*), menthol, pepper and spear mints (*Mentha* spp.), ambrette seed (*Abelmoschus moschatus*), basil, rosemary, patchouli (*Pogostemon cablin*), vanilla, bergamot (*Citrus aurantium* subsp. *bergamia*), cananga (*Cananga odorata*), etc.

(c) **Oils from the processing of raw material:** ajowan *(Trachyspermum ammi)*, anise (*Pimpinella anisum*), caraway (*Carum carvi*), cardamon (*Elettaria cardamomum*), celery, coriander, cubebs (*Piper cubeba*), cumin, curcuma oleoresin, (*Curcuma* spp.), dill, fennel, fenugreek, garlic, hops, nutmeg, etc.

(d) **By-products of other industries:** cinnamon leaf, clove leaf, lemon, mandarin and tangerine oils, orange peel, etc.

In India the most profitable cultivated aromatic medicinal plants are lemon grass, vetiver (*Vetiveria zizanioides*), ginger grass, sandalwood, agarwood, ginger, pepper, jasmine, rose, mint, citronella (*Cymbopogon* spp.), cedar wood, bergamot, spear mint, lavender, patchouli, geranium, basil, eucalyptus, etc. Their cultivation is backed by agricultural research.

ALKALOID-CONTAINING PLANTS

The following plants have a considered market for industrial application: *Atropa belladona, Catharanthus roseus, Cinchona* spp., *Datura* spp., *Rauvolfia* spp., *Hyoscyamus* spp., *Duboisia* spp., *Cephaelis impecacuanha*, etc.

GLUCOSIDE-CONTAINING PLANTS

Those plants with a considered market for industrial applications include: *Digitalis* spp., *Dioscorea* spp., *Glycyrrhiza* spp., ginseng (*Panax ginseng*), etc.

ENZYME-CONTAINING PLANTS

Carica papaya is one of the leading plants for enzyme application. Pineapple (bromotin), figs (ficin), etc., are also marketable.

STEROGENIC PLANTS

For agalactia and reproductive disorders, etc., the plants with the most commonly used sterol contents are *Withania somnifera* and *Asparagus racemosus*. *Solanum myricanthum* (syn. *S. khasianum*) is the high sterol yielding plant used as the basic raw material in the manufacture of the contraceptive pill; it is easily cultivated. There are a number of other such interesting plants.

Dioscorea spp. are important sources of steroids and have a world market, similarly for *Solanum* spp. Some *Agave* spp. and *Yucca* spp. which yield hecogenin are new steroid sources.

LAXATIVE PLANTS

Vegetable based laxatives represent a substantial part of the retail market and is valued at $500 million.

(a) **Stimulant laxatives:** these include *Aloe, Senna alexandrina, Cascara sagrada (Rhamnus purshiana)* and species of rhubarb (*Rheum* spp.).

(b) **Bulk forming laxatives:** psyllium husk (*Plantago arenaria*) forms the major part of this market.

VALEPROTRAITE-CONTAINING PLANTS

Valeriana spp. has the established world market in this field.

OTHER PLANTS

Current research on indigenous drugs has started to generate a market for various other plants, examples of which are given below.

(a) **Antibacterial:** *Azadirachta indica, Annona squamosa, Ocimum* spp., *Vitex negundo*, etc.

(b) **Pediculocidal, insecticidal and larvicidal:** *Annona squamosa* seeds, *Vitex negundo*, etc.

(c) **Aphrodisiacs, sex stimulants and spermatogenics:** *Sida cordifolia, Mucuna pruriens, Cardiospermum halicacabum*, etc.

(d) **Urinary disorders:** *Crateva nurvale, Boerhavia diffusa, Saxifraga callosa subsp. callosa* (syn. *S. ligulata*), *Hygrophyla auriculata* (syn. *H. spinosa*), *Hyoscyamus niger*, etc.

(e) **Antiamoebic, gastric and liver disorders:** *Holarrhenia antidysenterica, Aegle marmelos, Berberis aristata, Cyperus* spp., *Phyllanthus emblica*, etc.

(f) **Respiratory and cardiovascular:** *Alpinia* spp. (syn. *Galanga* spp.), *Ocimum* spp., *Rauvolfia* spp., etc.

CONCLUSIONS

The passage of time depicts the increasing use of medicinal plants as scientifically proven, patent remedies without adverse side-effects and toxicity. This will require the increasing involvement of farmers to include such plants in their cropping, either through cultivation, hedge plantations, agroforestry or ornamental horticulture. The pattern will differ according to regional ecological conditions.

The continental organization of medicinal plant banks by harnessing the natural resources should lead to the development of an intercountry network of rural based new resources of herbal drugs, food and industry, thereby creating potentials for rural employment and multifaceted industrial development. This will benefit the small farmer and relieve the suffering of man and his animals.

REFERENCES

Anjaria, J. V. (1986) *Traditional veterinary medicine. The modern trend – traditional drug research. A brief review.* Paper presented with the second interim progress report on Traditional Veterinary Medicine Project of SL-ADB-LD Project: 58–87. Sri Lanka, Veterinary Research Institute, Gannoruwa, Mimeo.

Chemexcil India (1986) *Report of Conference on Chemical and Pharmaceuticals, 10th February, 1986.* Basic Chemicals, Pharmaceuticals and Cosmetic Export Council, Bombay and UK Trade Agency for Developing Countries.

FAO (1984a) *Traditional (indigenous) systems of veterinary medicine for small farmers in India.* FAO/UNO Regional Office for Asia and the Pacific, RAPA-80 by Anjaria, J. V., Bangkok.

FAO (1984b) *Traditional (indigenous) systems of veterinary medicine for small farmers in Nepal.* FAO/UNO Regional Office for Asia and the Pacific, RAPA-81 by Joshi, D. D., Bangkok.

FAO (1984c) *Traditional (indigenous) systems of veterinary medicine for small farmers in Thailand.* FAO/UNO Regional Office for Asia and the Pacific, RAPA-82 by Buranamanus, P., Bangkok.

FAO (1986) *Traditional (indigenous) systems of veterinary medicine for small farmers in Pakistan.* FAO/UNO Regional Office for Asia and the Pacific, RAPA-83 by Maqsood, M., Bangkok.

Gujarat Export Bulletin (1987) Volume 21–22, No. 11–12, Statistical Number. Gujarat Export Corporation Ltd., Ahmedabad, India.

NIMEXE (1980) *Analytical tables of foreign trade.* CST Vol. 1. European Communities Statistical Office, Luxembourg.

UNCTAD/GATT (1982) *Markets for selected medicinal plants and their derivatives.* International Trade Centre Publications, Geneva.

WHO (1976) WHO Handbook. Report of UNICEF – WHO Joint Committee on Health Policy. Alternative approaches to meeting basic health needs of developing countries. Fifth meeting, 17th September, 1976. WHO, SEA/RC.20/Min-5, Geneva.

WHO (1977) Health for all by the year 2000. WHO, Geneva.

World Drug Market Manual (1982–83) Yorkhouse, 37 Queen Square, London.

9

The Potential Value of Under-exploited Plants in Soil Conservation and Land Reclamation

J. Smartt

INTRODUCTION

The subjects of soil conservation and land reclamation are both vast and can be dealt with from the points of view of numerous specialisms. The present treatment is perhaps from a novel standpoint, that of the plant breeder. Obviously no definitive treatment of either topic can be contemplated but the breeder can make a valid if belated contribution to the discussion of these subjects.

There is an ever-present hazard in the introduction of exotic species to a new geographic area in that, although they may perform their allotted tasks exceedingly well, they may outstay their welcome. Possibly the best example of this is kudzu (*Pueraria lobata*) an introduction from Japan to the United States. It has been grown as a cover crop for erosion control, a function it has performed admirably, but it has invaded areas where it is unwelcome and has thus become a serious pest in the south-eastern United States. The reason why this particular introduction gave rise to a problem was that the question of post-establishment management was neglected. It is obvious that we need plants which will grow where we want them, when we want them, and which we can dispose of easily when their function has been fulfilled. Essentially we wish to treat them in precisely the same way as we treat our crop plants. We need therefore to look at plants which appear to have potential for use in soil conservation and reclamation, not only from the point of view of the attributes they need to possess for use in soil conservation and land reclamation, but also the qualities which make for ease of management. These attributes can be found in a range of plant species, which can for convenience be called under-exploited plants. This group includes notably wild plants but also weeds, minor crop species, and superseded land races and cultivars of major crops.

SOURCES OF PLANTS FOR USE IN CONSERVATION

It is probably a common experience for agriculturalists to encounter plants which have qualities that seem to cry out for exploitation but are almost invariably coupled with other qualities which may for the present preclude any such exploitation. The experience which set me thinking on this subject was in Central Africa where I was involved in groundnut breeding and agronomy

(Smartt 1961). In the course of my work I came across some very vigorous and impressive weed species which seemed to have characteristics of great potential value. The possibility of exploiting their good points and circumventing their drawbacks seems on reflection to be worth exploring.

There were in particular two noteworthy grass species which excited my special interest, *Rottboellia cochinchinensis* (syn. *R. exaltata*) and *Eleusine indica*. The first species is a tall, fast growing annual grass which commences growth immediately after the first rains fall and, when present at sufficient density, effectively suppresses the growth of all competitors. The reason why it is able to do this is that the seeds lack dormancy; as soon as conditions are favourable for growth this occurs very rapidly. Since all the seed germinates with the first rains, none remains dormant in the soil. It is therefore not persistent under an effective system of cultivation. This quality could be useful in systems designed to restore the organic matter of soils; it could be a valuable green manure crop.

The second species mentioned, *Eleusine indica*, has different but equally interesting properties. In southern Africa it is known as ox-grass and with good reason. It is a very palatable annual grass which produces abundant seed with well developed dormancy. This has unfortunate consequences for the farmer who uses farmyard or kraal manure since he thereby introduces an abundance of a very effective weed to his land. When it is established it is a difficult plant to remove. This characteristic is of great interest from the standpoint of soil conservation; it rapidly produces a very dense tussock of tillers and a remarkable and equally dense fibrous root system. This root system is extremely effective both in soil binding and in developing a very good crumb structure in the soil. These qualities, its palatability, its ease of establishment, its soil binding and its soil improving qualities invite exploitation, but this is obviously not straightforward. What then are the stumbling blocks to actual exploitation? The key perhaps lies in the dormancy of its seed. If a non-dormant genotype could be developed then the problem might well be overcome. I am indebted to Dr. G. E. Wickens of the Royal Botanic Gardens, Kew, for bringing to my attention the fact that the robust African form of *E. indica* is apparently tetraploid ($2n = 4x = 36$) and confined to Africa (Ivens 1967). Interestingly, when occurring as a weed in crops of its relative the finger millet *E. coracana* hybridization apparently occurs with free recombination of characters. From such hybrid populations it could well be feasible to select genotypes which retained the vigorous growth of *E. indica* with the non-dormant seed of *E. coracana*. This segregant could have value in both soil stabilization and building and also as a useful grass for short term leys.

While I should not advise agrostologists to drop all else and initiate crash programmes on *Rottboellia cochinchinensis* or *Eleusine indica*, what I am suggesting is that the local flora should not be neglected in the search for useful plants; there is scope for 'talent spotting' in the local flora. In the general enthusiasm for plant introduction local 'talent' has often been ignored. Great effort can be expended on selection and acclimatization of exotic species when perhaps less effort expended on indigenous species could be more productive. Weed species are probably worth looking at initially, especially grasses and legumes. Appropriate simple selection regimes for desirable variants could produce rapid advances in a relatively short time. It would not be the first time in crop plant domestication that weeds have become crops.

It is axiomatic that weeds and crop plants have many common characteristics. The major difference being that weeds do not produce anything that we want, or if they do we are not prepared to accept the concomitant disadvantages. There is a substantial group of plants which exist in a kind of limbo between crops and weeds and these are commonly lumped together as underexploited crop plants. Some of these are discarded crops, such as the former oilseed crop *Camelina sativa*, which have become very subtle and resourceful weeds, if one may be permitted the anthropomorphism!

CHARACTERISTICS OF CROP PLANTS USEFUL IN CONSERVATION

Harvest index

Probably the major character determining success or failure of a crop in modern cultivation, second only to yield, is the harvest index. Attitudes to crop production can become more and more specialized, perhaps excessively so, particularly with regard to grain crops where grain yield alone becomes the sole economic yardstick of agricultural success. With cereals the value of straw can be discounted, with the legumes the haulm may not be valued. After passage through modern combine harvesters for example, legume vines and haulm may be totally pulverized and consequently of no practical value thereafter. Pea vines and groundnut haulm are both useful stock foods and it is regrettable that the post-harvest value of these is largely lost. A high harvest index in cereals can obviously reduce the amount of straw produced and its contribution to livestock feeding and its suitability for thatching. This is not to imply that a high harvest index is a bad thing but it is not a universally good thing. It behoves us to anticipate, as far as we can, the consequences of changes which can be effected by successful selective breeding.

Vegetative morphology

The improvement in harvest index is but one of a whole series of changes which has been brought about by selection in plant breeding programmes. Selection for changed morphology has been practised over the centuries in fact, with the primary objective of producing manageable growth forms, such as the production of dwarf bush variants of wild rampant forms. Selection for high harvest index is, in part at least, morphological selection. For use in soil conservation or land reclamation work, an efficient vegetative canopy reflected in a high harvest index is almost certainly not required, rather the reverse in fact. The growth pattern of the root system is also important. Individuals in populations of highly sophisticated crop cultivars (especially those of self-pollinating species) and F1 single cross hybrids tend to be or are genetically homogeneous. Aggressive competitive ability is undesirable, an efficient but not overly competitive root system is required. By contrast a vigorous and aggressive growth of both root and shoot system is appropriate, rather than otherwise, in plants required for conservation and reclamation.

Gigantism

Perhaps the most remarkable and consistent feature of domesticated species is gigantism, most especially of the part used by man. This is most noticeable in grain and fruit crops, where size increases have been considerable. This kind of change is not desirable in species to be used for the purposes under discussion. Seed size should be large enough to favour efficient establishment of the seedling. Excessively large seeds are usually not optimal as such seeds can unduly attract both predators and pathogens to the detriment of establishment. Quite a small seed size may be optimal, which in fact enables a given quantity of dry matter to go further in establishment of a new stand.

Loss of seed dormancy

Loss of seed dormancy is a *sine qua non* of successful crop plants and can also be of value for uses we are now considering. Management is greatly facilitated if seed dormancy is lacking and rapid germination of cover crop and green manure seed can be readily obtained. However, there are some conditions in which seed dormancy could be useful. In long-term projects, a cover may have to be maintained for lengthy periods, during which there may well be a turnover in the plant population and a need for replacements. The presence of a dormant seed bank in the soil would help maintain the species as long as desired, until it is either replaced in the course of a natural or semi-natural succession by a crop. This obviously leaves the problem of the dormant soil seed bank, if a subsequent crop is to be taken.

Loss of dispersal mechanisms

Equally characteristic of successful crop plants is the suppression of the mechanisms for seed dispersal. Here we have a parallel with the seed dormancy situation in that the primitive state of this character (i.e. with an active seed dispersal system) favours the strategy of a self-sustaining plant community rather than the advanced state, in which the character is totally suppressed. Perhaps an intermediate state, in which a partial suppression of the dispersal mechanism was established, could be optimal in that there was an option, if needed, for the conventional harvesting seed; any seed unharvested could then be left to be distributed by the still partially functional dispersal mechanism.

Reduced efficiency of dispersal mechanisms could well provide a substantial part of the solution to management problems in restricting the crop specifically to those areas where it was needed.

Presence or absence of toxins

A major problem in the utilization of some crops is the presence of toxic materials which limit, if not prevent, their use as human and livestock food. The lupins are such an example, where the presence of alkaloids in both the vegetative parts and the seeds has created detoxification problems for those attempting to exploit the plant for food. However, in certain circumstances, the presence of alkaloid toxins can be of positive advantage. A good demonstration of this was brought to my notice by Mr. George D. Hill, a well known New Zealand legume agronomist. Reclamation of a particular area was being hampered by a plague of marsupials, which apparently devoured most types of plants. Sweet lupins and those with reduced alkaloid contents went the way of the rest. However the situation was retrieved when alkaloid-rich Russell lupins were introduced, no doubt other alkaloid-rich forms and species of lupins would have been equally suitable. Alkaloids are still a problem in lupin cultivation, but the cause of this problem has become the solution of a different problem!

The presence of toxic and distasteful principles may be of prime importance where grazing could be a hazard. It may be necessary, where the option is open, to forgo the possibility of exploiting cover crops for grazing or fodder by establishing genotypes with protecting toxins or, alternatively, by deliberately excluding the livestock.

ADDITIONAL EXPLOITATION OF PLANTS USED IN CONSERVATION

We have heard in the past a great deal about the exploitation of leaves as a source of proteins for food manufacture. Apart from the use of soya bean meal as a protein source for the production of textured vegetable protein, very little has been heard of exploiting protein-rich seeds in general. Seeds have obvious advantages over leaves as a protein source, their dry matter content is so much higher and the material itself is less bulky and probably easier to harvest and handle. Problems with toxic materials are likely to be no more serious with seeds than with leaves, but will obviously still have to be solved satisfactorily. Many plants which are of value in conservation work produce a bulk of seed which is largely unexploitable at the present time due to toxic concentrations of a range of materials, such as alkaloids, non-protein amino acids and cyanogenic glycosides, etc. It might well be possible to devise simple detoxification procedures to produce a protein concentrate material of value for livestock feeding or further processing for the food industry. Such processing might be carried out either by an industrial- or cottage-based industry. Detoxification is of course a standard procedure in the utilization of bitter cassava and is not a totally foreign procedure in such agricultural economies.

If a fully effective and economic technology is developed, there is no reason why crops, primarily produced for soil conservation or used in land reclamation, could not also be a source of the basic raw material for such an industry, provided that this could be done without prejudice to the primary objective. If such materials are unacceptable as human food they could well find use in formulating livestock rations.

LAND RECLAMATION

Land dereliction

Dereliction of land can arise from three main causes, uncontrolled erosion, nutrient exhaustion in the broadest sense, and from the accumulation of toxic materials. The first question which has to be decided is whether the land is to be rehabilitated, or whether it can be exploited in its present state. In cases of extreme nutrient depletion, soils can be reduced to a skeletal state. In high silica soils reclamation may involve a phase of near hydroponic culture, some initial input of nutrients is essential. The question becomes how much is economic, if a cash return is possible in the short or medium term it may well be worthwhile to initiate such a programme of reclamation. This could be something comparable to the very effective marram – lupin – pine succession used on unstable sand-dunes. In different geographic areas, different successions would have to be devised. A keen eye should be kept on weeds, wild plants and underexploited crops and their potential utility assessed. It may well be possible to initiate artificially controlled successions which could restore land to productivity over a shorter time scale than would have occurred naturally. There would seem to be scope for such artificially accelerated successions.

Salinity

Salinity and alkali accumulation are two difficult problems, the former perhaps the more tractable. While considerable efforts have been made to breed salt-tolerant genotypes of cereal and other crops we should not lose sight of the fact that some crops have evolved from strand plants, e.g. the beets and brassicas and a level of salt tolerance is a feature of the biological species to which these crops belong. Any loss of salt-tolerance consequent on domestication and selection for cultivation on non-saline soils could probably be restored by appropriate crosses and selection. Where salinity is combined with a high water-table it might be possible to develop halophyte communities with potential for exploitation by grazing animals, in much the same way that salt marsh is used in the British Isles. In drier saline environments the potential for exploitation is being evaluated for jojoba, a Mexican desert plant tolerant of both salinity and alkalinity (CAST 1984). Another species also being closely examined for potential exploitation in arid lands is the buffalo gourd (NAS 1975). It may become possible to exploit certain salt lakes if it ever proves possible to cultivate eel-grass (*Zostera marina*) in them. Natural eel-grass communities are exploited for their grain on the west coast of Mexico. Saline and alkaline water bodies could also be exploited, in some cases at least, for the production of *Spirulina*, an alga which has a multiplicity of uses as a food protein source.

The problems of reclamation have already been touched upon, these involve the rebuilding of the soil's physical structure, restoring the nutrient status chemically and ultimately re-establishing the soil biota. This full range of objectives may not always be achieved. In areas subject to processes of desertification, totally new systems of environmental use may have to be devised. There is obviously scope for developing culture of those grass species, for example, which have been used as a food source in the Sahel (J. M. J. de Wet 1986, unpublished). For this to be done efficiently may necessitate devising some system of extensive production, a form of 'crop-ranching'. Such a system of production would of course be geared to subsistence needs. In these desert margin areas there is obviously scope for developing a wide range of forage and browse plants. Deep-rooting legume trees such as *Acacia* species could be used and a suitable mixture could provide forage for much of the year, if not year round.

In many ways the winning of new land from the sea has parallels with the reclamation of saline land. In the former instance, under a good rainfall regime the weather can eventually be expected to reduce salinity; it may then be helpful to accelerate the ecological succession by the timely and deliberate introduction of species with an appropriate level of salt-tolerance. In areas where salinity is man-induced, it may be possible to use irrigation water to wash salt from the soil profile if adequate drainage can be provided. If this proves possible subsequent cultivation of crops should be restricted to those with the most efficient water economies (Nabhan 1979), in such a situation *Phaseolus acutifolius* would be preferred to *P. vulgaris* for example.

Toxic materials

A great deal of dereliction of land arises incidentally from spoil dumps from mines of all types. Heavy-metal mining probably produces the most intractable problems. Although it may be possible by use of heavy metal tolerant genotypes of grass species (Bradshaw 1960) to produce a vegetation cover over such areas, full restoration of biological activity in the soil is unlikely to be achieved in the short term. It is possible that the activity of toxic elements could be reduced by chemical treatments to produce insoluble salts or chelates. The most intractable problem of such lands is that any products by tolerant genotypes tend to contain unacceptably high contents of the toxic heavy metals in question. This clearly sets a limit to the extent of reclamation that may be possible.

In general terms physical reclamation is easiest to achieve, it may necessitate some civil engineering work in addition to the establishment of plant species. Reclamation from chemical dereliction is usually much more problematical and may be intractable to a greater or lesser degree.

CONCLUSION

A conclusion which can be drawn from this all too brief survey is that for most usage in conservation and reclamation work, plants fairly close to their natural state may be the most effective for the purpose. While most changes which have occurred under the selection pressures prevailing under domestication render plants less suitable for such use, some of these changes can greatly facilitate management. The key management factors are loss of seed dormancy and inactivation of seed dispersal mechanisms. Where these are not apparent in species populations at large, a mass selection programme may be sufficient of itself to establish them. Where the plant in question (e.g. *Eleusine indica*) has a congeneric cultigen it may be possible to introgress the character. Failing this it may be worthwhile to attempt artificial mutagenesis prior to a selection programme. Some under-exploited crop plant species may have a suitable combination of characters for immediate exploitation.

REFERENCES

Bradshaw, A. D. (1960) Population differentiation in *Agrostis tenuis* Sibth. III. Populations in varied environments. *New Phytol.* 59: 92–103.
CAST (Council for Agricultural Science and Technology) (1984) *Report No. 102. Development of New Crops: Needs, Procedures, Strategies, Options.* Ames, Iowa.
Ivens, G. W. (1967) *East African Weeds and Their Control.* Oxford University Press, Nairobi.
Nabhan, G. P. (1979) Tepary beans, the effects of domestication on adaptations to ariel environments. *Arid Lands Newsletter* 10: 11–16.
NAS (1975) *Underexploited Tropical Plants with Promising Economic Value.* National Academy of Science, Washington, DC.
Smartt, J. (1961) Weed competition in leguminous grain crops in Northern Rhodesia. *Rhodesia Agric. J.* 58: 267–273.

10

Green Biomass of Native Plants and New, Cultivated Crops for Multiple Use: Food, Fodder, Fuel, Fibre for Industry, Phyto-chemical Products and Medicine

R. Carlsson

INTRODUCTION

Primary production by photosynthesis in green leaves means that in temperate countries 20 tonnes of dry matter and 3 tonnes of leaf protein or more per hectare can be produced per annum, while in tropical countries, due to better access to solar energy, 80 tonnes of dry matter and 6 tonnes of protein per ha can be produced. In the latter case the so called C4 plants (e.g. tropical grasses and species of Amaranthaceae) are more efficient than 'normal' so called C3 plants.

GREEN BIOMASS AND ITS WET-FRACTIONATION

Research and development on green biomass for wet-fractionation is in progress in more than 70 countries, of which the majority are in the tropics. In industrialized countries, wet-fractionation factories exist in about ten countries, including France, USA, USSR, Denmark, Japan, Spain, Italy and New Zealand. The process capacity of such factories is often between 50 to 120 tonnes processed biomass per hour. Such factories sell leaf protein concentrates for feed as fibre-rich/sugar-rich pellets for ruminants. On the other hand, small-scale production of leaf nutrient concentrate for human consumption is in progress in ten or more tropical countries. By the use of appropriate technology in the local villages, up to 100 kg of leaves per hour is processed using various 'mincer-presses'. Most of the human food consumption of leaf nutrient concentrates is connected with a society called Find Your Feet Ltd., with the main office in London. In both cases described there is a need to investigate more fully the locally available plant species that may be useful for wet-fractionation and with multiple uses of the product obtained.

Wet-fractionation of green crops produces as a first step a fibre-enriched pressed crop and an expressed green juice (Fig. 10.1). The pressed crop can (due to the extraction efficiency of the soluble substances and type of plant material) have a different composition and use. Traditionally the pressed crop, often with

Figure 10.1: Multiple use of Green Biomass by Wet-fractionation

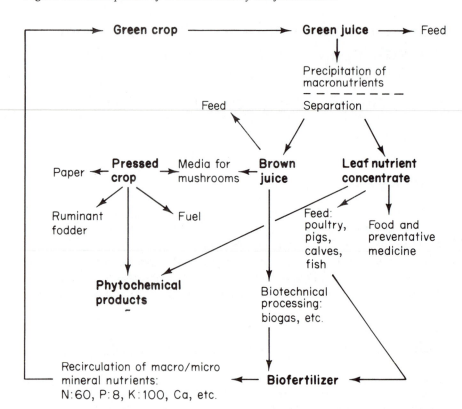

added brown juice (Fig. 10.1), is used as ruminant feed, even though the protein content is still rather high. In other cases there is a trend to use the pressed crop for biogas production, or as a source of fibre for the paper and fibre industries, or for the production of solid and liquid fuel. In the latter case the pressed crop is 'cleaned' of other substances such as proteins. As both the fibre and fuel industry can accept large quantities of the pressed crop, this aspect of its use is receiving greater attention in industrialized countries today where there is an over-production of food and feed grains. There is also a minor interest in the use of the pressed crop as a substrate for the local production of mushrooms.

The green juice can be used direct as in traditional diets such as Kola Kanda in Sri Lanka, and as a health food drink. However, the largest use is in the raw state as a liquid feed for swine.

The second fractionation is made by heat-coagulation or acid-coagulation of the proteins in the green juice. The acid-coagulation can be by lactic acid fermentation of the sugar in the green juice to a final pH of about 4. By separation of the coagulated leaf protein and other macromolecules, a leaf protein/nutrient concentrate (LPC/LNC) and a protein-free brown juice are obtained. The LNC is rich in high-quality protein, pigments (chlorophylls, carotenoids), polyunsaturated fat (linoleic acid), starch and minerals such as calcium and iron. The brown juice is rich in sugars, amino acids and soluble materials such as potassium, phosphorus, nitrogen (as nitrates), and other macro-minerals and micro-minerals, which make it a useful biofertilizer.

In large scale processing LPC is used as non-ruminant feed for poultry, swine, calves and sometimes fish. For more or less direct consumption, LNC is used in selected villages as a food-supplement, rich in pro-vitamin A, protein, calcium and iron. The xanthophylls of the carotenoids are used as feed/food colourants, either directly or by conversion by the consuming animal. The possible anti-carcinogenic effects by carotenes such as beta-carotenes is also being investigated. Since all the valuable constituents in the LPC/LNC come from the green leaf, it should be emphasized the dark-green leafy vegetable should be eaten as a first choice before making LNC. On the other hand many leaves contain much anti-nutritive substances such as phenolics/tannins, saponins, phyto-oestrogens, goitrogenic and cyanogenic substances, alkaloids, nitrates and oxalic acid as soluble constituents, and proteins such as anti-trypsins and phyto-haemagglutinins. These substances are however digested as part of the normal metabolism and are often used as defence against microbial and insect attacks. Fortunately, the production of LNC reduces the amounts of these substances to about one to ten per cent of the original level, and/or inactivates the proteins by coagulation. The LNC is therefore advantageous for consumption where a concentrated diet is necessary, e.g. children and pregnant mothers, and where a diet largely free of anti-nutritive substances is required.

The brown juice can be used as a liquid feed for ruminants. However, the main interest today is in making biogas from the soluble carbohydrates or in making ethanol, in both cases as energy sources. The final residual juice, after a fermentation, is suitable for use as a biofertilizer.

Phytochemicals can also be produced from the wet-fractionation of green biomass. Cellulose, hemicellulose, pectin and lignin are obtained from the pressed crop. Proteins such as ribulose-diphosphate-carboxylase and other soluble proteins, glycoproteins and phytohaemagglutinins are part of the LPC, and can be fractionated off during LPC production if necessary. Lipids such as phospholipids, galactolipids, sterols (glycosterol, tocopherol, cholesterol), phenols (plastoquinone, ubiquinone), carotenoids (carotenes, xanthophylls, chlorophylls, etc.) are also found in the LPC. If needed these lipophilic substances can also be extracted. Part of the pharmacologically active substances mentioned above are also found in brown juice constituents.

Reviews on wet-fractionation of green plants and production of LPC, have been given by Pirie (1971, 1986), Costes (1981), Singh (1982), Telek & Graham (1983), and in the Proceedings of the 2nd International Leaf Protein Research

Conference (Tasaki 1985) and of the 15th International Grassland Congress (1985).

PLANT SPECIES USED FOR WET-FRACTIONATION

Among the species investigated for LPC production, those with a high dry matter and protein content and which give a high yield of high-quality leaf protein as LPC are the most desirable (see references above). Such species often belong to such plant families as: Leguminosae subfamily, Papilionoideae, Chenopodiaceae, Cruciferae, Solanaceae, Amaranthaceae and Cucurbitaceae.

(a) Leguminosae species suitable for wet-fractionation and multiple use of the wet-fractionation products are, e.g. *Medicago sativa* and species of the genera *Trifolium, Lupinus, Vicia, Pisum* and *Phaseolus* among temperate species, and *Vigna, Canavalia, Clitoria, Desmodium* and *Psophocarpus* species among the tropical genera. The advantage of leguminous species is that no nitrogen fertilizer is needed; on the other hand much potassium fertilizer is required. The yield of extractable leaf protein vary from 1000 kg to 3000 kg per ha.

(b) Centrospermae species such as those belonging to the Chenopodiaceae and Amaranthaceae are grown as leafy vegetables or spinaches in temperate and tropical countries (*Chenopodium, Atriplex, Beta, Spinacia, Amaranthus, Celosia*), plus the saltbushes *Atriplex, Kochia*), have given leaf protein yields between 1000 to 1500 kg per ha.

(c) Cruciferae species include those of *Brassica* (rape, kale, cabbage, mustard), *Raphanus* (radish) give yields similar to the Centrospermae species mentioned above.

(d) Solanaceae species used for the production of LPC are potato (*Solanum*) and tobacco (*Nicotiana*).

(e) Cucurbitaceae species also gives good quality LPC, but no yield data are available.

From other plant families one need only mention individual species such as *Helianthus annuus* (sunflower) and *H. tuberosus* (Jerusalem artichoke), *Urtica dioica* (stinging nettle), green cereals and *Tithonia rotundifolia* (syn. *T. tagetiflora*).

The species mentioned above are easy to wet-fractionate and the quality of their LPCs is good. Many tropical grasses, when fertilized, also give high yields of dry matter and of protein. One such species used for LPC and fibre production is *Pennisetum purpureum* (elephant or Napier grass) and can yield 3000 kg extracted leaf protein per ha. The energy consumption for wet-fractionation of such C4 grasses, however, is relatively high.

Another source of plants for biomass production and wet-fractionation are those growing in marshes or as aquatic plants. Such plants can also purify water at the same time as providing products for feed or energy. Some species worthy of special mention are *Eichhornia crassipes, Ipomoea aquatica, Nasturtium officinale, Nymphea* spp., *Pistia stratiotes, Polygonum* spp. and *Arundo donax, Phragmites australis* (syn. *P. communis*), *Typha latifolia* and some *Atriplex* spp. (Carlsson 1982).

PLANT SELECTION AND BREEDING

Preliminary studies to select especially suitable cultivars, provenances, breeding lines (strains) and even individual plants from established agricultural crops and some promising under-exploited plants have been carried out to increase the yield and quality of LPC from plant shoots (Carlsson 1982). Cultivars of *Brassica* spp., *Helianthus annuus, Medicago sativa*, and *Solanum tuberosum* have been investigated. Provenances of *Amaranthus, Atriplex* and *Chenopodium* species have also been studied, as well as individual plants of *Medicago sativa* and *Trifolium pratense*.

EFFECTS OF PLANT MATERIAL ON LPC QUALITY

Both yield and quality of LPC from a plant material is affected by the specific composition of substances other than those of protein and other macromolecules in such LPC. Species and cultivars contain specific secondary plant substances such as phenolics or tannins and their oxidation products, which reduce the extractability of the protein content by coagulation (unless a reducing sulphite is added), or react with amino acids such as lysine and cysteine, thereby reducing the proteins' nutritive value. (Table 10.1).

Table 10.1: *Protein Efficiency Ratio (PER) of LPC from Different Species*

Species	Non-Treated LPC	Sulphite-Treated LPC
Amaranthus hypochondriachus	1·8	1·9
Atriplex hortensis	1·5	2·0
Brassica hirta	2·1	2·1
B. napus	2·1	2·0
Chenopodium quinoa	1·9	2·0
Helianthus annuus	1·5	2·0
Medicago sativa	1·5	2·0
Casein control	2·5	—

The composition and amount of secondary substances varies with the physiological stage of development of the plant; this can affect the nutritive value of the protein in the LPC (Table 10.2).

Table 10.2: *Net Protein Utilization (NPU) of LPC from Species at Different Physiological Stages*

Physiological Stage	*Atriplex hortensis*	*Chenopodium quinoa*
Vegetative	55	—
Bud forming	63	55
Flowering	57	60
Seed setting	51	61
(Casein control: NPU=69)		

SOME PHYTOCHEMICAL SUBSTANCES OF WET-FRACTIONATED GREEN BIOMASS

In spite of the multitude of nutrients in green LPC, white leaf protein products as such have a special interest for use in different industries. One reason for this is their high nutritive value; another is their advantageous functional properties; a third is that due to its extreme purity and its complete digestibility, the crystalline ribulose-diphosphate-carboxylase can be used in medicine as a dietary protein for kidney patients.

Because of its anatomy C3 species contain 30 to 45 per cent more crystalline white protein than C4 species, and 10 to 15 per cent more soluble leaf proteins. Such a high proportion of soluble leaf protein indicates the feasibility of producing the crystalline fraction. A high proportion of soluble protein, compared to membrane-bound protein, is found in species of the Chenopodiaceae, Cruciferae and Solanaceae, plus individual species of Leguminosae. Crystalline protein has mainly been associated with Solanaceae species, e.g. *Nicotiana* spp. However, with a new technique the same type of crystalline protein has been produced from *Spinacia oleracea* and *Medicago sativa*. Complete data are given on white protein products in a review by Carlsson (1986).

Carotenoids have a wide application for industry. A major use is as food colourants and as pigments in animals' diets to cause colouring effects on the muscles of fish, the skin of broilers and egg yolk. In other cases, carotene (beta-carotene) is essential for fertility and normal pregnancies of animals. In many countries synthetic carotenoids have been given a restricted use or have even been forbidden in animal diets, consequently there is a strong interest in natural carotenoids. LPC, and especially the green fractionation of LPC, contains 10 to 20 times the amount of carotenoids as the original leaf. The LPC from some species may contain the equivalent of about 800 kg lipids per ha.

Different species and plant families contain different secondary plant substances. Solanaceae species, for example, contain alkaloids that could be used as biological pesticides and as a base for production of different medicines; these alkaloids are found in the brown juice. Chenopodiaceae and Amaranthaceae species have saponins in the brown juice, likewise *Medicago sativa and Trifolium* spp. produce a brown juice containing plant oestrogens, etc.

106

The pressed crop, a bulk product from the wet-fractionation of green crops, is creating considerable interest in Sweden as a basis for the production of short fibres for the paper industry. Plants such as *Medicago sativa, Urtica dioica, Miscanthus sinensis* and *Salix* spp. may be possible sources for a combined production of protein for feed and fibre for paper. Green crops apparently offer possibilities for both the production of bulk phytochemicals and for production of the special products for the pharmaceutical industry.

CONCLUSION

The growing interest for green biomass and the different valuable products that can be obtained from the same crops emphasize the need to investigate all green plants for multiple uses, and to save the genetic pool of the green plants. As a specific example one can emphasize the need to save plants from the genetic pool of the pseudocereals of the Amaranthaceae and Chenopodiaceae.

REFERENCES

Carlsson, R. (1982) Trends for future applications of wet-fractionation of green crops. In: *Forage Protein Conservation and Utilization. Proceedings of a Seminar in the EEC Programme of Coordination of Research on Plant Protein*, T. W. Griffith and M. F. Maquire (Eds), pp. 241–247. Grassland Research Institute, North Wyke, Okehampton, Devon.

Carlsson, R. (1986) White leaf protein products for human consumption. – A global review of plant material and processing methods. In: *Proceedings of the 1st International Congress on Health and Food*, Section 'Tobacco Protein Utilization Prospects', P. Fantozzie (Ed.), pp. 125–145. Salsomaggiore Thereme, Italy,

Costes, C. (Ed.) (1981) *Proteines Foliaires et Alimentation*. Guathier-Villars, Paris.

Pirie, N. W. (Ed.) (1971) *Leaf Protein: its agronomy, preparation, quality and use*. Blackwell Scientific Publications, Oxford.

Pirie, N. W. (1986) *Leaf Protein and Other Aspects of Fodder Fractionation* (2nd Ed.). Cambridge University Press, London.

Proceedings of the 15th International Grassland Congress, August 1985, Kyoto, Japan. The Science Council of Japan and the Japanese Society of Grassland Science, Nishi-nasuno, Tochigi-ken, pp. 329–327.

Singh, N. (Ed.) (1982) Progress in Leaf Protein Research, *Proceedings of the 1st International Leaf Protein Research Conference, October, 1982, Aurangabad, India*. Today and Tomorrow's Printers and Publishers, New Delhi.

Tasaki, I. (Ed.) (1985) *Proceedings of the 2nd International Leaf Protein Research Conference, August 1985*. Nagoya & Kyoto, University of Nagoya.

Telek, L. and H. D. Graham (Eds) (1983) *Leaf Protein Concentrates*. AVI Publishing Co. Inc., Westport, CT.

11

Strategy for Development of a New Crop

L. Lazaroff

INTRODUCTION

The International Council for the Development of Under-utilized Plants (ICDUP) has been attempting since its formation in 1978 to encourage the development of *Psophocarpus tetragonolobus*, popularly known as 'winged bean', as the first of what it hopes will be a number of selected plants with outstanding potential for humankind which it intends to support. Its experience in the determination of a strategy – a set of goals and the means of reaching them – for the purpose of encouraging that development may be of interest. This is, of course, not to say that the strategy for winged bean should be followed to guide the development of other plants. But it may be useful to recognize the dimension of the problem as we found it and the effort to focus on those issues considered most important, what these were, how they emerged and the results.

The impetus for the formation of ICDUP was the summary report first published in 1975 by a special panel of the US National Academy of Sciences (NAS 1975). Almost immediately, on publication of the report, a small group attempted to get things moving. It adopted a tentative set of goals, based to an extent on the Academy panel's recommendations and, with support from a private American foundation, proceeded to get the views of the principal winged bean researchers identified in the Academy report on whether and how this might be done. The result was the holding of an international meeting of a much broader body of researchers numbering about 200, from 26 countries and six continents, from developing and developed countries, for the purpose of bringing information on the winged bean up to date and for charting the direction of further development. It was at the conclusion of this meeting and for the purpose of encouraging and guiding continued development, that ICDUP was created.

THE STRATEGY

The unique method followed by the Council, on the recommendation of the assembled researchers, was to secure the cooperation of an international network of skilled scientists and practising agriculturists who would work on aspects of the problem in their home institutions in various parts of the world and who would share their results. No large, central research institution with heavy overheads was contemplated. The Council was to coordinate their work under an agreed-upon system of research priorities. This has worked.

The Council itself, which is non-profit making, is directed by an international board of distinguished agricultural scientists. Present members of the Board include Nobel Laureate Dr. Norman Borlaug, Dr. M. S. Swaminathan, Director General of IRRI, and Dr. C. C. Tan of the People's Republic of China. Until their death, Dr. Jose Drilon, Director of SEARCA in the Philippines and Professor J. P. M. Brenan, former Director of Kew Gardens, were active members. In addition, a Winged Bean Steering Committee, composed of winged bean researchers, was appointed to provide professional advice and guidance on Council grants.

With funds secured from interested supporters, the Council provides grants for research, principally by research workers in the developing world, for practical experimentation and field studies, for advanced training for younger researchers from developing countries to carry on the work, and for the application of research results to small farmers and processors.

The strategy to be pursued, as it evolved through discussion at the first international winged bean conference, meetings of the Board of Directors and of the Steering Committee, and refined more sharply at an even larger second international meeting of winged bean researchers in 1981 (over 200 from 36 countries), and the broad participation of researchers must be emphasized, was designed to accomplish three things:

(a) Early domestication and commercialization of the plant.

(b) Encouragement of research and application of that research by ICDUP and by others.

(c) Preparation and distribution of seeds and information on latest findings to farmers and policy makers.

It was emphasized repeatedly that the overall purpose of the work must be to benefit humankind and that the research results must reach farmers and policy makers.

Priorities were established to govern stimulation and support of research, action and training programmes.

These priorities may be of interest in considering the development of other under-utilized plants. The taxonomic work having been completed by Dr. B. Verdcourt and Miss P. Halliday (1978), the priorities now governing the Council's work are as follows:

(a) *Establishment of a central germ plasm collection:* Establishment of a centralized, well-organized gene bank for documenting, maintaining long-term storage, testing, propagating and distributing lines of all nine species in the genus.

(b) *Improvement of the crop*

    (i) Accelerated breeding programmes developed against carefully defined objectives, and taking into account the various end-uses of the winged bean. (Includes purification of existing lines; development of high-yielding varieties with non-shattering pods; development of light-coloured seeds, easily dehulled and of high nutritional value.)

    (ii) Research on development of the tuber since this part of the plant has been largely neglected. (Includes methods of propagating in different environments, nutritional content, harvesting, storing and alternative uses.)

    (iii) Further development of the winged bean as a green vegetable among subsistence farmers for small markets and for use as animal forage.

    (iv) Exploration of winged bean as a cover crop under plantation crops such as oil palm, coconut, rubber and coffee.

    (v) Development of self-supporting, determinant varieties that are day-neutral this was judged to be of highest importance. It was suggested that these could possibly be found growing naturally, or be developed through hybridization, tissue culture or genetic recombination. Such varieties could then be improved on lines already suggested for easier harvesting.

(c) *Reducing environmental and biological stress:* Identifying environmental and biological stress, including the effects of temperature, waterlogging, the interactions of genotype and environment and devising methods of reducing them and protecting the plant against disease.

(d) *Agronomy:* Various aspects of the agronomy require urgent research and study designed to achieve proper crop management and to integrate winged bean into existing agricultural systems.

(e) *Processing and product development:* Research on the processing and testing of the ripe seed for milk, edible oil, and of the residues for animal feeds; methods of deriving protein from the leaf; uses of the tuber for food and feed; uses of the ripe seed hull; research on processing that can be done on the farm (pickles, sauce, etc.) and commercially.

(f) *Marketing:* Methods of marketing winged bean products so as to ensure adequate production for commercial processing through contract farming among smallholders.

(g) *Harvesting and storage:* Harvesting and storage of ripe seeds, tubers, and immature pods.

(h) *Informational needs:* Preparation and distribution of relevant information as required for farmers and policy makers.

What is not stated under (h) above, but has been emphasized in plans for expanding uses of winged bean in new areas, is the importance of devising methods of encouraging uses of winged bean parts (immature pods, unripe and ripe seeds, leaves, flowers, shoots and tubers) into traditional local dishes and, for young children in particular, methods of preparing the leaves which are extraordinarily rich in vitamin A.

Two principles undergird the above priorities: (1) that cooperation with and involvement, if possible, of local authorities and institutions must be considered essential, and (2) that ICDUP strategy and priorities be treated as flexible and continuously reviewed.

In order to assure adherence to these agreed upon priorities, a rather strict set of 'Guidelines for Consideration of Winged Bean Research Proposals' was published in the Council's newsletter, *The Winged Bean Flyer* (Anon 1981). The full text of these guidelines, which indicate even more fully the shape of the Council's intentions, appears as Appendix A to this paper. It details a number of 'research thrusts', each based on the priorities with objectives given under each and a suggested strategy for achieving them. (A format is also given suggesting how the proposals should be organized, including a breakup of the anticipated expenditures. Each proposal for support must indicate the 'other sources of support' for the project and their nature, whether in cash and/or (estimated) in kind, it being indicated that such support is expected.)

Early on, during the discussion of strategy, a polar disagreement emerged over which cultivators' needs would be emphasized in the research, i.e. those of the subsistence farmer and smallholder, or those of the commercial firm.

For the winged bean is now known to be a vine crop that can grow into a bushy mass up to four metres tall. It is indeterminate and must be firmly staked for highest yield of immature pods and ripe seeds. It is an ideal crop for the subsistence farmer. The immature pod, the part most commonly used, can be harvested as a green vegetable every few days over a two to three month period, while the green leaves and the dried haulm can be fed to his animals. Traditionally, the smallholder will grow only a few plants, largely for household use and animal feed and for sale of immature pods and ripe seeds in the local bazaars or market towns.

But it is labour intensive and expensive to grow on a large scale. The wooden stakes, essential for highest yield, are costly, should be treated if they are not to sprout themselves in the moist tropical climate, and must be replaced at least every three years. Labour costs even in developing countries with large populations have risen.

Should the Council concern itself primarily with improving the winged bean as a household vegetable, soil improver and animal feed suitable as it is now to the moist tropics with its short days; should it defer or reject engaging in the difficult struggle to change its architecture, render it determinate and make its cultivation possible outside its present climatic zone?

There were many who strongly believed that the needs of the subsistence farmer and smallholder in developing countries for a protein-rich food and perhaps additional income were greatest, and that research to develop a plant that would meet the needs of large-scale cultivation, the commercial firm and the international market would capture greatest interest and dominate the effort. It was not readily recognized that developing improved varieties for the smallholder would not necessarily result in expanded use. For, as Professor Brenan put it, use of winged bean by the subsistence farmer and smallholder as a green vegetable and animal forage represents only a part of the potential of the plant and a part whose development is likely to be 'slow and protracted' and that 'the impact will be narrow'. Farmers will expand production of an existing plant or introduce a new one only if they are certain they can make additional money by doing so and are not likely to alter their farming systems because a plant may be more nutritious.

Within two years it had become clear that being concerned primarily with the needs of the subsistence farmer would severely limit prospects for the crop reaching its potential as a source of human benefit and it was agreed that we must work on both targets.

A number of researchers experimented with ways of processing winged bean parts. Among other things, work began on methods of easily removing the hard coat from the ripe seed by adapting existing rice hulling machinery. Flour, some fat-free, was easily made from the ripe seed and from the tuber and successfully introduced into bread, cakes and snacks. Nutritious weaning foods from the ripe seed were successfully tested in Vietnam, Ghana and Czechoslovakia. The European firm that had developed machinery for making liquid vegetable milk from the soya bean was able to adapt the machinery to make milk from ripe winged bean seed. The resulting milk tasted better than soy milk, proved to be as rich and was sterilized to provide long shelf life. It was expected, with good reason, that the meal resulting from the processing of the oil and milk, enriched with methionine, could be sold as feed for poultry and pigs. Definitive tests on the effect of feeding the enriched meal to piglets, chicks and calves are now being planned to demonstrate weight gains.

Commercial firms were prepared to discuss the possible production of liquid milk or oil, but only if an adequate supply of good quality seed at a good price could be expected to be available on a regular basis to keep the mills going. How to accomplish this when the winged bean was too expensive to grow on a large scale and the seed could be obtained only in small amounts from smallholders?

We believe we have the answer. Sufficient volume of seed to meet the commercial demand can be achieved, given the plant's present architecture, if farmers with small plots were to plant under a contract farming system. Such systems, employed in developing countries such as the Philippines and Sri

Lanka for the production of fresh fruit or tobacco, provide participating small-holders with required inputs and extension information and arrange for the collection of harvest at regular intervals from farmers in the scheme under agreement guaranteeing the price. Instead of growing only the few winged bean plants that has been traditional, farmers could be encouraged to grow it more extensively, provided they could be convinced it would be profitable for them to do so. This means expanded cultivation of winged bean must be fitted into exist-ing farming systems without prejudicing existing earnings but rather resulting in greater overall returns.

In a project we are discussing for one developing country, it is proposed that farmers who would be involved in the production of winged bean under a contract farming system would grow it on only 0·1 ha of their land.

For such a contract farming system to work successfully, however, the local group responsible for managing it must have achieved the trust of farmers, a reputation for fair dealing and a record of efficient, dependable operation. Farmers considering taking part in such a scheme must be able to rely on the guaranteed price as well as on the availability of inputs and appropriate extension information at the time and in the amounts required.

An ICDUP project for the production of liquid vegetable milk has been submitted to an African nation and to potential funders for their consideration. This would be the first project that the Council would itself be administering.

One aspect of this plan reflects our awareness that in introducing or even expanding the production of winged bean where it is already grown, that we treat the problem as site specific. Thus in this project (and indeed in all proposals made to us by commercial firms proposing, for example, to grow winged bean for feed for their animals), we insist on the preliminary testing on the site of between 20 and 30 varieties of winged bean selected for the particular features desired and most suited to the areas proposed for cultivation. The elite varieties selected would be tested on a relatively small, sample plot over two and preferably three growing periods and progressively narrowed down to the five to eight varieties judged to be outstanding. (Details of the rather rigorous testing procedure proposed, including the preliminary information which would need to be gathered on the local environment, etc., are available for examination if desired.)

A programme for the development and introduction of winged bean must always be politically sensitive. Headquartered as we are in the United States we must be alert to the political effect of our work on supporters of the American cultivation of soybean. What lack of interest or indeed opposition are we likely to encounter in encouraging development of a self-standing determinant crop, capable of being harvested mechanically, and perhaps relatively insensitive to length of day? If we are successful in developing such varieties, as we expect we will be, what hazards can we expect to run? We have been attempting to establish close relationships with researchers in the US who have long been conducting significant work on soybean and we plan to meet with

113

representatives of the soybean industry. In our strategy, we do not see winged bean as in competition with, but rather complementary to, the US soybean industry.

CONCLUSIONS

It may be useful to close this paper with a few observations on strategies for a 'new crop' based on our experience with winged bean. Although we have no illusions about the transferability of such experience, we suggest the following for the reader's rejection:

(a) Any plan designed to increase cultivation of a given plant or to introduce a new one must be developed hand-in-hand with a programme to increase the market for it. The two efforts must be concurrent and complementary or either will dissipate.

(b) Local political, administrative, scientific and popular support must be developed to encourage cultivation of a new crop or expansion of a known crop whose cultivation is being neglected. Three points might be emphasized in this connection: Be aware of existing prejudices against a neglected crop. Recognize the role of women in the region as farmers, housewives and sometimes income managers in the field and on the hearth, design the programme accordingly and enlist them as allies. Finally, give a few leading farmers in the areas roles as partners in introducing the new crop, letting them share seeds and advice with friends and neighbours.

(c) Wherever possible, encourage the preparation by a local institution (another potential ally) of clear, easily understood and simply illustrated materials in local languages for farmer and/or extension agent guidance or, if literacy is limited, illustrated handouts.

(d) Keep local officials closely informed, listen to what they have to say and encourage them to think of themselves as partners in the enterprise.

(e) Based on ICDUP's experience, we would strongly recommend that any programme be tailored to local realities, the selection and testing of varieties suitable to the local ecology and the development of the agronomy appropriate to local environments and customs, local farm size, existing prices and availability of inputs, the effectiveness of existing extension systems, marketing structures, etc. Careful effort must be made to incorporate the new crop into existing farming and marketing systems.

REFERENCES

Anon (1981) Guidelines for Consideration of Winged Bean Research Proposals. *The Winged Bean Flyer* 3, 2: 42–47.

NAS (1975) *The Winged Bean. A High Protein Crop for the Tropics*. National Academy of Sciences, Washington, DC.

Verdcourt, B. and P. Halliday (1978) A revision of *Psophocarpus. Kew Bull.* 33, 2: 191–227.

Appendix A

GUIDELINES FOR CONSIDERATION OF WINGED BEAN RESEARCH PROPOSALS (Anon 1981)

RESEARCH THRUSTS

1. Germplasm Collection

    A. Objectives

       (i) Determine the range of variability within three species of *Psophocarpus*, i.e. *tetragonolobus, scandens, palustris.*

       (ii) Make materials available to plant breeders.

    B. Strategy

       (i) Appoint Germplasm Collection and Screening Committee, to provide leadership and continuity in this effort, which must extent to Asia, SE Asia, Africa and Oceania.

       (ii) Establish a centre for collecting, cataloguing, testing, evaluating, storing and maintaining germplasm.

       (iii) Make materials available to plant breeders.

2. Crop Improvement

    A. Objectives

       (i) Homozymous lines for greatly reduced segregation (variability) within lines.

       (ii) Specialized lines and combinations for green pods, dry seeds, storage roots, leaves, for food, feedstuff, oil, forage, soil improvement and ground cover.

       (iii) Dwarf lines – short internodes – determinate – self supporting for single destructive harvest, and lines with minimum or no photosensitivity.

       (iv) Lines with highest nutritional content and lowest proportion of anti-nutritional factors of various plant parts.

       (v) Low pod wall fibre so that pods of large size remain edible and shattering of dry pods is reduced.

      (vi)   Seed coats more permeable to water and easier to remove (dehull).

B.   Strategy

    (i)   *Indeterminate habit* (climbing or trailing) through –

      (a)   Selection and breeding.

      (b)   Hybridization and variety synthesis.

    (ii)   *Determinate habit* (dwarf type) through –

      (a)   Worldwide search for naturally-occurring dwarf types to be encouraged by cash reward.

      (b)   Mutation breeding coupled with tissue culture technology and genetic engineering.

3.   Reducing Environmental Stress

   A.   Objectives

      (i)   Efficient production (input : output).

      (ii)   Local adaptation.

   B.   Strategies

      (i)   Rhizobium studies.

      (ii)   Cultural practices, studies, such as: crop establishment, supports and vine training, irrigation, fertilization, plant density and arrangement, use of growth substances.

      (iii)   Studies of effects of moisture, drought, environmental stress.

4.   Reducing Biological Stress

   A.   Objectives

      (i)   Control of insects, nematodes, diseases, weeds, rodents.

   B.   Strategies

      (i)   Resistant, vigorous, well-adapted varieties.

      (ii)   Chemical protectants.

      (iii)   Introductions of predators and parasites that feed on harmful insects and creation of conditions favourable to these beneficial insects.

      (iv)   Alteration of soil conditions to favour beneficial rhizobia and discourage nematodes.

      (v)   Detailed studies of insects, diseases and nematodes affecting winged beans with special attention to viruses that might infect other food plants in areas where winged beans are not now grown... special attention to *Phaseolus* beans.

5.   Harvesting and Handling

   A.   Objectives

      (i)   Uniform senescence.

      (ii)   Non-shattering pods.

      (iii)   Uniform seed and seed coat maturation and moisture content.

6.   Post-Harvest Physiology

    A.   Objectives

        (i)   Prevent loss of dry seeds, pods and edible root tubers to insects and diseases.

        (ii)   Prevent loss of food value.

        (iii) Prevent (retard) loss of seed germination.

        (iv) Understand conditioning of seed for planting or for food and feed uses.

    B.   Strategies

        (i)   Chemical treatment of seeds.

        (ii)   Mixing of winged bean seeds with those of other legumes to reduce insect infestation.

        (iii) Controlling temperature and humidity for dry seed storage.

        (iv) Determination of optimum moisture content for stored seed.

        (v)   Determination of methods for conditioning stored seed prior to planting and using for food.

7.   Product Development

    A.   Objectives

        (i)   Development of wholesome and palatable food and feed products such as: flour, milk, TVP, oil, protein isolates, etc.

        (ii)   Combination of winged bean products with more conventional products.

    B.   Strategies

        (i)   Biological and physical analyses of plant parts, plant products and extracts.

        (ii)   Determination of available nutrients coupled with animal feeding trials.

        (iii) Screening of selections for special useful attributes.

        (iv) Development of methods of removing beany smell and flavour.

        (v)   Development of machinery appropriate for small-scale processing of seeds for oil, flour, milk, etc. at the village level.

8.   Economics and Marketing

    A.   Objectives

        (i)   Develop model *production and distribution* enterprise analyses for different locations to estimate resources required for intensive and extensive production.

        (ii)   Develop model processing and enterprise analyses to estimate the amount of raw product required for efficient manufacture of various products.

    B.   Strategies

        (i)   Develop cost of production estimates for various locations, production and distribution systems.

        (ii)   Develop cost of manufacture estimates for large and small-scale processing units.

9.   Outreach Thrusts (Extension and Education)

    A.   Objectives

        (i)   Popularize winged bean products and increase skills in production and use.

        (ii)   Create an understanding of the benefits from eating winged beans.

        (iii)   Create an awareness of the financial benefits from growing and selling winged beans.

        (iv)   Increase production and distribution of winged bean seeds of recommended varieties.

    B.   Strategies

        (i)   Demonstrations and tests (at no risk to growers).

        (ii)   Publications – production guides and recipes.

        (iii)   Local workshops to provide information to outreach personnel.

        (iv)   Film with soundtrack in different languages (such as film produced in Thailand).

OUTLINE FOR FUND REQUESTS

Ten copies of your application should be forwarded to:

        Mr. Louise Lazaroff
        Director General
        International Council for Development of Under-utilized Plants (ICDUP)
        18, Meadow Park Court
        Orinda, CA 944563
        U.S.A.

All requests for funds should be organized as follows:

1.   Title of the project

2.   Responsible institution and department/branch/section

3.   Immediate research or project supervisor(s)

4.   Expected duration of project

5.   Description of project

      (i)   Introduction and background information
      (ii)   Purpose
      (iii)   Theoretical considerations

    (iv)   Methodology and plan of work

    (v)   Experimental plan for the first year

    (vi)   Anticipated impact of this project

6. Budget for annual expenditure (Please give full details under each item):

    (i)   Salaries

    (ii)   Operating Expenses

    (iii)   Materials and supplies

    (iv)   Travel/transportation

    (v)   Other

    TOTALS:

7. Other sources of support available for the project; counterpart support (nature and amount in cash and/or in kind in U.S. dollar equivalents

8. Winged bean research being undertaken with other resources.

9. References.

10. Signature of Head of Institution.

CROPS – REGIONAL ASSESSMENTS

12

New Crops for Food and Industry: the Roots and Tubers in Tropical Africa

B. N. Okigbo

INTRODUCTION

Root and tuber crops consist of all carbohydrate-rich underground storage structures that are consumed by man and/or domestic animals. Strictly speaking, from the botanical point of view, root crops are the edible underground storage structures developed from modified roots, while tuber crops are those crops in which the edible carbohydrate-rich storage organs are developed wholly or in part from underground stems. Edible underground storage organs thus consist of diverse variously modified roots such as cassava (*Manihot esculenta*), sweet potato (*Ipomoea batatas*), and carrot (*Daucus carota*), stems or tubers as in yams (*Dioscorea* spp.), Irish potato (*Solanum tuberosum*) and the Jerusalem artichoke (*Helianthus tuberosus*), corms, cocoyams (*Xanthosoma* spp., *Colocasia* spp.) and rhizomes, e.g. ginger (*Zingiber officinale*). Sometimes included under root crops are storage structures such as tubers or bulbils of the aerial yam (*Dioscorea bulbifera*) borne on stems and the edible swollen leaves of onions (*Allium cepa*). It is no wonder then that roots and tuber crops are found in at least 20 diverse families of plants, including for example, the Araceae, Aponogetonaceae, Bassellaceae, Cannaceae, Compositae, Cyperaceae, Dioscoreaceae, Euphorbiaceae, Labiatae, Marantaceae, Oxalidaceae, Leguminosae subfamily Papilionoideae, Solanaceae, Umbelliferae and Zingiberaceae.

There are two main groups of roots and tubers found in Africa according to their centre of domestication, those of African origin and others that are exotic.

Indigenous African roots and tubers

These consist of roots and tuber crops that were independently ennobled on the African continent, and include various species of yams such as the white Guinea yam (*Dioscorea cayenensis* subsp. *rotundata*), yellow yam (*D. cayenensis*), the bulbil-bearing or aerial yam (*D. bulbifera*) and the three-leaved yam (*D. dumetorum*), the yam bean (*Sphenostylis stenocarpa*), which is also grown for its edible protein-rich seeds, the tigernut (*Cyperus esculentus*), the Hausa potato (*Solenostemon rotundifolius*, syn. *Coleus dysentericus*) and risga (*Plectranthus esculentus*, syn. *Coleus esculentus*), both of the mint family Labiatae, and very minor roots and tubers which are only consumed at the time of famine, such as the false yam (*Icacina senegalensis*) and the elephant yam (*Amorphophallus* spp.) (see Table 12.1).

Table 12.1: *Indigenous Root and Tuber Crops of Africa*

Scientific Name	Common Name	Distribution and Status
Amorphophallus spp.	elephant yam	Humid and subhumid tropical Africa – wild.
Dioscorea bulbifera	aerial yam	Tropical Africa – cultivated and wild[3]
Dioscorea cayenensis	yellow Guinea yam	Humid tropical areas of West and Central Africa[2]
Dioscorea dumetorum	three-leaved yam	Tropical Africa, cultivated in Eastern Nigeria
Dioscorea praehensilis	bush or forest yam	Humid West Africa – wild and cultivated
Dioscorea rotundata	white Guinea yam	Humid and subhumid areas of West Africa – cultivated[1]
Icacina senegalensis	false yam	West and Central Africa – wild
Plectranthus esculentus	risga	Tropical Africa, cultivated[3]
Solenostemon rotundifolius	Hausa potato	Tropical Africa, cultivated[3]
Sphenostylis stenocarpa	African yambean	West and Central Africa, cultivated.
Stylochiton lancifolius syn. *S. warneckei*	Warnecke's ground arum	Tropical Africa – wild

1. Of major regional or local importance in West Africa. 2. Of limited local importance in West Africa. 3. Of minor local importance in parts of West Africa. Source: Okigbo (1983) and Kay (1973).

Exotic roots and tubers in Africa

These are root and tuber crops of varying nutritional and economic importance which have been introduced into different parts of Africa. They consist of the water yam (*Dioscorea alata*) and Chinese yam (*Dioscorea esculenta*) from central Asia, tannia (*Xanthosoma sagittifolium* and other *Xanthosoma* spp.) from Latin America, the cassava (*Manihot esculenta*) of Latin American origin, the Irish potato (*Solanum tuberosum*) from South America as is also the sweet potato and miscellaneous minor root crops, including the carrot, ginger, radish (*Raphanus sativus*), Queensland arrowroot (*Canna edulis*) and turmeric (*Curcuma longa*) (see Table 12.2).

This paper deals with the processing and utilization of some of the important roots and tubers grown in Africa, i.e. yams, cassava, cocoyam and sweet potato.

IMPORTANCE OF ROOTS AND TUBERS IN AFRICA

In general, the importance of roots and tubers is in their use as sources of food, feed and miscellaneous products required by man. When sold, roots and tubers are also sources of local income but, since their products are rarely exported, they contribute little to balance foreign exchange. African and world production of cassava, yams, sweet potato and cocoyam are shown in Table 12.3.

Table 12.2: *Exotic Root and Tuber Crops Grown in Tropical Africa*

Scientific Name	Common Name	Origin	African status and location
Beta vulgaris	garden beets	Mediterranean region & Europe	Trop. Africa (highlands)
Canna edulis	Queensland arrow root		Trop. Africa[3]
Colocasia esculenta	dasheen, edo	SE Asia	Trop. Africa[2]
Curcuma longa	turmeric	SE Asia (India)	Trop. Africa
Cyperus esculentus	chufa	Mediterranean	Trop. Africa
Daucus carota	carrot	Mediterranean	Trop. Africa[2]
Dioscorea alata	water yam	SE Asia	Trop. Africa[2]
Dioscorea esculenta	Chinese yam	SE Asia	Trop. Africa[3]
Helianthus tuberosus	Jerusalem artichoke	N America	Trop. Africa
Ipomoea batatas	sweet potato	C&S America	Trop. Africa[2]
Manihot esculenta	cassava, manioc	C&S America	Trop. Africa
Raphanus sativus	radish	Europe & Asia	Trop. Africa
Solanum tuberosum	Irish potato	S America	Highland tropics[2]
Xanthosoma saggitifolium	tannia	Trop. America	Trop. Africa[2]

1. Of major importance in Africa and worldwide. 2. Of minor importance. 3. Of much limited local importance. Source: Kay (1973).

Most of the roots and tubers are grown in the humid and sub-humid tropics of West and Central Africa, with Nigeria and Zaire accounting for 33·2 and 34·9 per cent respectively of the total production during 1982-1984. West Africa accounts for over 95 per cent of the yams produced in Africa, of which Nigeria alone produced 75·5 per cent. Central Africa (Rwanda and Burundi) and similar highland areas in East Africa produced 40·2 per cent and 32·4 respectively of the sweet potato crop grown in Africa. As for the cocoyam crop, Nigeria, Ivory Coast and Ghana accounted for 43, 21 and 16 per cent respectively of the area and 58, 7 and 18 per cent of the production during 1982–1984. The calorific contribution of root crops to the nutrition of African populations is shown in Table 12.4.

Table 12.3: *Statistics on Area, Production and Yield of Roots and Tubers in Africa in Comparison to those in the World in 1985*

Root crop	Area (000 ha)			Production (000 t)			Mean Yield	
	World	Africa	% of world	World	Africa	% of world	kg/ha	% world average
Cassava	14195	7491	53	136532	56527	41	7546	78
Yams	2470	2334	94	25860	2475	96	10609	101
Sweet potato	8003	11043	14	111438	6299	6	5706	41
Cocoyam (taro)	993	769	77	550	3422	61	4450	80
Root Crops Global Totals	46775	13151	28	582110	9954	17	7569	61

Source: FAO Production Year Book (1985).

Table 12.4: *Estimates of Number of Consumers of Major Staples in Tropical Africa*

	Per Capita calorie consumption	Number of people (millions)	Major Consumer Countries
Cassava	600	40	Zaire, Mozambique, CAR, Congo, Gabon
	200–600	120	Nigeria, Tanzania, Ghana, Uganda, Cameroon, Madagascar, Angola, Zaire, Ivory Coast, Rwanda, Guinea, Kenya, Burundi, Benin, Mozambique, Togo, Liberia
Yam	200	60	Nigeria, Ivory Coast, Ghana, Benin, Togo, Cameroon
Sweet potato	200	15	Uganda, Rwanda, Burundi, Tanzania
Cocoyam	200	10	Nigeria, Ghana, Cameroon, Ivory Coast

Source: Barker & Dorosh (1986).

Similarly, cassava in Africa produced annually 441 000 tons of protein as compared to 258 000 metric tons and 65 400 tons for yams and maize respectively. The overall value of cassava produced in tropical Africa per annum during 1982–1984 amounted to about US$2979 million as compared to US$4665 million and US$451 million for yams and sweet potato respectively.

Parts of roots and tuber crops and their waste products can be fed to livestock and they are of potential industrial importance if they are not being used as such already. For various uses of roots and tubers grown in Nigeria reference should be made to Table 12.5.

For example, cassava finds uses in industry as a source of starch used in the manufacture of textiles and in the distillation of alcohol for use either as a beverage or gasohol. Some species of sapogenin-bearing yams found in Asia and Central America are used for the manufacture of corticosteroid drugs and contraceptive pills.

PROCESSING OF ROOT AND TUBER CROPS

Roots and tubers are highly perishable since they are succulent living plant parts with over 70 per cent water content. They are therefore not easily stored for relatively long periods of time, unlike cereals and grain legumes. Since they are living and respiring at a higher metabolic rate than seeds, they lose a lot of dry matter when stored for long periods of time. When they are injured they are invaded by rot organisms which may result in total loss.

The demand for roots and tubers generally declines with rising income unless they are processed to render them into forms with a longer shelf life and able to meet the demand for convenience foods in urban areas. Processing of roots and tubers becomes imperative when there are anti-nutritional factors or toxic substances involved, such as the presence of cyanide in cassava. Processing is also necessary where there is need to blend several products in certain food preparations.

Table 12.5: *Main Food Preparations and Uses of Root and Tuber Crops Grown in Tropical Africa*

Crop	Food Preparations	Other Uses/Remarks
Cassava (*Manihot esculenta*)	Fufu, tapioca & farinha flour (lafun), gari. Boiled cassava (sweet varieties). Bread and composite flour. Leaves used as vegetables. Flakes and chips. Biscuits.	Starch for industrial use. Chips and pellets used for livestock feed. Alcohol.
Yams (*Dioscorea* spp.)	Boiled yam, fried and roasted yam, fufu or pounded yam, chips and flakes, yam flour cakes, mashed yam	Livestock-feed. Contains sapogenin as diosgenin used in corticosteroid drugs and contraceptive pills (African yams not rich in these). Alcohol also manufactured (not attractive).
Sweet potato (*Ipomoea batatas*)	Chips, tubers, boiled and canned, fried and roasted or baked, frozen, chilled and made into pruic. Leaves used as green vegetables.	Used for animal feed. Starch (not economical). Alcohol. Syrup.
Cocoyams (*Xanthosoma saggitifolium*)	Tuber boiled, roasted and eaten	Leaves yield fibre. Starch.
Dasheen (*Colocasia esculenta*)	Tuber, leaves and flowers boiled and eaten	
Hausa potato (*Solenostemon rotundifolius*)	Cooked and eaten as source of carbohydrate similar to yams and potato; may be curried and eaten with rice or boiled, baked and fried into chips	Sometimes used as drug plant for treatment of dysentery and eye disorders.
Risga (*Plectranthus esculentus*)	Tubers eaten raw or boiled, sometimes pickled	Claimed to give higher root yield than any other Labiatae
African yambean (*Sphenostylis stenocarpa*)	Storage roots eaten after boiling	More often cultivated for its edible seeds. Roots with 6–14% protein.
Chufa (*Cyperus esculenta*)	Corms eaten raw, baked as a vegetable; roasted and eaten or grated and used for ice cream, sherbets, and as a milky beverage 'horchasta de chufas' in Spain	Used in animal feed and in confectionery as substitute for almonds; ground corms used as substitute or adulterant for coffee and cocoa. Minor products include oil similar to olive oil used for soap; also yields starch, flour, alcohol and paper from leaves.

For example, in the use of cassava flour for baking bread, it is necessary to render the raw material both dry and powdery in order to facilitate mixing. Some form of processing is also necessary if some constituents are to be extracted for industrial use or when surpluses are produced and there is need to avoid wastage. Lastly, processing facilitates handling and transportation of the produce in packaged forms.

The nature of processing to be carried out is usually related to the end use, in addition to the texture and structure of the product. The range of uses of the major root and tuber crops, and their methods of processing and any problems encountered, are discussed below.

USES OF CASSAVA

Cassava roots contain about 62–80% water, 35% carbohydrates, 1–2% protein, 0·3% fat, 1–2% fibre and 1% minerals (Onwueme 1978). The leaves of cassava contain proteins of high biological value and are only deficient in methionine. Major uses of cassava include food, feed and industrial uses.

Processing cassava for food

Apart from processing to isolate the edible part of cassava and make it more appealing and easy to eat, cassava roots for food must be processed in order to remove the poisonous hydrocyanic acid which is abundant in bitter varieties. The bitter cultivars have high levels of cyanide in both the flesh and bark as compared to sweet varieties which have high levels in the bark or peel but with negligible amounts in the flesh. The various food preparations and the operations involved in the processing and preparation for food are listed below:

(a) *Fufu:*

    (i) Fresh sweet cassava varieties may be washed, peeled, boiled and pounded into a dough-like consistency and then eaten with soup.

    (ii) Fresh roots washed, fermented, debarked or peeled, sieved, boiled with one or two changes of water pounded and then eaten. A modification of this involves washing, peeling, fermenting, sieving, boiling and pounding.

(b) *Gari:* A fine gritty flour made by peeling, washing, grating, bagging, pressing, fermenting, macerating, sieving and roasting over the fire in shallow pans or pots with or without adding palm oil. The gari can then be taken with milk as cereal or prepared into fufu by treating it with hot water (Kreamez 1986).

(c) *Farinha:* A Brazilian product resembling gari is made by washing, peeling, grating, pressing, sieving and toasting in a shallow basin over a low fire. Farinha can be taken with water, milk, sugar or coconut, as is done with gari.

(d) *Boiled cassava flakes or chips:* Prepared by washing, peeling, boiling, slicing, soaking in changes of water and eaten moist either alone or with coconut. It may be dried and stored until required, then soaked before eating either alone or with the addition of palm oil, fish, salt, pepper, etc.

(e) *Cassava flour:* Prepared by washing, peeling, chipping and drying in the sun, then artificially stored and/or milled into flour, sieved and ready to use. Sometimes it may be partially fermented prior to drying.

(f) *Cassava starch:* Details of processing are described below under industrial processing, but the product may be used to prepare a fufu-like dish that is eaten with soups.

(g) *Cassava leaf processing for a vegetable:* Entails chopping, washing, boiling and then used for soups after leaching out the cyanide with many changes of water.

(h) *Cassava chips and pellets for animal feed:* These are prepared by washing, peeling, and cutting into thin slices 3–6 cm long and then dried either in the sun or artificially before bagging, storage and/or transportation. The size and shape of pellets determines the length of time for sun drying. Thin slices dry in about eight hours while thicker or roughly chipped cassava may take several days. Pelleting is better for material to be shipped since it reduces space requirements by 25–40 per cent (Muller 1977).

(i) *Feeding of cassava leaves or whole plants:* Cassava leaves can be prepared as pellets, leaf meal or leaf protein concentrates and used in feed formulations. Details of the operations used in the preparation of whole plant are shown in Muller (1977).

(j) *Cassava as substitute for single cell protein:* Cassava can be used as a substrate for growing of several species of filamentous fungi which are then used for animal feed.

More details about the various aspects of the use of cassava for feed are given in Nestel & Graham (1977). Fresh cassava roots and dried pulverized leaves can also be fed to livestock as are the peelings produced during processing.

Industrial uses and processing of cassava

The most important industrial product from cassava is the starch, which is prepared by washing the fresh roots, peeling, crushing or rasping to produce a mash or pulp which is formed into a suspension in water and finally screened to separate the fibre from the starch 'milk'. Some sulphur dioxide is added to the water to prevent microbial infestation and fermentation. After separating the dirt and removing the water the starch is dried; it is then used in the commercial production of glucose, textiles and confectionery. Cassava starch also finds uses in the manufacture of adhesives, cosmetics, sizing of textiles, laundering and paper making.

Cassava starch may be used to prepare tapioca by smearing with oil before drying; the grains are then toasted over the fire where they undergo partial dextrinization. Tapioca flakes are prepared in a special way which involves cooking a thin layer of starch for two minutes in a pan. Tapioca pearls are prepared by granulating the starch, gelatinizing and roasting for about 15 minutes in coconut oil before drying. For more details see Onwueme (1978).

Cassava is also used for the brewing of beer and the manufacture of alcoholic beverages and gasohol for use as fuel.

Developments in cassava processing in Africa

Until recently most of the cassava processing work was done manually and the processes of fermentation and changes that the products undergo were not well known. But at the Federal Institute of Industrial Research, Oshodi near Lagos, Nigeria, and elsewhere, progress has now been made in the development of mechanical cassava processing machines for production of gari (Akinrele *et al.* 1971). Prior to this, local small-scale cassava grating machines have been in existence for over a decade, but there was no facility that could handle the cassava from peeling to the final product. The identification of the micro-organisms responsible for fermentation of cassava as *Leuconostoc Streptococcus* yeasts may lead to the selection of more efficient strains for the faster fermentation of higher quality material.

FOOD AND FEED USES OF YAMS

The yam is a major source of carbohydrates and energy in the 'yam zone' of West Africa. Yam tubers usually contain 60–80% water, 15–30% carbohydrates, 0·2–0·3% fat, 0·5–2·6% minerals, 1–3% crude protein, 0·2-15% crude fibre and 0·5–0·6% sugar.

The yam is usually consumed after boiling or roasting of the peeled or unpeeled tuber. Yams may also be fried or baked and then eaten either alone or with oil and some kind of sauce. The boiled yam may also be mashed, or more traditionally pounded in a mortar into fufu or foofoo. It is then eaten with a variety of soups, such as okra and melon soups made with fish, meat, spices, vegetables and palm oil. Yams are also mashed or grated and made into balls and eaten after frying. Yams may be processed into yam chips similar to Irish potato chips. Yam is sometimes processed into flour, or more recently into flakes from which yam fufu, similar to pounded yams, are produced. For more details about the various way of preparing yams reference should be made to Coursey (1967) and Onwueme (1978). Yam peels, tops and damaged tubers unfit for human consumption are often fed to goats, sheep and other livestock.

A recent development in yam food processing is the discovery that the browning reaction is mediated by polyphenal oxidase (Ngoddy & Onuoha 1985). The character and activity of this enzyme have been studied and various ways of inhibiting the browning action are being investigated. Sodium benzoate, sodium metabisulphate and thiourea have been shown to be effective in inhibiting the browning reaction. Ngoddy & Onuoha (1985) also report rheological studies and textural characterization of yam products that can be used in comparable assessment of different processing methods.

Yams contain steroidal sapogenins, especially in the form of diosgenin, which is a key building block for cortisone and corticosteroid drugs used as sex hormones and oral contraceptives. Unfortunately, West African yams are low in sapogenins and the best varieties for commercial production of these drugs are *Dioscorea opposita* from Mexico and *D. elephantipes*. One of the most recent developments in yam processing for drugs is the discovery of *D. sylvatica* in South Africa.

Yams also contain alkaloids for which purpose some species are used as sources of fish and arrow poisons, insecticide and various pharmaceutical preparations. Yams also feature in traditional medicines and in traditional, cultural and religious ceremonies and rituals.

SWEET POTATO PROCESSING AND UTILIZATION

The chemical composition of sweet potato tuber consists of 50–81 per cent moisture, 8–29% starch, 0·5–2·5% reducing sugars, 0·5–7·5% non-starch carbohydrates, 0·9–2·4% protein, 1·8–6·4% ether extract and 0·9–1·38% mineral matter (ash). Sweet potato is rich in carotene, especially in the orange fleshed varieties. It is also a good source of vitamin C and those of the B complex. In most of sub-Saharan Africa the starchy and less sweet varieties of sweet potato are preferred.

Sweet potato processing for food and feeds

Sweet potato tubers can be boiled, baked and candied before consumption. They may be made into a syrup or used as purée in water or soups. They are sometimes also manufactured into flour and mixed with wheat flour for bread baking. The less-starchy, orange-fleshed varieties are not preferred in West Africa; they are better for canning. Sweet potato chips made by washing, peeling, slicing and frying are used and eaten as snacks.

The leaves are used as vegetables and contain up to 27% true protein, 10% minerals, 8% starch, 4% sugar and 56 mg/100 g carotene. The leaves are usually boiled and used in soups and stews.

Tubers of sweet potato can be washed, cut and fed directly to animals. They may also be washed, shredded, or sliced, treated with sulphur dioxide and dried in the sun or artificially dried at 80°C. In this form they can be fed directly to livestock or first ground and used in rations for cattle, pigs, sheep and poultry. It is estimated that in the US about 33% of the sweet potato crop is fed to livestock. The leaves may be either fed directly or in the form of silage.

Industrial uses

Sweet potato starch is known to be suitable for many industrial uses but at present it is not economically attractive. The starch is prepared from tubers in the same way as for cassava. It is used in the manufacture of textiles, and for the production of alcohol and syrup.

COCOYAMS PROCESSING AND UTILIZATION

The proximate composition of taro (*Colocasia esculenta*) consists of 63–85% moisture, 13–29% carbohydrates (mainly starch), 1·4–3% protein, 0·46–1·36% fat, 0·6–1·3% ash and 0·6–1·2% crude fibre. Similar values for tannia (*Xanthosoma* spp.) are 70–77% moisture, 17–26% carbohydrate, 1·3–3·7% protein, 0·6–1·3% ash, 0·2–4% fat. For details on the carotene and vitamin contents see Onwueme (1978).

Food and feed uses

The corms and cormels of taro and tannia are usually eaten alone or with other ingredients, i.e. meat, fish, etc., after boiling, roasting, baking and frying. Many cultivars of *Colocasia* have raphides and crystals of calcium oxalate which irritate the throat when they are roasted or fried and are therefore best eaten after boiling, with one or more changes of water. Cocoyams are sometimes boiled, pounded into fufu alone or mixed with yams. In Hawaii and Polynesia a product of taro called poi is prepared by pressure cooking, washing, peeling, mashing, passing through a series of strainers and stored to undergo fermentation by *Lactobacillus* spp. before being consumed or canned. Coconut products may be added to improve the flavour (Chandra 1979). The fresh or pre-cooked taro and tannia may be peeled, dried and ground into flour. Slices of tannia or taro may be soaked in water overnight, blanched and dried before grinding. The inflorescence or its spathe and spadix are also eaten after boiling. Leaves may be used for wrapping food, while the petioles may be dried and used as fibre. The cocoyams are not much use for animal feed or in industries in tropical Africa.

THE FUTURE AND POTENTIALS OF ROOTS AND TUBERS IN AFRICA

In tropical Africa today, unlike southeast Asia and Latin America, the root and tuber crops feature largely in human food and only their waste products are used for animal feed. It is not likely that they will be much used for industrial or animal feed until the demand for food is satisfied. However, during the last decade considerable progress has been made in their improvement and emerging technologies, such as tissue culture and genetic manipulation, plus the possibilities of applied biotechnology, promise to result in bumper yields sooner rather than later. When this happens, there is no doubt that Africa will join Brazil and other tropical countries in exporting cassava for feed in Europe and the processing roots and tuber for miscellaneous products. Likely developments in root and tuber improvements that will affect processing and utilization include:

(a) Use of biotechnology to produce high yielding varieties of cassava, making it possible to satisfy demand for food and provide surpluses for use in feed and industrial products. It may also be possible to develop more suitable varieties of cassava, sweet potato and yams for manufacture of sweeteners, as in maize.

(b) Use of biotechnology to produce varieties that possibly can fix nitrogen, are non-toxic and can be washed, peeled and dried without rigorous processing. Also possible are (i) new varieties of cassava free from cyanide, (ii) new varieties of cocoyam without crystals or raphides of calcium oxalate, and (iii) selecting and culturing improved micro-organisms for more effective fermentation of cassava.

(c) There are possibilities for advances in the mechanization of all stages of roots and tuber production and processing, thereby reducing the high cost of labour and drudgery involved. Even before the increasing demand for roots and tubers as food is satisfied it is very likely that biomass technologies would have been perfected and improved to the extent that the residues and woody plant constituents of cassava could be used alone or with trimmings of alley cropping hedges for the manufacture of a range of organic chemicals. Biomass technologies offer opportunities for (i) achieving regular processing of agricultural products for food and other products for which a given crop or animal is produced, (ii) by-products processing for food and feed, and (iii) waste product processing for food, feeds, fuel, manure and any other products of marketable value (Rutkowski 1983).

REFERENCES

Akinrele, I. A., M. I. Ero and F. O. Olatunji (1971) Industrial specifications for mechanized processing of cassava into gari. Federal Institute of Industrial Research Technical Memo No. 26.

Barker, R. and Dorosh, P. (1986) Relative importance and future prospects of crops in sub-Saharan Africa. Paper prepared for the IITA Board of Trustees, Nov. 1986.

Chandra, S. (1979) Handling, storage and processing of root crops in Fiji. In: *Smallscale Processing and Storage of Tropical Root Crops*, D. L. Plunknett (Ed.), pp. 53–63. Westview Press, Boulder, Colorado.

Coursey, D. G. (1967) *Yams*. Longman, Green & Co., London.

FAO (1986) *1985 Production Yearbook*. FAO Statistical Series No. 70. Rome

Kay, D. E. (1973) *Crop and Product Digest, No. 2: Root Crops*. Tropical Products Institute, London.

Kreamez, R. G. (1986) *Gari processing in Ghana; a study of entrepreneurship and technical change in tropical Africa*. Agricultural Economics Research 86–30, Dept. of Agric. Econ., Cornell University, Ithaca.

Muller, Z. (1977) Improving quality of cassava root and leaf product technology. In: Cassava as animal feed. Proceedings of a Workshop held at the University of Guelph, 18–20 April, 1976, B. Nestel and M. Graham (Eds). IDRC, Ottawa

Nestel, B. and M. Graham (Eds) (1977) Cassava as animal feed. Proceedings of a Workshop held at the University of Guelph, 18–20 April 1976. IDRC, Ottawa.

Ngoddy, P. and C. C. Onuoha (1985) Selected research problems in yam processing. In: *Advances in Yam Research*, G. Osuji (Ed.), pp. 295–318. Biochemical Society of Nigeria in collaboration with Anambra State University of Technology, Emegu. Tropical Products Institute, London.

Okigbo, B. N. (1983) Root and tuber crops in human nutrition. IITA, Ibadan (mimeo).

Onwueme, I. C. (1978) *The Tropical Tuber Crops, Yams, Cassava, Sweet Potato, Cocoyams*. John Wiley & Sons, Chichester.

Rutkowski, A. (1983) Conversion of plant and animal waste to food. In: *Chemistry and World Food Supplies: The New Frontiers*, L. W. Shermilt (Ed.), pp. 337–348. Pergamon Press, Oxford.

13

New Plant Sources for Food and Industry in India

R. S. Paroda and Bhag Mal

INTRODUCTION

Agriculture in today's context is one of the important sources of renewable wealth in the world. It is estimated that so far man has used about 3000 plant species for food and other purposes all over the world (Zeven & Zhukovsky 1975)[1]. However, over the centuries, the practice had been to concentrate on fewer and fewer species and today most of the people in the world are fed by about 20 crops (Anon. 1975). Hence, in spite of this rich flora that offers such great promise, we seem to depend on only a very small fraction of the global plant wealth. With ever increasing population pressure and fast depletion of natural resources, it is necessary that we explore the possibilities of using newer plant resources in order to meet the growing need of food, clothes and housing for human populations.

India with its varied climate, soil and agro-ecology possesses immense plant diversity, with over 15 000 species of higher plants. Both our Indian civilization as well as our diverse tribal heritage have gone a long way in conserving the wild weedy species, native land races and primitive cultivars. The Hindustani (or Indian) Gene Centre is endowed with rich flora especially with regard to several less known yet economically important plants, ca. 160 cultivated species of economic plants, plus 56 species of lesser known cultivated food plants. Furthermore there are ca. 320 species of wild and weedy economic types (Arora & Nayar 1984). Hence, the region serves as a reservoir for a large number of useful plant species which are yet to be domesticated and utilized for the benefit of mankind. In the present paper, an attempt has been made to list such important plants along with wild relatives occurring in India, which could in future serve as new crops, especially in stress conditions or in emergency situations. The results of recent efforts on under-utilized and under-exploited plants made under the All India Coordinated Programme and by the National Bureau of Plant Genetic Resources, New Delhi, towards collection, evaluation, utilization and conservation/maintenance have also been highlighted.

---

1. Over 12 650 species are recorded as having been eaten by man. See Kunkel, G. (1984) *Plants for Human Consumption.* Koeltz Scientific Books, Koenigstein (GEW).

WEALTH OF LESS-KNOWN PLANTS

Food plants

India is richly endowed with a wealth of relatively little-known native and naturalized cultivated food plants which have the potential to be commercially utilized. Information on the plant genetic resources of these less-known cultivated food plants and their wild relatives is well documented (Arora & Nayar 1984; Arora 1985). The 81 important little-known cultivated food plants awaiting exploitation in India are listed below:

(a) **Seed/Nut Crops:** important species requiring attention are: *Borassus flabellifer, Coix lacryma-jobi, Digitaria cruciata* var. *esculenta, Dolichos uniflorus, Echinochloa colona, Euryale ferox, Mucuna pruriens* var. *utilis* (syn. *M. capitata*), *Panicum sumatrense, Parkia javanica, Paspalum scrobiculatum, Vigna aconitifolia, V. trilobata, V. umbellata* and *V. angularis*.

(b) **Vegetable Crops:** there are two distinct categories, (i) plants whose leaves/young shoots are eaten cooked or used in making soups, and (ii) plants whose tender fruits/pods are consumed as vegetable after cooking. The important species are: *Amaranthus lividus* (syn. *A. polygonoides*), *Bambusa tulda, B. spinosa, B. vulgaris, Canavalia virosa* (syn. *C. polystachya*), *Corchorus capsularis, Crambe cordifolia, Dendrocalamus asper, Emila coccinea* (syn. *E. sagittata*), *Houttuynia cordata, Lactuca indica, Moringa oleifera, Mucuna pruriens* var. *utilis* (syn. *M. cochin-chinensis* and *M. utilis*), *Rorripa indica* (syn. *Nasturtium indicum* var. apetala var. *apetala*) and *Wolffia arrhiza*.

(c) **Root and Tuber Crops:** most of these are consumed after boiling, although occasionally tubers are consumed raw. The useful species are: *Allium tuberosum, Alocasia cucullata, Asparagus sarmentosus, Coleus forskohlii, Colocasia esculenta, Curcuma angustifolia, Eleocharis tuberosa, Flemingia procumbens* (syn. *Moghania vestita*) and *Nelumbo nucifera*.

(d) **Fruit Crops:** important less-known cultivated fruit species identified are: *Aegle marmelos, Artocarpus lakoocha, Carissa congesta, Citrus* spp., *Elaeocarpus floribundus, Phyllanthus emblica* (syn. *Emblica officinalis*), *Garcinia pendunculata, G. tinctoria, Grewia asiatica, Limonia acidissima, Malpighia coccigera, Morus alba, Pereskia grandifolia, Phoenix sylvestris, Phyllanthus distichus, Rhodomyrtus tomentosa, Rubus lasiocarpus, R. ellipticus, Syzygium cumini* and *Ziziphus mauritiana*.

(e) **Spices, Condiments and Beverage Plants:** the identified species are: *Amomum aromaticum, A. xanthioides, Anethum graveolens* (syn. *A. sowa, Areca triandra, Caryota urens, Euterpe edulis, Kaempferia galanga, Madhuca longifolia* (syn. *M. indica*), *Osmanthus fragrans* and *Piper longum*.

Plants for Extreme Environmental/Emergency Situations

There are large areas of India which suffer from one kind of extreme or other where the inhabitants often face many problems regarding their food and other requirements. However, there are a number of indigenous potentially useful plant species which can support life during these extreme environmental/emergency situations. The information on the little-known plant species used as famine foods in the Indian hot desert has been well documented (Gupta & Kanodia 1968; Bhandari 1974; Saxena 1979; Shankarnarayan & Saxena 1987; Shankar 1987). The life support species used for food, fodder, medicine, fibre,

Table 13.1: Food Plants for Emergency Situations

Situation and Usage	Species
Hot dessert:	
– Grain	*Setaria verticillata, Haloxylon salicornicum, Citrullus colocynthis, C. lanatus, Echinochloa colona, E. crusgalli, Abutilon indicum, Eleusine compressa*
– Leaves and tender shoots	*Amaranthus gracilis, A. hybridus, A. spinosus, Chenopodium album, Boerhavia diffusa, Euphorbia caducifolia, Moringa oleifera, Portulaca oleracea, Aloe barbadensis*
– Fruits	*Grewia tenax, Ziziphus nummularia, Cordia sinensis* (syn. *C. gharaf*), *Coccinia grandis, Cucumis callosus, Momordica dioica, Salvador oleoides, S. persica*
– Tubers	*Cyprus rotundus, C. bulbosus, Asparagus racemosus, Ceropegia bulbosa, Butea monosperma, Portulaca tuberosa*
Cold dessert:	
– Grain	*Fagopyrum esculentum, F. tataricum.*
– Leaves	*Allium leptophyllum, Aconitum heterophyllum Cicer songarium, Polygonum viviparum, Sedum tibeticum, Scorzonera mollis*
Semi arid region:	
– Leaves	*Boerhavia diffusa, Bryonopsis laciniosa, Triumfetta rhombifolia, Cocculus hirsutus, Amaranthus viridis, Leptadenia reticulata*
Saline area:	
– Fruit	*Salvadora oleoides, S. persica*
– Foliage	*Salicornia brachiata*
– Leaves	*Chenopodium album, Portulaca oleracea, Trianthema portulacastrum*
Marshy lands/Flood areas:	
– Vegetable	*Alternanthera sessilis, Hygroryza aristata, Monochoria hastata*

fuelwood and other purposes under extreme hot and cold climates, semi-arid situations, saline areas, marshy land and flood areas have also been dealt with in considerable detail by Singh & Gupta (1987). The important plant species which can be conveniently used under harsh environmental conditions are listed in Table 13.1.

At the International Workshop on Maintenance and Evaluation of Life Support Species in Asia and the Pacific Region held at the National Bureau of Plant Genetic Resources, New Delhi in April 1987, a number of potentially economic species were identified as priority species for further research. The important among these are: *Chenopodium album, Fagopyrum tataricum, Flemingia procumbens, Digitaria cruciata, Bambusa tulda, Capparis decidua, Ziziphus nummularia, Prosopis cineraria, Citrullus colocynthis, Indigofera cordifolia, Tecomella undulata* and *Salvadora* spp.

Industrial Plants

The various important plant species which could be used for industrial purposes are indicated below.

(a) **Dye yielding plants:** *Butea monosperma, Kochia indica, Wrightia tinctoria, Morinda citrifolia* and *Anogeisus pendula.*

(b) **Tannin yielding plants:** *Acacia nilotica* and *Cassia auriculata.*

(c) **Plants used as detergents:** roots and leaves of *Euphorbia thompsoniana, Silene indica* (syn. *Lychnis indica*) and *Silene griffithii.*

(d) **Gum, wax and resin plants:** important sources are: *Acacia senegal, A. nilotica, Butea monosperma, Commiphora wightii, Prosopis juliflora, P. cineraria, Moringa oleifera* and *Salvadora oleoides.*

(e) **Timber and fuel plants:** important species include: *Acacia jaquemontii, A. nilotica, A. senegal, Calligonum polygonoides, Capparis decidua, Dalbergia sissoo, Azadirachta indica, Cordia dichotoma, C. sinensis* (syn. *C. gharaf*), *Moringa oleifera, Tecomella undulata* and *Boswellia serrata.*

(f) **Fibre plants:** important species are: *Leptadenia pyrotechnica, Crotolaria burhia, Saccharum bengalensis* (syn. *S. munja*), *Calotropis procera, C. gigantea, Cordia dichotoma, Acacia leucophloea, Butea monosperma, Sida cordifolia, S. carpinifolia* and *S. rhombifolia.*

RESEARCH EFFORTS ON UNDER-UTILIZED PLANTS OF ECONOMIC IMPORTANCE

Recognizing the potential of new plant species for use as food, fodder, energy and industrial crops, work on collection, introduction, domestication and utilization was initiated in the 1960s at the Indian Agricultural Research Institute, New Delhi. This activity was later extended to other research centres in the country. In order to strengthen further research in this direction, an All-India Coordinated Research Project on Under-Utilized and Under-Exploited Plants was initiated in 1982 with its headquarters at the National Bureau of Plant Genetic Resources, New Delhi, with 15 main centres and 10 cooperating centres in different agro-climatic zones of the country (Fig. 13.1). The project embraces research work on selected food plants (winged bean, rice bean, amaranth, buckwheat, chenopods), fodder plants (*Leucaena leucocephala, Albizia amara, A. procera, Cassia sturtii, Hardwickia binata, Dichrostachys glomerata* (syn. *D. nutans*), *Colophospermum mopane*), energy plants (bamboo, sugarcane, sweet potato) and hydrocarbon and industrial plants (guayule, jojoba, *Jatropha curcas, Euphorbia* spp.). Recently, a few more plants, i.e. bambara groundnut, cuphea, rumba (*Citrullus colocynthis*) and *Simarouba glauca* have been included in the programme. The salient research achievements over the past 5 years are highlighted below.

Food Plants

WINGED BEAN (Psophocarpus tetragonolobus)

Winged bean, immensely rich in protein and oil, offers good promise as a multipurpose crop. In India, it is largely cultivated in the humid subtropical parts of North-eastern Region, less common in the Bengal plains and is of sporadic occurrence in the Western Ghats, Maharashtra and in Central/Eastern Peninsular Region. Sizeable germplasm collections have been assembled from Papua New Guinea, Ghana, Indonesia, Thailand, Philippines, Australia, Sri Lanka and India. A wide range of genetic variation was observed (Chandel *et al.* 1984). A sizeable germplasm bank has also been built up at the Indian Institute of Horticultural Research, Bangalore and the University of Agricultural Sciences, Bangalore. A promising dwarf type has been developed which does not need staking. Based on multi-locational testing, four selections, namely EC 38821, EC 38821 B, EC 38955 and IIHR – 13 were identified as promising and recommended for pre-release multiplication. Amongst the new introductions, high yielding accessions included Blue Course (EC 27884), Blue Fine (EC 28864) both from Ghana, Molk (EC 38855 ex Papua New Guinea), EC 116886 and EC 121818 (ex Nigeria). Promising early maturing, high grain yielding types were EC 121296 (ex Nigeria), Makura (EC 38755 A), EC 38823 A, EC 27886 A and EC 38825 (all from Papua New Guinea).

New Plant Sources in India

Figure 13.1: All-India Coordinated Research Project on Under-Utilized and Under-Exploited Plants.

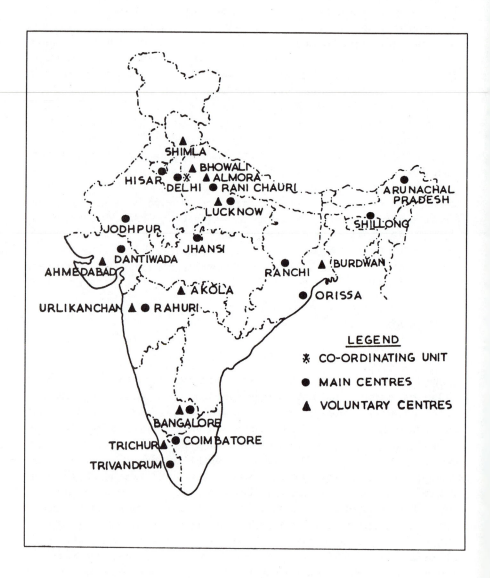

Staking, non-synchronous pod ripening and prolonged crop duration are the major problems which restrict the use of winged bean to the small farmer. Research efforts are under way to develop short duration varieties with bushy growth habit and synchronous maturity.

RICE BEAN *(Vigna umbellata)*

This is an important pulse crop, rich in protein, calcium, iron and phosphorus, awaiting exploitation. Over 500 germplasm collections from Nepal, China, Sikkim and India (North-eastern, Central, East Peninsular Regions) were made and evaluated at NBPGR, New Delhi for various growth characters. A wide range of genetic variability was observed. Hybridization between an early variety from China and bold seeded variety from Mysore and yellow seeded variety from Nepal led to the development of several promising lines combining earliness and high yield. A promising selection $C \times M12 P_2$-3 capable of up to 25 q/ha grain (2500 kg/ha) under Shimla conditions holds good promise for cultivation in the hills. The variety 'Rajmung' has been developed in Punjab. Promising introductions which give consistently superior performances under north Indian conditions included EC 93452, EC 101887, PI 247685 and PI 247693. Promising cultivars giving 9–17 q/ha grain (900–1700 kg/ha) at Shillong included RB 44, RB 53 and RB 56. Good genetic diversity in *Vigna umbellata* and its wild relatives *V. pilosa, V. vexillata* (syn. *V. capensis*) and *V. radiata* var. *sublobata* occurs in North-eastern Region.

AMARANTH *(Amaranthus spp.)*

Amaranth, a pseudo-cereal, has a tremendous potential for use as vegetable as well as a grain crop. The grains are rich in protein, fat and carbohydrates and are comparable to wheat, rice and oats. The grains are milled into flour and used as a staple food throughout the entire Himalayan region. In the North and South Indian Plains, it is consumed in the form of sweet balls (ladoos). Information on its origin, distribution, breeding, cultivation, pests and diseases is well documented by Singh & Thomas (1978). Three domesticated species of grain amaranth, i.e. *A. hypochondriacus, A. cruentus* and *A. caudatus* are important for cultivation, of which the former is the most popular.

Extensive collections of over 3000 accessions have been built up from different sources; germplasm suitable for hill regions and the plains are being maintained at NBPGR Regional Stations, Shimla and Akola, respectively. A wide range of genetic variation was observed. Based on multi-locational trials of promising entries, selection IC 42258–1 was identified as the best and was released as 'Annapurna'. This variety possesses drought tolerance and wide adaptability and is capable of producing 20–25 q/ha (2000–2500 kg/ha) grain with about 15% protein. Two other accessions, IC 5564 and an exotic introduction of *A. caudatus* cv. *edulis* (syn. *A. edulis*) from Taiwan were also observed to be very promising.

In view of its wide use, there is a need to develop high yielding varieties for the plains. Germplasm collecting efforts have been intensified and a sizeable collection from the cold arid tract, eastern UP, southern Bihar, western Rajasthan, Himachal Pradesh and tribal areas of Bastar in Madhya Pradesh have been collected.

BUCKWHEAT (Fagopyrum spp.)

In India, buckwheat is grown entirely in the temperate parts of the Himalayan range and on the southern Indian hills where it can withstand poor, infertile and acidic soils. Two species, *F. esculentum* and *F. tataricum* are domesticated. It is a potential pseudocereal with varied uses. It is used both as a food grain and as a leafy vegetable. The flower produces nectar, a food source of the honey bee. It is also used as a medicinal plant, the active principle being the glucoside rutin. The origin, distribution, breeding, cultivation, uses, pest and disease aspects have been reviewed by Singh & Thomas (1978). Germplasm collections comprising over 300 accessions have been built up at NBPGR Regional Station, Shimla and evaluated for different characters. Based on multi-locational testing of selected accessions under the All India Coordinated Programme, IC 13374 has been found to be the most promising in yield stability.

The major problems with this crop are lodging, shattering of grains at maturity, poor response to fertilizers and poor market availability for its products. Efforts are under way at NBPGR Regional Stations at Shimla and Bhowali to collect diverse germplasm and to develop lodging resistant, non-shattering and fertilizer responsive cultivars. Adaptive trials are also being conducted in non-traditional areas to extend its area of cultivation. Suitable husbandry practices are being developed to raise the level of productivity.

Chenopods (Chenopodium album)

Of the four domesticates *C. album* is the most widely distributed species found in the Himalayan region. It is an important crop for the hill region and is generally consumed mixed with other cereals. In addition to serving as food source, the grains are also used extensively in local alcoholic beverages (Partap & Kapoor 1985) and in times of scarcity, the dried stem is used as fuel. Agronomically, the crop is most suited to a mixed farming system, particularly with a multiple cropping pattern (Partap & Kapoor 1987). The germplasm collection efforts have been rather meagre; 84 accessions were evaluated at Shimla and are being maintained. The preliminary evaluation studies revealed good variability in plant height, flowering and maturity period, leaf and inflorescence size and seed yield. The promising types are now being evaluated in multi-location trials at Shimla, Delhi, Bangalore and Bhubaneshwar. An accession NC 58613 was identified as the most promising. *C. quinoa*, an exotic introduction, has also shown promise.

Bambara groundnut (Vigna subterranea)

It is an important legume crop with underground pods and is consumed in various ways. It contains high amounts of carbohydrate and is particularly rich in lysine, one of the more limiting amino acids. It can thrive well under dry arid situations. So far 192 accessions have been introduced from Nigeria and other African countries and evaluated for different traits. Adaptability tests under the coordinated programme are under way to find the best area for its cultivation. More rigorous efforts are needed in order to isolate and develop high yielding types.

Fodder Plants

LEUCAENA LEUCOCEPHALA

This is a fast growing, multi-purpose tree with a very high coppicing ability. It is expected to find place in the existing farming systems due to its versatility for producing out of season green forage, firewood and small timber without affecting crop yields. It has tolerance to drought, pests and diseases. In India, it assumes special significance in agroforestry. Over 500 germplasm collections have been built up from diverse sources and evaluated at different locations. Hybridization work involving Hawaii, Salvador and Peru types as well as *L. pulverulenta* is in progress at the Indian Grassland and Fodder Research Institute, Jhansi for developing fast growing, high yielding varieties with low mimosine content. K-8 (EC 124343 ex Philippines) and El Salvador (EC 123866 ex Australia) proved superior for fodder and fuel in Gujarat and Kerala States. Silvi-4 and K-28 were observed to give the best above ground biomass yields under Jhansi conditions. Adaptability studies are in progress at Jodhpur, Bangalore, Trichur and Hisar.

OTHER FODDER TREES

A few potential fodder species, *Hardwickia binata, Albizia amara, A. procera, Cassia sturtii, Dichrostachys glomerata, Colophospermum mopane* have been included in the coordinated programme; feasibility trials have been planned to be conducted in Jodhpur, Jhansi, Hisar, Dantiwada, Hyderabad, Delhi and Almora.

Energy Plantation Crops

BAMBOO (Bambusa spp.)

Bamboos are fast growing multipurpose trees which have a tremendous potential for cottage and paper industries as well as for fuelwood. Rich diversity exists in the north-eastern region and about 45 germplasm collections belonging to 8 species (*Bambusa tulda, B. pallida, B. nutans, B. arundinacea, B. balcooa, B. multiplex, Dendrocalamus sikkimensis* and *D. giganteous*) have been established at Basar in Arunachal Pradesh and are being evaluated. Further collections of wild and cultivated types are being planned.

CASUARINA spp.

These are fast growing tree species having a good potential for fuelwood. Over 20 accessions of *Casuarina equisetifolia, C. cunninghamiana* and *C. cristata* have been assembled at Mettupalayam, Tamil Nadu and are being evaluated. An accession of *C. equisetifolia* from the Philippines was observed to be very promising. Studies conducted at the Soil Salinity Research Institute, Karnal revealed that *C. equisetifolia* showed 90–95 per cent survival using both surface planting and channel planting methods under high salinity levels (Tomer & Gupta 1985).

TUBER CROPS

Exploratory studies for alcohol production from tuber crops are in progress at the Central Tuber Crops Research Institute, Trivandrum, Kerala. Evaluation of 100 accessions of *Dioscorea cayenensis* subsp. *rotundata* revealed that starch content varied from 65–90%, the highest being recorded for Boki-9. Sweet potato and *Dioscorea esculenta* were observed to have a higher amino acid content than *D. alata* and *D. cayenensis* subsp. *rotundata*; cassava and sweet potato varieties had a higher fibre content than *Dioscorea* spp.

SUGARCANE (Saccharum officinarum)

The work on sugarcane for energy production is in progress at the Indian Institute of Sugarcane Research, Lucknow. Sugarcane genotypes COLK 8001, COLK 7001 and Co 1148 were observed to show the best potential for bioenergy and ethanol production.

Hydrocarbon and Industrial Plants

GUAYULE (Parthenium argentatum)

This desert shrub from north-central Mexico and south western United States is drought hardy and can thrive well under 250–300 mm rainfall. It is a good source of rubber. About 80 accessions have been introduced and evaluated at Jodhpur, Hisar, Mettuapalayam, Dantiwada, Urlikanchan and Delhi. Promising accessions have been tested in multi-locational trials. The best performances were observed for Arizona-2 at Jodhpur, G-4 at Dantiwada and HG-8 and HG-9 at Hisar. The cultivar Arizona-2 (ex US) with good quality latex and 6–8% rubber content was recommended for release following the 1986 Annual Workshop. Analyses at the Bio-centre, Ahmedabad revealed that EC 148913 had the maximum rubber content (10·3%) and resin content (9·1%); HG-9 produced 8·9% rubber.

Recently, 23 new accessions from the US and Mexico have been introduced and will be evaluated for growth and other characters. Efforts are also being made to hybridize guayule with *P. incanum, P. tomentosum* var. *tomentosum* and var. *stramonium* (syn. *P. stramonium*) at the Central Salt and Marine Chemicals Research Institute, Bhavnagar. The hybrids with *P. tomentosum* vars. *tomentosum* and *stramonium* show considerable promise (Patel *et al.* 1986). A joint programme on guayule was initiated in 1985 with the Indian Council of Agricultural Research and the Department of Science and Technology collaborating in research on cultivation and other basic problems. The agronomic studies revealed that the best growth performance was obtained with 75 × 45 cm spacing at Jodhpur.

JOJOBA (Simmondsia chinensis)

A hardy shrub, native to north Mexico and south western United States and is valued for its seed oil, which is a good substitute for sperm whale oil. The oil (liquid wax) is used as a lubricant in high pressure machinery, and when mixed with alcohol, yields an excellent diesel fuel. It is a drought and salt tolerant species that can be grown economically on arid lands and coastal wastelands. The plant tolerates extreme desert temperature and can thrive well on lands unsuitable for traditional agriculture.

Over 75 accessions have been introduced at NBPGR, New Delhi. A wide variation has been observed for different traits amongst the cultivars studied. It has been tried in Gujarat, Rajasthan and Uttar Pradesh. One variety EC 33198 (ex US) has been identified as promising and recommended for large-scale cultivation. Three accessions, EC 124381, EC 99690 and EC 99691 were also observed to be promising. Breeding work done at NBPGR Regional Station, Jodhpur has shown that early fruiting can be achieved. Agronomic studies to standardize the spacing are underway. Standardization of micro-propagation techniques using tissue culture is also in progress at NBPGR, New Delhi.

EUPHORBIA spp.

Adaptability studies of different latex bearing species of *Euphorbia* have been conducted at different centres. The germplasm comprising 37 accessions of *E. caducifolia* were evaluated at Krushinagar and the genotypes, Sabarmati-1, Vijapur, Keva-2 and Jesanpura were high dry matter producing types. Evaluation of 15 accessions of *E. tirucalli* both from indigenous and exotic sources evaluated at Krushinagar revealed that Chandisar-1 was the best in growth performance. *E. antisyphilitica* is a good source of candellila wax and has established well under Jodhpur conditions. Germplasm of *E. lathyrus* has been introduced from US, USSR and France and their evaluation studies are in progress. Other species commonly growing on rocky soils which can be exploited for hydrocarbon are *E. neriifolia, E. nivulia* and *E. trigona*.

JATROPHA CURCAS

This species is adapted to marginal lands and can grow on gravel, sandy, clayey and eroded lands. It produces a semi-drying oil which can be used in fuel mixtures, as an illuminant and for making soaps and candles. Ten local types collected from Banaskantha and Mehsana Districts of north Gujarat were evaluated at Krushinagar. Agronomic studies revealed that wider spacing of 2 × 2 m and 2 × 1·5 m were far superior to other spacings.

Other Potential Plants

SIMAROUBA GLAUCA

This introduced tree species has a high potential for its edible oil. It has established well under Maharashtra conditions. Adaptability trials are planned to be conducted at Jodhpur, Rahuri, Mettupalayam, Ranchi and New Delhi.

CUPHEA

This is an oil yielding plant with a high degree of drought resistance. It shows promise for cultivation in the north-eastern hills. Seven exotic collections were evaluated at NBPGR Regional Station, Shimla. EC 133506 was observed to provide the best yield, giving 5·6 q/ha (560 kg/ha) seed yield.

FAIDHERBIA ALBIDA (syn. Acacia albida)

This leguminous tree is known for its high foliage and fruit production during the dry season; it is leafless during the rainy season. Its leaves and pods, relished by all livestock, are often the only fodder available at that time. This is an excellent tree species for use in agroforestry systems. The cultivar EC 133772 (ex Senegal) is doing well at Jodhpur (Rajasthan).

CITRULLUS COLOCYNTHIS

This species, which has a tremendous potential, occurs throughout India and grows wild in the warm arid and sandy areas of north-west, central and southern India as well as in coastal areas. This species is found in abundance in most arid districts, namely, Bikaner, Barmer and Jaisalmer Districts, and is also found growing wild in Jodhpur, Nagaur, Churu and Sikkar Districts (Paroda 1979).

The dried pulp of the unripe but fully grown fruit (excluding the rind) constitutes the drug colocynth. The roots have purgative properties and are used in treating jaundice, rheumatism and urinary diseases. The seed has a brownish-yellow oil which contains an alkaloid, a glucoside and a saponin (Anon 1960). The germplasm comprising 20 entries is being evaluated at Jodhpur. Multi-locational evaluation trials are being planned.

GENETIC RESOURCE CONSERVATION OF NEW CROP PLANTS

The National Bureau of Plant Genetic Resources, New Delhi, maintains active collections of various under-utilized and under-exploited plants at Headquarters as well as at Regional Stations located in different agro-climatic zones of the country (Table 13.2). In addition, the main centres of the All-India Coordinated Research Project have also been entrusted with the responsibility for germplasm maintenance of particular plant species. The base seed collections are kept under controlled conditions of temperature and humidity for both long- and short-term storage. Recently, a self-contained, portable cold store module, obtained under the British Aid Programme, has been commissioned and 50 000 accessions of

Table 13.2: Germplasm Assembled and Evaluated in Important Under-Utilized Crops

Crop	No. of accessions	Source country
Winged bean	319	India, Ghana, Papua New Guinea, Nigeria, Indonesia, Philippines, USA, Thailand, Sri Lanka
Rice bean	500	Nepal, China, India, Indonesia, Brazil, Nigeria
Amaranths	3000	India, Nepal, Malawi, Zambia, Poland, Taiwan, USA
Buckwheat	331	India, USSR
Chenopods	92	India, Poland, USA, Italy, Hungary, USSR
Bambara nut	192	Nigeria, India, Australia, UK, USA
Cuphea	7	USA
Leucaena	574	Philippines, Australia, USA, Syria
Guayule	80	USA, Mexico
Jojoba	75	USA, Israel, UK, Mexico
Euphorbia	56	USSR, USA, Philippines
Jatropha	10	India
Casuarina	65	Philippines, Australia, Tanzania
Bamboo	45	India
Thumba	34	USA

various agri-horticultural crops, including under-utilized and under-exploited plants, are being stored. Four cold store modules capable of running at −20 °C have also been procured. Efforts are in progress to commission these modules and, with this, the National Repository will be in a position to accommodate about 0·25 million germplasm accessions.

Future Thrusts

(a) Germplasm collections from indigenous and exotic sources of high priority species for food, fodder, energy, hydrocarbon and industrial plants need to be increased. Specific exploration programmes to collect genetic variability need to be undertaken for specific crops according to priority.

(b) Many of these less wellknown yet economically important species require greater publicity. Also appropriate research and development efforts are required to improve further some of the selected species.

(c) Adaptability studies need to be undertaken in order to find out the most suitable agro-climatic condition for each species. Standard practices for their efficient cultivation are also necessary.

(d) Concerted efforts are required to examine the possibility of introducing new crops into existing cropping/land use systems in order to help promote their use as supplementary rather than as substitute crops.

(e) For the crops which produce raw materials for industry, simultaneous arrangements for setting up the necessary infrastructure for their processing and marketing are required.

(f) Concerted plant breeding efforts are also required to develop genetically superior strains of these economically important species in order to increase their productivity.

(g) The entire genetic diversity built up is required to be maintained by appropriate long- and short-term conservation measures.

REFERENCES

Anonymous (1960) *Medicinal Plants of the Arid Zones*. UNESCO Publication No. NS. 59/111–17A, Paris.

Anonymous (1975) *Under-exploited Tropical Plants with Promising Economic Value*. National Academy of Sciences, Washington DC.

Arora, R. K. and E. R. Nayar (1984) *Wild relatives of crop plants in India*. NBPGR Science Monograph No. 7, New Delhi.

Arora, R. K. (1985) *Genetic resources of less known cultivated food plants*. NBPGR Science Monograph No. 9, New Delhi.

Bhandari, M. M. (1974) Famine foods in Rajasthan desert. *Econ. Bot.* 28, 1: 73–81.

Chandel, K. P. S., K. C. Pant and R. K. Arora (1984) *Winged bean in India.* NBPGR Science Monograph No. 8, New Delhi.

Gupta, R. K. and K. C. Kanodia (1968) Plants used during scarcity and famine periods in dry regions of India. *J. Agric. Trop. Bot. Appl.* 15, 7 & 8: 265–285.

Paroda, R. S. (1979) Plant resources of Indian arid zones for industrial uses. In: *Arid Land Plant Resources,* J. R. Goodin and D. K. Northington (Eds), pp. 261–281. International Centre for Arid and Semi-Arid Land Studies, Lubbock, Texas.

Partap, T. and P. Kapoor (1985) The Himalayan grain chenopods. 1. Distribution and ethnobotany. *Agric. Ecos. Environ.* 14: 185–199.

Partap, T. and P. Kapoor (1987) The Himalayan grain chenopods. 3. An underexploited food plant with promising potential. *Agric. Ecos. Environ.* 19, 1: 71–79.

Patel, K. M., E. R. R. Iyengar and P. M. Sutaria (1986) Guayule (*Parthenium argentatum*) for bio-products on wastelands. In: *Proceedings Bio-energy Society Second Convention and Symposium, 13–15 Oct. (1985),* R. N. Sharma, O. P. Vimal and V. Bakthavatsalam (Eds), pp. 130–136. Bio-energy Society of India, New Delhi.

Saxena, S. K. (1979) Plant foods of western Rajasthan. *Man and Environment* 3: 35–43.

Shankar, V. (1987) Life support species for emergencies in the Thar desert, India. Paper presented at the *International Workshop on Maintenance and Evaluation of Life Support Species in Asia and the Pacific Region. 4–7 April 1987* New Delhi, India.

Shankarnarayan, K. A. and S. K. Saxena (1987) Life supporting arid zone plants in famine periods. Paper presented at the *International Workshop on Maintenance and Evaluation of Life Support Species in Asia and the Pacific Region. 4–7 April 1987* New Delhi, India.

Singh, H. B. and T. A. Thomas (1978) *Grain Amaranths, Buckwheat and Chenopods.* ICAR, New Delhi.

Singh, P. and J. N. Gupta (1987) Unique life support species and their populations used by local people in India under extreme environmental conditions. Paper presented at the *International Workshop on Maintenance and Evaluation of Life Support Species in Asia and the Pacific Region. 4–7 April 1987* New Delhi, India.

Tomer, O. S. and R. K. Gupta (1985) Performance of tree species in saline soils. In: *Annual Report 1985.* Central Soil Salinity Research Institute, Karnal.

Zeven, A. C. and P. M. Zhukovsky (1975) *Dictionary of Cultivated Plants and their Centres of Diversity.* Centre for Agriculture Publishing and Documentation, Wageningen. (Ed. 2 by A. C. Zeven and J. M. C. de Wet 1982).

14

Potential New Food Crops from the Amazon

D. B. Arkcoll and C. R. Clement

THE NEED FOR NEW CROPS

Research on new crops is usually justified by the need to diversify from the very small number of species that provide the vast majority of our diet. The spectre of sudden appearance of an uncontrollable disease in one of these is particularly pertinent in these days of AIDS and it is worth remembering that a few historical examples of crop diseases have been just as devastating. A further interest in new crops results from the increasing desire for novelty in our diets. This is well demonstrated by the proliferation of ethnic restaurants and the variety of foods available in modern supermarkets. New agricultural systems may also require new crops. Examples are the need for shade tolerant species in agroforestry systems; deciduous self pruning species for alley cropping, and plants with high dry matter productivity for biomass fuels. Nutritional recommendations and a need for new chemical compounds with desirable characteristics have also resulted in the introduction of several new crops. Examples include crops like the Evening Primrose (*Oenothera* spp.) (Wolf *et al*. 1983) to supply gamma-linolenic acid and several plant sources of natural sweeteners, colourants and rubber.

Expanding the area farmed into new regions, often considered marginal for conventional crops, may also require new crops. Brazil, like many developing countries, faces the problem of improving the lives of its rapidly expanding population. It sees the migration of its underprivileged from crowded areas in the south and northeast to exploit the natural resources of the empty areas of the north as a partial but important solution to the county's economic and social problems. The colonization that has taken place so far has often been chaotic and unsuccessful partly because the traditional diet of cassava and hunted meat and fish do not respond well to intensification. One consequence is that much of the region's food has to be imported (Giugliano *et al*. 1978). There are also serious problems and restrictions that limit the expansion of major crops (rubber, cocoa and the African oil palm) that are most appropriate for producing income in these hot wet climates with very poor soils. Large capital investments are needed, there is a long delay before a return is obtained, markets are volatile, and the two species native to the Amazon suffer from devastating local diseases (Fig. 14.1).

Figure 14.1: *The Need for New Crops.*

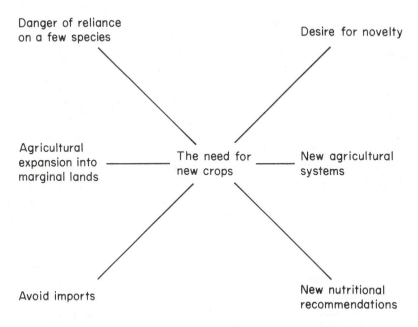

Thus expanding the area farmed, and reducing imports, are important reasons for developing new crops and systems to produce both food and income in such regions. Unfortunately, the loss of tropical forest taking place there at the moment is rapidly reducing the potentially useful germplasm available for developing appropriate new crops. So this needs to be studied and collected urgently if it is not to be lost (Clement & Arkcoll 1979; Clement & Chavez 1983).

The African oil palm owes its relatively recent success not only to its adaptation to soils and climates where few alternative crops are available, but also to the very high and competitive yields that have resulted from successful breeding. High yields in some crops have contributed to gluts and surpluses, especially when subsidized, that may also help create a demand for new or alternative crops. An example is the cereal surplus in Europe, which, together with the desire to reduce oil imports and the breeding out of anti-nutritional factors has led to a spectacular increase in the production of oilseed rape in the last decade. This in turn has helped lower world oil prices, forcing oil exporting countries also to look for alternatives.

SELECTION CRITERIA

After the initial enthusiasm and bewilderment at the vast number of little known fruit encountered in the region, one is faced with the difficult task of deciding which really have any commercial future. Much of the detail needed for such a decision is usually missing. Firstly, one finds that the vast majority of fruit are collected from the wild so that no yield or agronomic data are available. Secondly, most are unexciting or even unpleasant to people unfamiliar with them although they may be very popular and provide a useful dietary supplement locally. Thirdly, the scientific literature on them is largely botanical with the occasional reference to the fruit's potential due to local popularity and good flavour (Correa 1926; Le Cointe 1947; Cavalcante 1976). There is sometimes a chemical analysis of the fruit which may be quite detailed if reasonable levels of oil are found because of the historic interest in finding different types of oil for various end uses (Bolton 1928; Pesce 1941; Eckey 1954; Hilditch & Williams 1964). However, in general, most of the selection criteria needed to estimate their potential as crops still remains to be obtained (NAS 1975).

These criteria include diverse agronomic characteristics and those that affect potential for both industrialization and marketing (Fig. 14.2). Amongst the desirable field requirements are: a good fit with current farm practice, rusticity, precocity, high yields and easy propagation and harvesting to give steady returns. Prolonged harvests can be useful to avoid large peaks in demand for field equipment, labour, transport, storage and processing facilities. Slow perishability, firm texture, allowing easy transport and long storage, are often prerequisites for industrialization and marketing of most fresh fruit. Uniformity of size and shape and a suitable texture may help sorting, cleaning, skinning, depulping, etc. in the processing factory. Processors also look for attractive fruit that do not deteriorate on processing, especially on heating. They should either have little waste or valuable by-products to improve profitability.

Marketing is much more complex with the easiest solution being to look for products like oil which already have large world markets (Arkcoll & Aguiar 1984). Novelty alone is rarely sufficient to sell the product more than once. It is generally agreed that new products must have some instant appeal, especially in shape, colour, texture and flavour (Arkcoll 1986). This is complicated by big differences in individual tastes. Products that require familiarity before they are liked must have some hidden value like 'status', 'rejuvenating properties', 'be good for health', etc., that can be successfully exploited by skilful advertising, if the initial barriers are to be overcome. An attractive price is also important and is dictated by field, processing and marketing costs and losses.

Criteria for new crops that are useful for the survival of smallholders and colonists will be different and often less rigorous. Rusticity, prolonged harvests, high nutritional value and familiarity or a bland flavour are specially important (Arkcoll 1984a, 1986).

Figure 14.2: *Some Selection Criteria for New Crops*

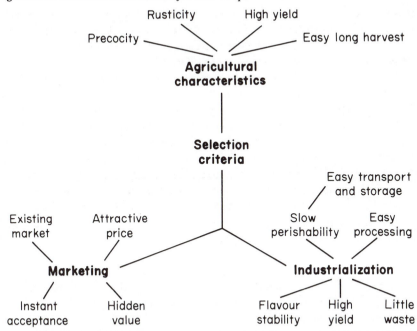

Current knowledge on some of the more interesting species from the Amazon region with potential as new food resources will now be discussed briefly in terms on these criteria. More detail on most of the species can be found in a recent FAO forestry paper on 'Food and fruit-bearing forest species' (FAO 1986).

CROPS FOR CARBOHYDRATE

The main staple in the diet of the whole region is cassava and, although well known as a subsistence crop, it is seldom treated as a serious agricultural crop. It is also prepared and eaten in an unusual form in the Amazon. Whole roots are peeled and then soaked in water for a few days during which they undergo a microbial acid ferment and become so soft that much of their moisture can be squeezed out easily before drying to a gritty sour flour on a griddle. These flours may be eaten dry or mixed with many other foods and drinks. Another novelty is that many of them are yellow in colour because they are made from varieties rich in carotenoids. Some varieties have been selected for this useful nutritional component as cassava eaters are often known to be deficient in vitamin A (Marinho & Arkcoll 1981). Another interesting variety is high yielding and precocious. Its roots start to swell after four months and can give up to 70 t/ha in 7 months (Arkcoll 1981). This and several others are used on the river banks during the short period between the annual floods; thus some also have good tolerance to a root rot (*Phytophora drechsleri*) common in such wet soils.

Some indian tribes use many different varieties and species of *Dioscorea*, and also *Pachyrhizus tuberosus* and *Calathea allouia* (Kerr & Clement 1980). Some of the yams obviously deserve further study and there is some interest in the other two because of their crunchy texture (Kay 1987).

Several palms produce starchy fruit that are widely used in the region, the most interesting being *Bactris gasipaes* (peach palm, pupunha, pejibaye). The N-free extracts (mainly starch) of the dry mesocarp of this fruit have been shown to vary from 14·5 to 84·8% (Arkcoll & Aguiar 1984). Its texture is usually like potatoes but may be as floury as maize or fibrous or even oily, and it has a bland agreeable flavour when boiled. Some indian tribes appear to have selected it for both high starch and size and use it as an additional staple to cassava, sometimes as a griddled dry flour, sometimes as a boiled fruit and sometimes as a fermented peach-flavoured drink after hydrolysing the starch with spittle (Clement, in press). This species has attracted considerable attention because of its small crown and potentially high yields that fit it well into our attempts to produce food ecologically with trees. Such 'food forests', in which the trees themselves produce the basic diet, are seen as a logical extension of the agroforestry systems that seem most suited to the wet tropics (Arkcoll 1978, 1979a&b, 1981, 1984a&b, 1985). *B. gasipaes* is undoubtedly the most interesting species in the region and it will consequently be discussed in more detail in a later chapter. Its true potential has never been realized partly because of excessive vegetative growth that makes harvesting difficult and partly due to a serious fruit drop problem now believed to be caused principally by an insect-introduced fungus. It is an excellent example of the difficulty of trying to grow a crop in its centre of diversity, so we are encouraging work in other tropical areas to avoid the problem of pest and diseases.

Babassu (*Orbignya* spp.) is another palm that has a very dry starchy mesocarp that is sometimes used to make a gruel in times of necessity in some very poor rural areas where large natural stands and nuts are exploited principally for their oily kernels. Commercial exploitation of the whole nut, to include the use of the mesocarp for industrial starch and alcohol production, has been examined. However, the logistics of collecting and transporting wild nuts of low overall value, and the expense of cracking and separating them, has not been a viable proposition to date (Mattar 1984). Several other Amazonian palms (*Mauritia flexuosa* and *Acrocomia* spp.) are reputed to accumulate a sago-like starch in their stems, although this has not been confirmed and there does not seem to be any obvious exploitation. The same can be said of tapping these palms for a sweet sap.

Several species of *Pouteria* have very starchy mesocarps that are eaten in some areas but they are not pleasant or productive enough to have attracted attention. Similarly, the starchy kernel of *Poraqueiba sericea* is used for washing but is said to be too bitter for eating. However the starchy by-product seeds of *Bixa orellana* are raising some interest now that yields of 2 t/ha are thought to be possible. They are exploited in several countries for the natural carotenoid colourant, bixin. This is found on the surface of the seeds and seldom passes 4% of their total weight (I. Guimaraes, pers. comm.).

CROPS FOR OIL

Plants with fruit that are rich in oil are extremely interesting as new crops because there are large world markets already in existence for both the oil and the residues if their prices are attractive. Consequently much attention has been paid to the various options available.

The major oils produced in the Amazon are lauric oils, chiefly from palm kernels collected from a fraction of the enormous wild stands of babassu (*Orbignya* spp.) found in the transition zone around the edge of the forest. About 150 000 t of oil are obtained annually from this genus and smaller amounts from *Astrocaryum vulgare* and *Scheelea martiana* (Table 14.1). There is a glut of lauric oils on the world market at the moment and this may well last because of the rapidly increasing plantations of African oil palm (*Elaeis guineensis*) in the Far East (Arkcoll 1988). However Brazil is very interested in raising its own production to avoid future imports as demands rise and the poorly rewarded collection from the wild dwindles. The large natural stands (over 12 million hectares), slow growth and low oil per bunch have helped inhibit the planting of native palms to date, but recent collection of both *Orbignya* and *Astrocaryum* have identified more useful germplasm and raised enthusiasm over their potential. A small amount of lauric oil was also obtained from *Astrocaryum murumuru* in the past and some is extracted from *Acrocomia* spp. in Paraguay and the south of Brazil. The latter are common in the Amazon and are being intensively studied in attempts to exploit their potentially high yields of mesocarp oil (palmitic-oleic type) (Wandeck & Justo 1982; Lleras & Coradin 1983). Predictions of over 6 t/ha of oil have been made for this species but confirmation has been hindered by difficulty in breaking the prolonged seed dormancy. Rapid putrefaction, with hydrolysis of oil extraction from the muci-

Table 14.1: *Government Data on Food Collection from the Wild in the Amazon*

Species	Product	Quantity (t/yr)
Astrocaryum vulgare	Oil kernels	1364
Scheelea martiana	Oil kernels	6304
Orbignya spp.	Oil kernels	204766
Bertholletia excelsa	Brazil nuts	40710
Euterpe oleracea	Palm hearts	94594
Euterpe oleracea	Fruit	92983
Couma spp.	Chicle latex	4786

Source: Fundação IBGE (1985).

laginous pulp, and slow growth, remain to be solved. This is a clear example of a need to consider factors other than yield when selecting new crops.

Cuphea spp. have also attracted our attention as they are being examined as potential sources of lauric oils that might be grown as arable crops in temperate climates and avoid the need to import coconut and palm kernel oil from the tropics (Graham *et al.* 1981; Hirsinger & Knowles 1984). Dehiscence is the main problem but there are many others to be overcome. There are over a hundred Brazilian species, some from the Amazon that are mostly unstudied and it is hoped that some may have useful characteristics. Species known to be rich in C8 and C10 fatty acids exist, and these may be a useful source of medicinally and nutritionally interesting medium chain triglycerides in the future (Arkcoll 1988; Bach & Babayan 1982).

Elaeis oleifera is being collected and used to produce hybrids with *E. guineensis*, the most important oil crop of the tropics. Although it is low yielding, and the hybrids have excessive vegetative growth, it may be useful to reduce height as the stem tends to creep along the ground; it also has some disease resistance and compositional differences of interest to breeders (Ooi *et al.* 1981).

Bactris gasipaes fruit are mainly starchy but some do exude oil on cooking. A recent study of several hundred fruit showed a tremendous variation in shape, size and composition (Arkcoll & Aguiar 1984). A few very oily fruit were found and it is hoped that this characteristic can be combined with the potentially high yields to produce a useful new oil crop for the wet tropics with a quality residue with over 20% protein. The enormous variation in oil content of the dry mesocarp, from 2·2 to 61·7%, gives an idea of how difficult it is to exclude a little-known species from a list of those with most potential, and also how rapidly one might make progress in a crop improvement programme.

A search for extenders or substitutes for high priced cocoa butter is currently in progress and has identified two interesting options from the Amazon. The first of these is the oil from the seeds of *Theobroma grandiflorum* that are now being produced as a by-product from the pulp market; this is expected to increase in the future (see exotic fruit section). Seed are said to be unsuitable for direct use as chocolate so uses for its oil are now being sought (Alvim, pers. comm.). It has been possible to separate about 65 per cent of a solid fraction from *Caryocar cuneatum* oil recently that is rich in 2-oleodipalmitin (Arkcoll & Jee, unpublished data). About 400 t of cooking oil per year are produced in the northeast of Brazil at the moment from the mesocarps and endosperms of these fruit. A more valuable oil might justify the collection of greater amounts from the millions of native trees or even justify the plantations of this *C. villosum*, tried in Malaysia some 60 years ago (Lane 1957).

Several palms are used to produce oily emulsions from their thin mesocarps often thickened with cassava flour. These may be important items of the diet in some areas. The most common is *Euterpe oleracea* (Table 14.1) but *Mauritia flexuosa*, *Astrocaryum vulgare*, *Jessenia bataua* and *Oenocarpus bacaba* are all widely used. Their flavour and texture are unusual and are rarely liked by those unfamiliar with them. Good quality table oil may be obtained from some of these by boiling and concentrating them and they are of use in remote areas where large natural stands exist (Balick 1981). At the moment these do not have much appeal as commercial plantation crops because of slow growth, low oil content and low yields, but it is felt that this could change if better germplasm is found. Other species that have been exploited from the wild to a small extent but have major defects as crops are *Pentaclethra macroloba* (dehiscence and low yield), *Virola* spp. (slow growth and low yields), and *Pachira aquatica* (toxic cyclopropenic fatty acid content).

CROPS FOR PROTEIN

Hunted animals, turtles and fish were, and in many areas, still are the major source of protein (Giugliano *et al.* 1978). The need to replace these is one of the main reasons for the ranching that is expanding in many areas. Low protein contents (1%) in the regions staple, cassava, led us to search briefly for better varieties and some with over 3% were found (Arkcoll & Aguiar, unpubl. data). This small increase could obviously be quite useful in a diet that may contain as much as 700 g a day of cassava flour (Giugliano *et al.* 1978).

Cowpeas (*Vigna unguiculata*), Lima beans (*Phaseolus lunatus*) and peanuts (*Arachis hypogaea*) are grown to a very small extent in the region and some interesting varieties have been noted. These include very small-seeded cowpeas that are held high above the crop to facilitate drying in a wet climate; numerous different shapes, colours and sizes of lima beans adapted to the wet tropics, cultivated by certain Amerindian tribes, and extremely large peanuts, with some individual seeds weighing over 2·5 g, used by others.

Some time has been spent looking for a tree or perennial with an edible bean to fit into the food forests described earlier, without much luck. Some tribes use the large seed of a *Swartzia* sp. from the forest but this did not grow well in Manaus (Arkcoll 1984a). *Parkia platycephala* produces copious amounts of an extremely useful fodder pod in drier regions at the edge of the Amazon forest but these, and most other species tried, are very fibrous or too bitter to be used by humans. This is true of several *Inga* spp. tried to date, however *I. paterno* is apparently eaten in El Salvador and Honduras. Most nuts are good sources of protein in remote regions and several species are eaten, often in quite large quantities when collected from the forest. Brazil nuts (*Bertholletia excelsa*) are too valuable as export items to make much impact on the diet in populated regions. However, many palm kernels are important in some areas. The residues after oil extraction from babassu and other species have about 20% protein and are exploited commercially as an ingredient of cattlecake. Mature kernels of most palms are too hard and fibrous to eat whole but it is common practice to eat them immature, when they are still soft. The relatively small kernels of *Bactris gasipaes* are also rather fibrous normally but some as soft and attractive as coconut have been noticed. However, mature kernels of several species, especially babassu are, often pounded with water to give a milk-like substance that is widely used in certain regions, especially for children. Similarly the purple, brown or orange milk-like emulsions from *Euterpe oleracea* (Table 14.1), *Jessenia* spp. and a few other palm mesocarps may contribute reasonable quantities of protein of apparently good value when consumed in sufficient amounts (Balick & Gershoff 1981). The large amount of mesocarp in most fruit of *Bactris gasipaes* may also be a useful source of protein (Arkcoll & Aguiar 1984). Protein contents vary from 3·1 to 14% in the samples studied, so some are higher in protein than most cereals, and are of sufficient quality to replace maize in chick diets. The residue after extracting oil from one of the best introductions had 22·6% protein and might be useful as a complete ration or protein concentrate for monogastric animals, especially after the removal of large fibres.

EXOTIC FRUIT AND NUTS

Many of the fruits found in local markets come from the wild or from smallholdings where they are beginning to be domesticated (Clement 1983). A surprisingly large number of them are considered uninteresting or even unpleasant by people not accustomed to them (Arkcoll 1986). Some are important in the local diet as mentioned in previous sections, and deserve attention for this restricted market. It may be possible to extend this market slowly as people nearby become more familiar with the fruits. However, there are quite a number of fruits that are much more appealing so most attention has been directed towards selecting these for their commercial potential. It is in fact extremely important to select and concentrate work on just a few of the very best fruits if one is ever to encourage commercial plantations of a species with no market and so break the vicious circle in which there are no markets because there are no plantations (Leakey 1968). Unfortunately it is not so easy to agree on a short-list because people have natural differences in their tastes and preferences. This problem has been exacerbated by the collection of fruit that sometimes seem fantastic miles up river, at the end of a long day's search, but may be considered quite dull, once back in the laboratory under the cold critical glare of a tasting panel. Indeed that fantastic fruit may only be re-examined several years later after being grown up from seed and then it may not breed true, adding a further query – was the original better? At the other end of the scale, rejected or low priority species take on a new appeal when one suddenly finds an amazing tree with vastly superior fruit to any seen previously.

The answer has been to collect and grow many of the interesting species in field banks initially and decide the best options after evaluating them in terms of the selection criteria discussed previously. Emphasis has been placed on those that have instant appeal that wears the test of time as assessed by tasting panels, newcomers and popularity in fresh juice and ice-cream shops in different towns (Arkcoll 1986). Simple field characteristics, like rusticity and precocity, are easily evaluated in the germplasm bank and have also helped reduce the list significantly. At the same time, new species or better germplasm of older species are continually being added.

Species that raised initial interest but have now been lowered in our priority list include *Pouteria caimito, Couepia bracteosa, Pourouma cecropiifolia, Talisia esculenta, Inga edulis* (not sufficiently appealing to create quick markets), *Couma utilis* (hard to harvest and too soft to transport), *Brysonima crassifolia, Astrocaryum aculeatum,* and *Platonia esculenta* (strong flavoured requiring familiarity). Interest in *Myrciaria dubia* because of its very high vitamin C content has been dampened by its poor flavour profile and low yields (Charley 1970). Similarly the future for the high yielding *Solanum topiro* is affected by its flavour being much inferior to that of the excellent fruit of *S. quitoensis*, now being commercially grown in the Andes. An initial interest in *Quararibea cordata* as a table fruit, because of its attractive flavour and thick skin which would allow long-distance transport, has now waned because of its demand for good soils and long delay to first fruit and susceptibility to drought and several pests (Clement 1982). Similarly *Annona muricata*, already grown commercially for its excellent juice in several parts of Latin America, has been low yielding and troubled by pests and diseases in Manaus (Ferreira 1985). Two other species with good yields and superbly flavoured juices are *Eugenia stipitata* and *Psidium angulatum*, although both have been attacked by fruit fly in spite of being very acid (Chavez & Clement 1983). *Bactris gasipaes* fruit with sweet flavour, and good texture (medium oil content, low fibre and edible skins) have been collected and are so much better than the normal fruit that a market may be created for them as table fruit. Its susceptibility to fruit drop has already been mentioned.

Although we always hope that resistance might be found to these various pests and diseases or that they can be controlled economically, they all draw attention to the difficulty of trying to grow and evaluate the real potential of many of the best species in their centre of diversity. This is especially true in the wet tropics where pests and diseases are common, multiply very rapidly, and are not broken by cold or drought as in other climates. The need to test them in other regions is obvious and we look forward to improved germplasm exchange schemes and an end to the unrealistic possessiveness sometimes encountered.

Several excellent fruit trees have not been troubled by pests and diseases so far. These include some very good introductions of *Spondias lutea*, superior in flavour to those usually found in the West Indies. They are both rustic and productive and a search is on for fruit with more pulp and easier harvesting. There is also considerable interest in *Theobroma grandiflorum* whose pulp is widely considered to be the best flavoured of the region for making juices and charlotte creme desserts. However it does not produce well either in the field or in terms of pulp and juice yield and is susceptible to Witches Broom disease.

The fruit must also be harvested off the ground within a day or two of dropping or they rot and unfortunately they do not ripen if harvested off the trees before fully mature. Consequently fruit are very expensive so efforts are in progress to overcome the problems. Fruit without seed and thus with higher pulp yields have been found. The pulp is best stored frozen as the very strong fruity aroma is partially destroyed by heating. Both this species and *Eugenia stipitata* have such interesting aromas that these might find markets of their own in fruit cocktails, yoghurts, perfumes, etc.

Hancornia speciosa and *Eugenia uniflora* are two other species with fruits of very attractive flavour much appreciated in other parts of the north of Brazil. These are both found in large natural stands in the drier areas at the edge of the Amazon forest, and clearly deserve much more attention than they have received so far. There are undoubtedly many other species that we have not had the chance to look at yet for which this is probably also the case.

The well-known commercial fruit that have their origin in the Amazon (cocoa, pineapple, passion fruit and guava are the principal ones) are collected by specialists from time to time so we have not paid them much attention. However interesting material that has been noticed include very large (15 kg) and red-skinned pineapples and orange-fleshed guavas.

Brazil nuts (*Bertholletia excelsa*) have a steady world market that could expand especially into the mixed nut and confectionery markets if they were cheaper. As current production is almost exclusively from the wild in Brazil (Table 14.1), a search is on for more precocious dwarfs, highly productive and rustic germplasm, with the aim of increasing the production in plantations (H. Muller, pers. comm.). *Lecythis usitata* is far less productive than Brazil nuts and also sheds its nuts when ripe so do not offer obvious advantages although of good quality. *Caryodendron orinocense* is being actively studied in Colombia (Martinez 1970) and is said to be very pleasant to eat although, however, fears have been expressed over its possible toxicity. We have paid some attention to various *Couepia* spp. and found that *C. longipendula* has an excellent flavour. Attempts are in progress to obtain precocious dwarfs and to reduce the thickness of shells which must currently be cracked with an axe. The nuts of *Caryocar nuciferum* were exported from Surinam to Europe in the 1920s but have not been studied since (Stahel 1935). One wonders if the crossing of this species with smaller ones from Brazil, that have useful mesocarp fats, might justify their joint exploitation. The nuts of *Pachira aquatica* are very common throughout Central America and Brazil and serve as a useful supplement to the diet in many regions. The species is very fast growing, rustic and takes easily from large cuttings so is useful for live fence posts and street shading. It produces large quantities of pleasant flavoured nuts in big pods, but unfortunately contains high amounts of cyclopropenic fatty acids in its oil, and these are known to be toxic and possibly even carcinogenic (Bruin *et al.* 1963; Tinsley *et al.* 1982). This raises the unfortunate question of the need for toxicological studies on new species before their widescale use is recommended.

CROPS FOR DIVERSE USES

Of considerable surprise to the newcomer to the region is the almost complete lack of little-known vegetables and the very low consumption of leafy spinaches. It seems that the abundance of fruit available in the forest are eaten in sufficient quantity to complement vitamin requirements. *Solanum topiro* is occasionally used in some areas as a sour tomato substitute in fish stews. A few are used as condiments and those of cassava are eaten as a speciality in Para State after prolonged boiling and spicing. Leaves are quite poorly used in the diet throughout the country because of the association with animal food, however there are three excellent spinaches used in a few rural areas that deserve much wider recognition. These are *Xanthosoma brasiliense, Talinium triangulare* and *Pereskia grandifolia*. The first two are relatively shade tolerant, so might be useful for agroforestry systems, and the latter is a rustic cactus. All three have about 20% protein in their leaves so make a significant contribution to the diet of some people.

Palm hearts are also rich in protein, being immature leaf meristem. Those of several species are eaten by rural communities but in general most palms are protected in the Amazon for their fruits. However, the elimination of much of the accessible *Euterpe edulis,* used for commercial canning in the south of Brazil in the past, has driven the canners up to the Amazon estuary where they are now exploiting wild *E. oleracea* stands at the rate of almost 100 000 t per year (Table 14.1). Other species used in a smaller way include *Astrocaryum jauari* and *Scheelea martiana*. Commercial plantations of *Bactris gasipaes* have not been able to compete with this wild exploitation. However this palm is able to produce excellent palm hearts on a sustainable basis (because they tiller), at six times the rate of *E. oleracea* in experimental plantations (Gomes & Arkcoll 1987). Such plantations may become viable in the future once yields are increased, uses are found for the 99 per cent of the tree that is presently unused, and the prices of hearts exploited from the wild increases as these are eliminated. Rational use of the wild stands allowing recovery of plants could halt this trend (A. Anderson, pers. comm.). Palm hearts are considered to have excellent export potential and over 200 000 t per year have been produced in the past to satisfy this growing market. Unfortunately, poor quality control by the numerous small firms involved in the canning of palm hearts, has led to the rejection of much material and the current stagnation. The product is often very variable because of the subjective decision used to eliminate outer fibrous and inedible leaves, and the need to acidify the pasteurized product to the correct level to avoid botulism.

Couma spp. are also being rapidly eliminated from accessible areas because of destructive tapping for a latex to produce chicle. About 5000 t of latex are exported each year (Table 14.1) and the possibility of plantations is now being examined. Trees of *C. utilis* are growing very fast and look quite promising as the fruit can also be sold as a valuable by-product for the local market.

161

An interesting cola-type drink made from the seeds of guarana (*Paullinia cupana*) has a large market within Brazil. Some 1000 t are produced each year from about 5000 hectares and the crop is expanding now that export markets are opening up. It owes much of its success to the stimulation produced by its high caffeine content and a widely held belief in its rejuvenating and aphrodisiacal properties. These are good marketing points although the drink is grim to pleasant, depending on formulation. In fact, several of the most popular brands on the market contained no real guarana at all until recent legislation made this obligatory.

The potential for finding other minor food ingredients like gums, colourants, etc., is large and little explored. *Bixa orellana* is a good example and is now grown commercially in several countries as a source of natural colourant. It is increasing in importance now that it is considered to be less of a health hazard than some of the synthetic alternatives. Several forest species exude large amounts of gum from damaged trunks and deserve attention.

CONCLUSIONS

A search for new food crops can be justified by our need to reduce our dependence on so few species, our desire for novelty and new useful compounds, and our interest in increasing crop returns and reducing imports and surpluses. New crops may also be useful for new agricultural systems and for expanding the productive area into marginal regions like the Amazon where few known crops are adapted, yet food and income are required by the rapidly expanding population.

It is obvious that means must be found to overcome the risks of early commercial plantations of interesting species with no established markets, if any of them are ever to make the large jump from botanical curiosities to useful crops. Careful selection of the most promising species will obviously help and some of the main criteria for selecting new crops have been outlined. They include many agronomic, industrial and marketing considerations. These criteria are being used to select the best Amazonian species for a variety of end uses. Some of the most interesting species that have been chosen so far are *Bactris gasipaes* as a starchy staple; *Astrocaryum aculeatum, Acrocomia aculeata, Elaesis oleifera, Bactris gasipaes* and *Caryocar* spp. as oil crops; *Annona muricata, Eugenia stipitata, E. uniflora, Hancornia speciosa, Psidium angulatum, Spondias lutea, Theobroma grandiflorum* and *Couepia longipendula* because their fruits or nuts have exquisite flavours; *Couma utilis* as a chicle substitute; *Paullinia cupana* as a cola drink and *Bactris gasipaes* as a source of palm hearts. Natural variation is extremely large and has helped enormously in the selection of these species. However, it has also made it difficult to eliminate those with problems that one hopes to overcome by finding better germplasm. There may well be other rare species with excellent germplasm and prospects that we have not seen yet that deserve equal or more attention.

Germplasm banks of most of the selected species are in various stages of formation and field trials and breeding programmes are in their early stages. These are providing the missing data needed for a more complete evaluation of their potential and the improvements needed to turn them into viable crops. This requires time and a multidisciplinary research team that has proved hard to form and harder to keep. Maintenance of live germplasm banks and breeding perennials is expensive and requires a long-term financial commitment and research philosophy. A lack of continuity in funding is often disastrous and banks of Amazon fruit in various stages of decay can be seen in Costa Rica (CATIE), Trinidad (University of the West Indies), Surinam (CELOS), Campinas (IAC), Belem (CPATU), Rio de Janeiro (km 47) and now in Manaus (INPA). Outside international help is needed to avoid the devastating effects of cyclical financial crisis often seen in developing countries. A way round the problem of attracting research workers, especially breeders, to the slow rewards obtained with tree crops must also be faced. Another problem is that the potential of many of the best species may be difficult to realize in or near their centre of diversity, because of native pests and diseases. Testing in other parts of the country, or in other countries, is clearly needed to evaluate fully many of the most promising identified to date. Improved exchange of new crop germplasm is an urgent requirement.

Attention is also drawn to interesting germplasm of a few well-known crops like cassava in which high yielding, high protein, high carotenoid content, precocity and disease resistance have been seen.

REFERENCES

Arkcoll, D. (1978) Food forests, an alternative to shifting cultivation. *Abstracts of XI International Nutrition Congress*, p. 300. Rio de Janeiro.

Arkcoll, D. (1979a) An evaluation of the agroforestry options for the Amazon. In: *Anais do Simposio sobre ciências básicas e aplicadas*, pp. 101–111. Academia de Ciências do Estado de São Paulo.

Arkcoll, D. (1979b) The production of food from trees and forests. In: *Proceedings of an International Symposium on Forestry Science and its Contribution to the Development of Latin America*, M. Chavarria (Ed.), pp. 171–173. Euned, Costa Rica.

Arkcoll, D. (1981) Some interesting varieties of cassava from the Amazon. *Acta. Amazônica*, 11: 207–211.

Arkcoll, D. (1984a) Nutrient recycling as an alternative to shifting cultivation. In: *Ecodevelopment*, B. Glaeser (Ed.), pp. 39–44. Pergamon, Oxford.

Arkcoll, D. (1984b) Some leguminous trees providing useful fruit in the north of Brazil. *Pesquisa Agropecuária Brasileira* 19 (s/n): 61–68.

Arkcoll, D. (1985) The production of food in the humid tropics. In: *Anais do II Encontro brasileiro de agricultura alternativa*, pp. 221–224. AEARJ/FAEAB, Rio de Janeiro.

Arkcoll, D. (1986) Some lesser known Brazilian fruit with unexploited potential. In: *Proceedings of the XIX Symposium of the International Federation of Fruit Juice Producers*, pp. 27–34. Juris Druck, Zurich.

Arkcoll, D. (1988) Lauric oil resources. *Econ. Bot.*, in press.

Arkcoll, D. and J. P. L. Aguiar (1984) Peach palm (*Bactris gasipaes* H. B. K.) a new source of vegetable oil from the wet tropics. *J. Sci. Food Agric.* 35: 520–526.

Bach, A. C. and V. K. Babayan (1982) Medium chain triglycerides, an up-date. *Amer. J. Clinical Nutr.* 36: 950–962.

Balick, M. J. (1981) *Jessenia bataua* and *Oenocarpus* Species: Native Amazonian Palms as new sources of edible oil. In: *New Sources of Fats and Oils*, E. H. Pryde (Ed.) pp. 141–155. American Oil Chemists Society, Champaign.

Balick, M. J. and S. N. Gershoff (1981) Nutritional evaluation of the *Jessenia bataua* palm: source of high quality protein and oil from tropical America. *Econ. Bot.* 35: 261–271.

Bolton, E. R 1928 *Oils, Fats and Fatty Foods*. Churchill, London.

Bruin, A., J. E. Heesterman and M. R. Mills (1963) A preliminary examination of the fat from *Pachira aquatica*. *J. Sci. Food Agric.* 14: 758–760.

Cavalcante, P. B. (1976) *Frutas comestíveis da Amazonia*. Instituto Nacional de Pesquisa da Amazônia, Belem.

Charley, V. L. S. (1970) Some tropical fruit juices. In: *Proceedings of a Conference on Tropical and Subtropical Fruits*.: 161–166. TPI, London.

Chavez, F. W. B. and C. R. Clement (1983) Considerações sobre a araçá-boi (*Eugenia stipitata*) na Amazonia brasileira. *Anais. Cong. Bras. Fruticultura.* 7: 167–177.

Clement, C. R. (1982) Observações sobre a sapota da América do Sul (*Quararibea cordata*). *Proc. Amer. Soc. Hort. Sci./Trop. Reg.* 25: 427–432.

Clement, C. R. (1983). Underexploited Amazonian Fruits. *Proc. Amer. Soc. Hort. Sci./Trop. Reg.* 27A: 117–142.

Clement, C. R. (in press) Domestication of the pejibaye palm: past and present. *Advances in Economic Botany*.

Clement, C. R. and D. B. Arkcoll (1979) Forestry politics and the future of fruit culture in the Amazon. *Acta Amazônica.* 9 (supplement): 173–177.

Clement, C. R. and F. W. B. Chavez (1983) Review of genetic erosion of Amazon perennial crops. *Plant Genetic Resources Newsletter* 55: 21–23.

Correa, M. P. (1926) *Dicionário das plantas utéis do Brasil*, 6 volumes. Imprensa Nacional, Rio de Janeiro.

Eckey, E. W. (1954) *Vegetable Fats and Oils*. Reinhold, New York.

FAO (1986) *Food and fruit bearing forest species No. 3: Examples from Latin America*. Forestry Paper No. 44/3. FAO, Rome.

Ferreira, F. W. C. (1985) A influência de dois tipos de consórcios de fruteiras na incidencia de broca do tronco (*Cratosomus* sp.) nas graviola (*Annona muricata* L.). *Acta Amazônica* 15: 3–11.

Fundação IBGE (1985) *Anuario Estatistico do Brasil* 1985. IBGE, Rio de Janeiro.

Giugliano, R., R. Shrimpton, D. B. Arkcoll and M. Petrere (1978) A study of food and nutrition in the Amazon. *Acta Amazônica* 8 (supplement 2): 1–54.

Graham, S. A., F. Hirsinger and G. Robbelen (1981). Fatty acids of *Cuphea* seed oils and their systematic significance. *Amer. J. Bot.* 68: 908–917.

Gomes, J. B. M. and D. B. Arkcoll (1987) Initial studies on the production of palm hearts by *Bactris gasipaes* in plantations. In: *Anais do I Encontro nacional de pesquisadores em palmito*. In press. EMBRAPA, Curitiba.

Hilditch, T. P. and P. N. Williams (1964) *The Chemical Constitution of Natural Fats*. Chapman and Hall, London.

Hirsinger, F. and P. F. Knowles (1984) Morphological and agronomic descriptions of selected *Cuphea* germplasm. *Econ. Bot.* 38: 439–451.

Kay, D. E. (1987) *Root Crops* Ed. 2, revised E. G. B. Gooding. TDRI, London.

Kerr, W. E. and C. R. Clement (1980) Prática agrícolas de consequências genéticas que possibilitaram aos índios da Amazônia uma melhor adaptação às condições regionais. *Acta Amazônica* 10: 251–261.

Lane, E. V. (1957) Piquia, a potential source of vegetable oil for an oil starved world. *Econ. Bot.* 11: 187–207.

Leakey, C. L. R. (1968) Problems in the introduction of new crops in developing countries. In: *Essential oils production in developing countries*, pp. 97–110. Tropical Products Institute, London.

Le Cointe, P. (1947) *Amazonia Brasileira III, Arvores e plantas utéis*, Nacional, São Paulo.

Lleras, E. and L. Coradin (1983) La palam macauba (*Acrocomia aculeata*) como fuente potencial de aceite combustivel. In: *Palmeras poco utilizadas de America Tropical*, pp. 102–112. FAO/CATIE, Costa Rica.

Marinho, H. A. and D. B. Arkcoll (1981) Studies on the carotenoids of some Amazonian cassava varieties. *Acta Amazônica* 11: 71–75.

Martinez, J. B. (1970) *El Inchi (Caryodendron orinocense Karst.)*, 1–52. Faculdade de Agronomia, Universidad de Nariño, Colombia.

Mattar, H. (1984) Industrialization of the babassu palm nut: the need for an ecodevelopment approach. In: *Ecodevelopment*, B. Glaeser (Ed.), pp. 71–87. Pergamon, Oxford.

NAS (1975) *Underexploited tropical plants with promising economic value.* NAS, Washington DC.

Ooi, S. C., E. B. da Silva, A. A. Muller and J. C. Nascimento (1981) Oil palm genetic resources – native *E. oleifera* populations in Brazil offer promising sources. *Pesquisa Agropecuária Brasileira* 16: 385–395.

Pesce, C. (1941) *Oleaginosas da Amazonia*. Of. graf. da revista da Veterinaria, Belem.

Stahel, G. (1935) De Sawarie-noot, *Cariocar nuciferum* en enkele andere in Suriname in het wild grociende noten. *Serie Overdrukken* No. 7: 1–32. Landbouwproef Station, Suriname.

Tinsley, I. J., G. Wilson and R. R. Lowry (1982) Tissue fatty acid changes and tumor incidence in C3H mice ingesting cottonseed oil. *Lipids* 17: 115–117.

Wandeck, F. A. and P. G. Justo (1982) Macauba, fonte energética e insumo industrial. *Vida Industrial*, October, pp. 33–37. São Paulo.

Wolf, R. B., R. Kleiman and R. E. England (1983) New sources of gamma linolenic acid. *J. Amer. Oil Chem. Soc.* 60: 1858–1860.

15

Potential Multipurpose Agroforestry Crops Identified for the Mexican Tropics

J. L. Delgardo Montoya and E. Parado Tejeda

INTRODUCTION

Lack of firewood, insufficient forage for cattle and low nutrient level of the soils are the three critical agricultural problems in most of the developing countries. As a consequence most of the tropical forests are removed at an unacceptable rate, with the object of satisfying the ever increasing demand for firewood, agricultural land and forest timber (Delgardo 1987).

For the past 12 years the Instituto Nacional de Investigaciones sobre Recursos Bioticos (INIREB) has dedicated a research group to identify and evaluate species of great potential value that would otherwise have been neglected, ignored or destroyed. All the species we recommend in this paper belong to the humid tropics.

Jatropha curcas belongs to the Euphorbiaceae. It is native to tropical America, from southern Florida to Argentina. In Mexico it occurs in both coastal areas, up to 1300 m asl. It prefers hot and humid climates. It is easily propagated by seed and cuttings. In the Caribbean area it is known as physic nut.

It is considered as an oleaginous crop, because of the high oil content of the seed. The physical and chemical properties of the oil are such that it is often used for food and medicinal purposes.

Brosimum alicastrum is a Moraceae tree, commonly called ramon. At present it is utilized to a very limited and local extent as forage for pack animals. However, its abundance in the tropical and subtropical forests of Mexico and its high nutritional value may possibly merit increased utilization.

Many fast growing leguminous tree species could serve as fuelwood and provide other products as well as help in the control of erosion. Another characteristic of these species is their capacity to fix atmospheric nitrogen, thereby increasing content of this valuable element in poor soils, or by providing additional sources of nitrogen in the form of green manure and in the production of high protein forage.

INIREB selected eight legume tree species to conduct an experiment which was also supported by the US National Academy of Sciences. Its goal was to evaluate their growth and development. The results show that *Acacia pennatula, Gliricidia sepium, Albizia lebbek*, and *Piscidia piscipula* (syn. *P. communis*) are outstanding in terms of firewood, forage and nitrogen-fixing crops and could be recommended for use.

RESULTS AND COMMENTS

Jatropha curcas

The high oil and protein content of *Jatropha* seed is shown in Table 15.1.

Table 15.1: *Nutritional Analysis of Jatropha curcas Seeds*

Component	g/100 g Seeds
Moisture	4·08
Ash	4·98
Crude fat	50·33
Crude protein	27·13
Crude fibre	5·12
Carbohydrates	8·36

Source: Cano & Hernández (1984).

Table 15.2 shows that the essential amino acid methionine content is high, especially when compared with other grains and seeds used in human nutrition.

Table 15.2: *Amino Acids Content of Jatropha curcas Seed Protein*

Amino Acid	g Amino Acids/100 Protein
Methionine	2·2
Valine	5·9
Lysine	2·5
Isoleucine	4·2
Leucine	7·1
Histidine	2·0
Phenylalanine	4·6
Threonine	4·0
Glutamic acid	17·4
Cystine	1·9
Tyrosine	2·8
Aspartic acid	11·5
Proline	5·5
Serine	6·0
Arginine	14·5
Glycine and Alanine	5·7
Tryptophan	0·83

Source: Cano & Hernández (1984).

Table 15.3 shows that the quality of *Jatropha curcas* seed protein is relatively similar to that of lentils and beans, lower than wheat but higher than corn.

167

Table 15.3: *Protein Quality of Jatropha curcas Seeds compared with other Foods. (In relation to limiting amino acids according to WHO 1973)*

Limiting Amino Acids	Limiting amino acids of the sample $\times$ 100 Limiting amino acids of reference protein					
	Eggs	*J. curcas*	Corn	Wheat	Beans	Lentils
Lysine	100[1]	45	36	50	91	100
Methionine + Cystine	100[1]	89	100	100	29	44
Tryptophan	100[1]	83	50	100	100	85

1. Indicates an apparent total greater than 100. Source: Cano & Hernández (1984).

The physical-chemical properties of *Jatropha curcas* seed oil compared with other edible oils is shown in Table 15.4. *Jatropha* was found to have a remarkably low fusion point.

Table 15.4: *Physical and Chemical Characteristics of Jatropha curcas Oil Compared with other Edible Oil.*

	J. curcas	Sesame[1]	Cotton[1]	Sunflower[1]	Olive[1]
Density 25°C	0·9172	0·916	0·917	0·917	0·912
Refractive Index	1·470	1·472	1·466	1·467	1·462
Iodine Index	112·51	109·5	106·0	130·5	84·0
Saponification Index	190·47	191·0	193·5	193·0	192·0
Fusion point °C	−8 to −6	20 − 25	11	17 − 26	17 − 26

Source: Cano & Hernández (1984); 1. Bailey (1964).

At present *Jatropha curcas* is grown as a living fence. Its seeds are employed for the elaboration of local dishes. Because of its easy adaptability to various soils and climates it should be employed as an intercropping species for agroforestry systems similar to *Vanilla planiflora* (syn. *V. fragrans*). Its management could provide an alternative seed crop for human consumption and supplementary forage as well as providing a new raw material for industrial oil.

Brosimum alicastrum

Brosimum alicastrum may be found in almost pure stands within the tropical forests. It occurs at altitudes varying from 50 to 800 m asl (Sosa *et al.* 1975) and forms part of the high evergreen and semi-deciduous forests; occasionally ramon is also found in moist ravines within the semi-arid zone.

Extensive high and medium tropical rainforests containing this species occur in more than one third of the Mexican states. See Pennington & Sarukhán (1968) and Puleston (1972) for distribution maps.

The time of flowering and fruiting varies according to its distribution within Mexico. The fruits ripen on the tree over an approximately two month period. Germination occurs during the rainy season several months after seeds have fallen and is probably triggered by moist conditions; approximately 85 per cent of the seeds germinate.

A unique aspect of the ramon tree relative to many other tropical plants is that all of its parts are usable (Pardo & Sanchez 1981). Leaves and seeds have been utilized as forage for cattle, while Puleston (1972) has described the important role ramon played as a subsistence food crop among the ancient Mayans. Roasted seed can be consumed like chestnuts, and when boiled used as a potato substitute. The latex and cortex may be used for pharmaceuticals.

In addition to animal forage, food and medicine, the wood can be used in the construction of bee hives, tool handles, packing crates, fuel and inexpensive furniture.

The results of a chemical analysis of ramon seed is shown in Fig. 15.1. The average crude protein compares favourably with that of conventional foods.

Figure 15.1: Comparison of the Crude Protein Content of Brosimum alicastrum with Seven Common Foods.

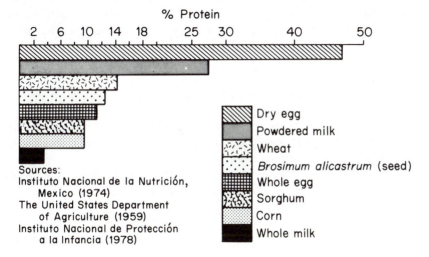

Sources:
Instituto Nacional de la Nutrición, Mexico (1974)
The United States Department of Agriculture (1959)
Instituto Nacional de Protección a la Infancia (1978)

Dry egg
Powdered milk
Wheat
Brosimum alicastrum (seed)
Whole egg
Sorghum
Corn
Whole milk

Source: Pardo & Sánchez (1981).

In addition, it can be seen that *B. alicastrum* seeds also compares quite favourably with sorghum, a major forage and important crop of Mexico (Table 15.5).

Table 15.5: Protein Quality of B. alicastum Seeds Compared with Other Foods (in terms of the quantity of each of the three limiting amino acids proposed by the WHO 1973)

Limiting Amino Acid	Limiting amino acid in the sample / Limiting amino acid in reference protein × 100					
	Eggs	Milk	*B. alicastrum*	Sorghum	Corn	Wheat
Lysine	100[1]	100[1]	42	47	36	50
Methionine + Cystine	100[1]	97	100[1]	92	100	100
Tryptophan	100[1]	100[1]	100[1]	97	50	100[1]

1. Indicates an apparent total greater than 100. Source: Pardo & Sanchéz (1981).

The potential benefits of using ramon foliage and seeds as a forage are quite high. For example, in Veracruz there are 238 800 hectares of mid-altitude tropical forests. Stand densities of *B. alicastrum* are as high as 50 trees/ha in the high altitude forests and can be up to 400 trees/ha in mid-altitude forests. If only 50 per cent of the total forest area contains 5 trees/ha (a conservative estimate) it is possible to calculate that: (1) The total population of ramon trees is approximately 5·2 million; (2) The annual seed production from these trees is about 83 000 tons, based on a known annual production of 16 kg/dry seed/tree; (3) This implies that more than 10 000 tons of crude protein are produced each year, assuming a crude protein content for the Veracruz sample of 12·5 per cent by dry weight, and (4) This amount of protein is sufficient to feed more than 96 000 steers with 600 g daily for the six months normally allowed in local commercial fattening operations.

Brosimum alicastrum is still a widely distributed and abundant tropical source; available literature suggests that it has a high potential for utilization as food, livestock forage, medicine and wood product. At present, our results indicate that the most practical and immediate application is the use of its seed for cattle feed. In addition, seed harvesting would create new jobs and provide a source of income in the rural areas. Used either as food or forage, the seeds are rich in crude protein, and in particular tryptophan, one of the major limiting amino acids in the Mexican diet.

Increased utilization of ramon for forage may also help deter its continued destruction since tropical forests are currently being destroyed at an alarming rate, ironically to create new pastures and grazing lands.

Recent studies in the Botanic Garden of INIREB show that an experimental crop of four year old trees, will fruit even when grown without shade.

Leguminous trees

The following species: *Piscidia piscipula*, *Gliricidia sepium* and *Acacia pennatula* are native to humid and semi-arid tropical America; *Albizia lebbek* is native to tropical Asia.

All these species were compared with *Leucaena leucocephala*. They are all fast growing tree species which will fix nitrogen, as well as being multiple-purpose trees with a high potential for soil restoration, fuel and forage.

Except for *Albizia*, they are widely distributed in the State of Veracruz, mainly in the coastal and central areas; *Albizia* is found in the semi-arid areas up to 1500 m asl.

A domestic study was carried out on the uses and exploitation of huizache, *Acacia pennatula*, a native legume which has established itself as an aggressive climber in secondary vegetation. The vines are browsed by cattle, who are also the main seed dispersal agent; the seeds passing through the digestive system are scarified, resulting in very high level of germination.

A laboratory analysis showed protein content of the vines of huizache (15·2%) to be higher than that for the shoots of *Prosopis juliflora* (13·8%). The huizache is also used for shade in coffee plantations, the wood for firewood, and in rural areas the trunks are used for construction purposes.

The growth and biomass production of *Piscidia piscipula*, *Albizia lebbek* and *Gliricidia sepium* were compared with *Leucaena leucocephala* over a four year period. *Leucaena* proved to have the best growth at all times, with *Piscidia* second and *Albizia* third.

For the lowland site the mean basal diameters differed for *Leucaena*, *Piscidia* and *Albizia* at 19, 25 and 31 months, but at 43 months they were similar (Table 15.6).

Table 15.6: *Mean Basal Diameter (cm); La Mancha Station*

	Age (months)				
	19 (Dec.)	25 (Jun.)	31 (Dec.)	37 (Jun.)	43 (Dec.)
Albizia lebbek	3·9 ± 1·7	3·8 ± 1·7	7·6 ± 2·5	7·2 ± 3·1	8·2 ± 4·6
Erythrina americana	3·6 ± 1·6	3·3 ± 1·9	4·9 ± 2·8	6·5 ± 3·2	9·9 ± 3·6
Enterolobium cyclocarpum	3·4 ± 1·5	3·5 ± 1·6	6·4 ± 2·5	7·5 ± 3·8	9·8 ± 5·0
Piscidia piscipula	5·2 ± 0·9	5·0 ± 0·9	7·7 ± 1·4	8·3 ± 2·6	9·6 ± 2·9
Leucaena leucocephala	5·9 ± 1·8	5·5 ± 1·9	7·8 ± 2·3	8·5 ± 3·0	9·2 ± 3·5

The highest dbh for the species in decreasing order were: *Leucaena*, *Albizia* and *Piscidia*, but the standard error for *Piscidia* was smaller (Table 15.7).

Table 15.7: *Mean diameter (cm) Breast Height (DBH) La Mancha Station*

	Age (months)		
	31 (Jan.)	37 (Jun.)	42 (Dec.)
Albizia lebbek	6·5 ± 2·8	6·8 ± 3·2	8·8 ± 4·0
Erythrina americana	2·9 ± 1·2	3·5 ± 1·1	5·6 ± 2·0
Enterolobium cyclocarpum	4·5 ± 2·3	5·2 ± 2·3	7·8 ± 3·0
Piscidia piscipula	5·8 ± 1·3	6·2 ± 1·5	8·1 ± 1·8
Leucaena leucocephala	6·7 ± 2·2	7·1 ± 2·4	8·0 ± 2·7

The production of total biomass at 29 months showed that *Leucaena* and *Piscidia* produced the best yield, the absolute value was similar but the standard error for *Piscidia* was lower than that of *Leucaena*. At 41 months *Piscidia*, *Leucaena* and *Albizia* gave the best performance (Table 15.8).

Table 15.9 shows that *Albizia* and *Piscidia* have similar yield in volume/ha, as well as in heat and specific gravity values.

The relation between biomass and energy production at 29 and 41 months is shown in Table 15.10. *Leucaena* and *Piscidia* have the best relation, followed by *Albizia*.

171

Table 15.8: *Biomass (kg/ha) from 2500 Trees/ha; La Mancha*

Species	Age (mths)	Leaves	Branches	Stems	Pods	Total	kg/ plant	SE	CV
Albizia	29	3998	—[1]	4945	124	9067	3·69	1·93	52·4
lebbek	41	4711	—[1]	17189	300	22200	8·88	5·12	58·4
Erythrina	29	154	—[1]	286	—	440	0·21	0·19	91·3
americana	41	1311	—[1]	3714	—	5025	2·01	2·18	108·4
Enterolobium	29	1948	858	2705	—	5511	2·20	1·65	75·0
cyclocarpum	41	3047	1301	10602	—	14950	5·98	5·22	87·2
Piscidia	29	3500	2562	5840	—	11902	4·76	2·05	43·2
piscipula	41	6251	4716	16687	—	27654	11·06	7·73	69·8
Leucaena	29	1104	2416	7824	639	11938	4·80	4·62	96·4
leucocephala[2]	41	2002	4458	15256	1512	23228	9·29	9·55	102·7

1. Stems and branches determined together. 2. Start of fruiting stage.

Table 15.9: *Firewood and Energy Production at 29 and 41 months from 2500 trees/ ha; La Mancha*

	Yield Estereos/ha[1]		Heat (Kcal/g)	Specific Gravity	Humidity %
Months:	29	41			
Albizia lebbek	15·4	104·1	4·2581	0·71	54
Erythrina americana	4·9	43·7	4·0587	0·37	80
Enterolobium cyclocarpum	18·1	72·6	4·1949	0·39	62
Piscidia piscipula	13·5	104·1	4·1974	0·75	61
Leucaena leucocephala	11·7	158·3	4·3954	0·60	47

1. 1 m^3 of logs = 1 estereos.

Table 15.10: *Biomass and Energy Production at 29 and 41 months; La Mancha*

La Mancha Months:	Biomass[1] (kg/ha)		Heat (Gcal/ha)	
	29	41	29	41
Albizia lebbek	4945 c	17189 c	21·0 c	73·1 c
Erythrina americana	286 e	3714 e	1·2 e	15·0 e
Enterolobium cyclocarpum	3563 d	11903 d	15·9 cd	49·9 d
Piscidia piscipula	8402 b	21403 a	35·3 b	89·8 a
Leucaena leucocephala	10240 a	19714 b	45·0 a	86·6 ab

[1] Stems and branches.

Table 15.11 shows forage production per species per density at 11, 12, 14, 23 and 26 months, also cumulative yields for replicated and non-replicated treatments. Cumulative yield shows that *Gliricidia* and *Piscidia*, both at 20 000 plants/ha, produced the best results with 19·3 tonne/ha and 15·1 tonne/ha respectively.

The proximate analysis of the forage samples on a dry weight basis shows the following crude protein content: *Gliricidia* 23·4%, *Albizia* 21% and *Piscidia* 16·2%. However, relating harvest yield with protein content, the best species for high cumulative crude protein are *Gliricidia* 4·5 tonne/ha and *Piscidia* 2·2 tonne/ ha.

Table 15.11: *Forage Production (kg/ha)*

Species	Plants/ha	Age harvested (months)					Cum.	DMS
		11 Jul.	12 Aug.	14 Oct.	23 Jul.	26 Oct.	Yield	5%
Replicated treatments:								
Piscidia piscipula	20000	–	–	4875	4479	5797	15151	b
Leucaena K-8	50000	1374	976	2852	2556	2966	10724	c
Gliricidia sepium	20000	–	–	5077	7049	7160	19286	a
Panicum maximum	control	3702	–	210	6044	1830	11786	bc
Non-replicated treatments, with data adjusted:								
Piscidia piscipula	10000	–	–	3655	3149	4925	11729	abcd
Gliricidia sepium I	40000	–	–	7228	6337	1570	15135	abc
Gliricidia sepium II	10000	–	–	4300	6696	5897	16893	ab
Albizia lebbek	20000	–	–	4147	9225	7734	21106	a
Enterolobium cyclocarpum	20000	–	–	154	1990	152	2296	d
Leucaena K-8 inoc.	50000	1357	702	1523	2810	2316	8708	bcd

1. Dates of harvest depend on the rate of growth.

The response to forage harvesting management was different between species; *Gliricidia* gave a very good response to forage harvesting its shoots were vigorous, productive and healthy. It is possible to conclude that *Gliricidia* is very well adapted species to the region and is amenable to management as a forage-cut crop. In most cases the woody legumes were generally more effective and productive than the control (*Panicum maximum*).

REFERENCES

Baily, A. E. (1964) Industrial Oil and Fat Products. Interscience Publications, New York.

Cano, A. L. and A. C. Hernández, (1984) *El Piñoncillo, Jatropha curcas. Recurso Biotico Silvestre del Tropico*. INIREB, No. 14, Xalapa, Veracruz.

Delgado, M. J. L. (1987) *Native FGNFTs in Upland and Lowland Sites as Source of Fodder, Fuelwood and Soil Enrichment*. INIREB 8730080. Xalapa, Veracruz.

Pardo, T. E. and M. C. Sanchez (1981) *Brosimum alicastrum. A Potentially Valuable Tropical Forest Resource*. INIREB No. 3. Xalapa, Veracruz.

Pennington, T. D. and J. Sarukhán (1968) *Arboles Tropicales de México*. Benjamin Franklin, Mexico, DF.

Puleston, D. E. (1972) *Brosimum alicastrum, as a subsistence alternative for the classic Maya of the Central Southern lowlands*. Master of Arts Thesis. University of Pennsylvania.

Sosa, V., A. Gomez-Pompa and A. Barrera (1975) Un arbol tropical de importancia economica, el ramon (*Brosimum alicastrum*). Manuscript.

WHO (1973) Energy and Protein Requirements. World Health Organization, WHO Tech. Rep. Ser. No. 522, Geneva.

16

Food, Fuel and Jobs from Sugar Cane and Tree Legumes

T. R. Preston

INTRODUCTION

There have been significant improvements in food supply in Third World countries during the last two decades. However, a close examination of the statistics reveals that this development has mostly taken place in Asia, and has been almost entirely confined to yield increases in two cereal crops – rice and wheat. There have not been comparable increases in the production of other food staples; and in the case of animal protein the per capita supply has decreased.

TECHNOLOGY TRANSFER

Why has technology development and transfer been so successful in the case of wheat and rice? The answer in part is that the research which led to the development of the technologies was appropriately carried out in the target environment – in the Philippine tropics (IRRI) in the case of rice and in the highlands of Mexico (CIMMYT) in the case of wheat. There has not been a similar concentration of effort for the other crop staples.

In the field of animal production, almost all the intentions have failed, primarily because effort was concentrated on transfer of existing technologies rather than on the development of new ones. Furthermore, the technologies which were transferred were not developed in the target regions; nor were they adapted to address the multiple facets of livestock development in Third World countries. For example, the pig and poultry production systems which have been introduced in tropical countries almost invariably are mirror copies of models which had their origin in the industrialized countries mostly situated in temperature latitudes where available resources – feed, environment, technological support – differ markedly from those encountered in the tropics of the Third World.

The result has been the opposite of what is required. The end products of the 'high technology' systems are costly to produce, putting them out of reach of that part of the population whose nutritional need is greatest. From a macro-economic viewpoint, the high dependency on imported inputs creates a sink for scarce foreign exchange. This puts a limit on expansion and adds further to production costs. The 'modern' systems, by seeking 'economies of scale', use little labour and become foci of pollution and wasteful use of transport. In other words, such systems have proved to be inappropriate when assessment takes into account the broader issues of overall socio-economic development.

It is obvious that more appropriate technologies must be developed. But to do this, it is necessary to make a more precise diagnosis of the problems and then to develop production systems which are in harmony with the realities of the environment in which they are to operate and are in line with the rural development strategy of the country concerned.

STRATEGIES FOR RURAL DEVELOPMENT

Food production is but one item in an overall rural development strategy. Equally important issues are the development of renewable energy sources, employment generation and protection of the environment. All these factors must be taken into account when planning livestock production systems.

It is in this context, of catering to multifaceted needs, that the issue of new crops and new methods of processing and utilizing them become relevant.

This paper describes a multipurpose system which is not just a viable alternative to traditional livestock production policies, current in the industrialized countries, but offers much more, namely, higher levels of productivity per unit of renewable resource input, lower costs, increased employment opportunity, self-reliance in fuel as well as food, and protection of the environment.

BIOMASS, INTEGRATED SYSTEMS AND SELF-RELIANCE

It is increasingly being argued (Preston 1980; Lewis & Slessor 1982) that the long-term substitute for fossil fuel is biomass derived from solar energy capture; and this is a more viable and desirable alternative than nuclear energy, when environmental and social issues are taken into account. Furthermore, it can be shown that there need be no conflict, indeed the prospects are for complementarity, in the use of biomass to satisfy both food and fuel needs.

Such a policy presupposes a series of conditions, principal among which are the following:

(a)   Crops and cropping systems must be chosen which permit:

   (i)   maximum capture of solar energy and its conversion into biomass,

   (ii)   optimum fixation of atmospheric nitrogen in relation to the nutrient needs of the selected crops and associated livestock systems, and

175

    (iii)   fractionation of the crops to satisfy dual needs of food/feed and fuel.

(b)   The livestock components of the system should address the complementary needs of monogastric and herbivorous animal species. The overall system should:

    (i)   be at least self-sufficient in, and preferably a net exporter of, energy,

    (ii)   not contaminate the environment,

    (iii)   not destroy natural ecosystems,

    (iv)   optimize employment opportunities, and

    (v)   promote a maximum degree of self-reliance.

Taking account of these guidelines, it is relatively easy to justify the selection of sugar cane and forage trees (mainly leguminous) as the source of the biomass. Both these crops are easily separated into fractions of low and high cell wall content, which facilitates satisfying the food needs of monogastric and herbivorous animals as well as providing an exportable surplus of fuel (Fig. 16.1). The sugar cane provides highly digestible carbohydrate (soluble sugars) which drives the system; the trees supply protein-rich foliage. Both are important sources of fuel which is derived from the residual fibre-rich structural components.

COMMERCIAL EXPLOITATION OF SUGAR CANE FOR PRODUCTION OF FOOD, FEED AND FUEL

The milling of sugar cane stalk to extract the sugar-rich juice is an old technology, which has been developed to varying degrees of sophistication – ranging from the artesanal production of 'gur' and 'panela' to the industrialized production of crystalline white sugar. These procedures, developed over centuries, are immediately applicable to the utilization of sugar cane as a source of animal feed and fuel (Fig. 16.2). They can be likened to the actual systems of processing cereals in temperate countries which also are fractionated into high-value (grain) and low-value (straw) components.

Since it was first proposed some seven years ago (Preston 1980), the concept of sugar cane fractionation for feed and fuel production has received increasing support by policy makers and is now officially endorsed by several Caribbean countries (FAO in press; IFS in press; both cited in Preston & Leng 1987).

There are three ways in which the system can be applied commercially. Two of these (Fig. 16.2) involve the integration of a livestock/fuel system into the actual commercial processes for industrial production of crystalline sugar and artesanal production of brown sugar (gur and panela). In the third method, no

176

Figure 16.1: *Flow Diagam of Integrated Systems Utilizing Sugar cane, Legume Trees and Aquatic Plants for Conversion of Solar Energy into Animal Products and Fuel.*

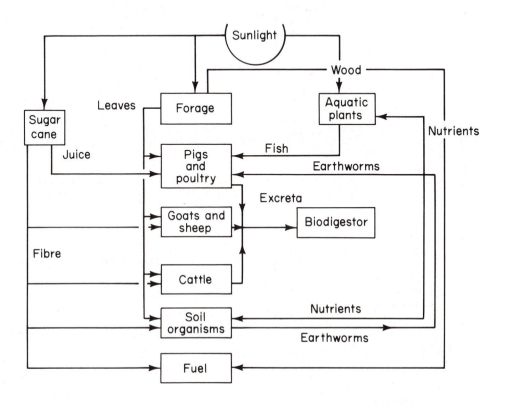

sugar is produced for human consumption, the end products being meat and/or milk and fuel.

In the industrial system changes are required both in the factory and in the field. In the factory, the process is terminated at the penultimate centrifugation so that the final products are a mixture of 'A' and 'B' sugar and 'B' molasses giving a sugar : molasses ratio of about 1:1 instead of the traditional 3:1 ratio when the aim is to maximize extraction of sucrose. The 'B' molasses has a sucrose content of ca. 56% and total soluble sugars of ca. 75%. Its metabolizable energy value for monogastric livestock is ca. 75% that for sorghum grain (Ly 1987).

Figure 16.2: *Model for Integrating Livestock Production into Existing Factory and Artesanal Sugar Enterprises.*

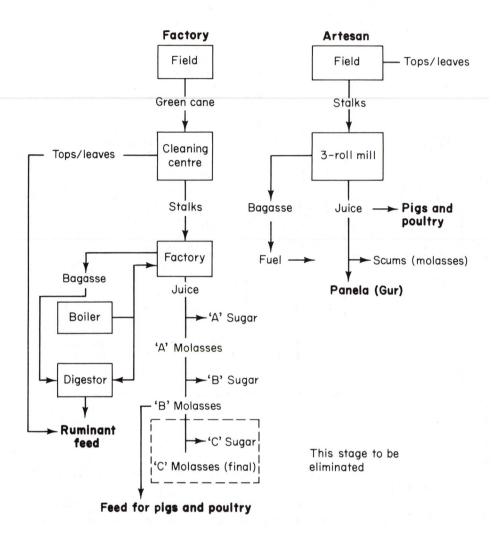

In the field, the objective is to utilize as ruminant feed, or fuel, the green leaves and dry trash presently either burned as a pre-harvest treatment, or left as mulch. The whole cane is harvested green either by machine or hand and delivered to intermediate collection centres where the clean stalk is separated for onward transmission to the factory. The residual fibre can be prehydrolysed (with steam, acids or alkalis) prior to feeding to ruminants, or converted into briquettes for subsequent use as fuel by direct combustion or gasification.

The modifications to the artesanal system involve concentrating the 'scums' to give a stable 'molasses' of ca. 50% total sugar content suitable for feeding to monogastric animals. The cane tops are fed to ruminants.

The third system uses a single two- or three-roll mill which extracts in a single pass about two thirds of the total juice present in the stalk (45–48% of the weight of cane). This is used as the basis of the diet for monogastric animals and small herbivores such as rabbits. The residual bagasse is a potential feed resource for ruminants, can be used as fuel or as a substrate for micro-organisms which later are harvested by earthworms, which in turn become a protein source for poultry.

Sugar cane juice and 'B' molasses as livestock feed

Performance date for pigs which are fed can juice as the basis of their diet are summarized in Table 16.1. It is apparent that the levels of performance are comparable with those normally achieved on grain-based feeds; and that there are interesting possibilities for saving on protein inputs. Pig feeding systems based on sugar cane juice are now in commercial use in the Dominican Republic, Colombia and Cuba (FAO in press; IFS in press).

Table 16.1: *Pigs Fed on Sugar cane Juice Require Less Protein (8 pigs per group; 16–110 kg liveweight)*

	Dietary protein (% in DM)		
	11	8	5
Liveweight gain (g/d)	850	820	800
Feed intake (kg/d)			
Cane juice	10	12	14
Protein suppl. (40% protein)	1	0·75	0·50
DM conversion	3·6	3·8	4·4

Source: Mena (1986).

Systems for feeding broilers and layers on sugar cane juice are presently being investigated (Posso 1987). The birds prefer a solid feed and must be encouraged to consume the juice by restricting the protein supplement. Performance rates are inferior to those achieved with grain-based diets, but already offer economic advantages to farmers for whom land and labour are more readily available resources than capital (Table 16.2).

Ducks, geese and turkeys take more readily to liquid diets and commercial systems for these species, using high-test and 'B' molasses, are already being used in Cuba (R. Perez, pers. comm.). Few problems are anticipated in developing cane juice feeding systems for these species.

Table 16.2: *Broilers Grow More Slowly on Sugar cane Juice than on a Cereal-based Concentrate, but Probably at Less Cost*

	Cereal grain	Cane juice
Liveweight (g)		
At 15 days	124	125
At 64 days	2110	11570
Daily gain	41	31
Feed intake (g/d)		
Cereal concentrate	93	
Cane juice		240
Protein suppl. (40% protein)		38
Feed conversion (DM basis)	2·06	2·61

Source: Mena (1986).

Research with rabbits began only recently and is still at an early stage of development (Solarte 1987). It has proved to be feasible to provide 50% of the diet dry matter of growing rabbits in the form of sugar cane juice preserved with sodium silicate (5 g/litre). The rest of the diet was fresh foliage of *Erythrina glauca*. Growth rates were slower than would normally be achieved on cereal-based diets but part of the problem may have been the presence of tannins in the *Erythrina* reducing protein digestibility.

Sugar cane bagasse for ruminants

Research in Mauritius (Wong *et al.* 1974) showed that high *pressure* steam treatment (200 °C) for 2–5 minutes raised rumen *in situ* digestibility from 30 to 60%. Subsequent work (Naidoo *et al.* 1977) emphasized the critical role of supplements providing bypass nutrients (protein and glucose precursors), in addition to urea-N and minerals for rumen microbes. In Colombia this technology is now being applied commercially for cattle fattening, using the foliage from the legume tree (*Gliricidia sepium*) as the main source of bypass protein (Table 16.3).

Table 16.3: *Fattening Cattle on Predigested Bagasse*

Liveweight (kg)		
Initial	265	
Final	310	
Daily gain	0·8	
Feed intake (g/d)		(DM)
Pre-digested bagasse	10	(4·0)
Molasses (with 10% urea)	1·5	(1·2)
Gliricidia leaves	5·7	(1·1)
Poultry litter	0·6	(0·46)
Rice polishings	0·6	(0·46)

High pressure steam at 200 °C for 5 minutes; n = 24
Source: CIPAV (1987).

FOLIAGE FROM LEGUME TREES

The value of the foliage from *Leucaena leucocephala* as well as protein supplement in carbohydrate-rich diets for ruminant is well documented (Preston & Leng 1987). It has also been used to replace partially the protein component in diets for growing chickens (D'Mello 1982).

More recent developments, which promise to be easier to apply commercially, are the findings that other legume trees, more tolerant to acid soils and easier to establish than *Leucaena*, are equally effective in stimulating animal performance (Tables 16.4–6).

Table 16.4: *N Content and in vitro digestibility of the foliage of Gliricidia sepium and Erythrina poeppigiana.*

	E. poeppigiana			G. sepium	
	Leaves	Stem	Bark	Leaves	Stem
N content (% of DM)	4·9	1·8	2·2	4·6	2·1
DM digestibility (%)	44	45	78	69	45

Source: CATIE (1982); Benavides (1983), cited by CIPAV (1987).

Table 16.5: *Growth Rates of Zebu Weaner Steers Fed a Basal Diet of King Grass Supplemented with Gliricidia Foliage (3 steers per group)*

	Level of Gliricidia (% fresh basis)				
	0	15	25	30	50
Liveweight gain (g/d)					
– Trial I (1985/86)	75		380		400
– Trial II (1986/87)	105	300		390	

Source: ICA (1986/87), cited by CIPAV (1987).

Table 16.6: *Milk Production of Goats Given a Basal Diet of King Grass, Reject Bananas and Leaves from Erythrina poeppigiana*

	Erythrina leaves (% LW, DM Basis)			
	0	0·5	1·0	1·5
Milk yield (kg/d)				
High yield group	0·38	0·7	0·82	0·96
Low yield group	0·25	0·5	0·51	0·64

Source: Esnaola & Rios (1986), cited by Preston & Leng (1987).

Conservation of these foliages is proving to be relatively simple using acetic acid derived from the sugar cane juice by aerobic fermentation (CIPAV, unpublished data).

ON-FARM FUEL PRODUCTION

Continuous flow plastic (PVC) bag digesters are proving to be a low-cost solution to the problem of deriving fuel energy from livestock wastes (CIPAV 1987). Investment in the plastic and accessories is of the order of US$6·00/m^3 of liquid digester capacity.

A number of technologies offer promise as a means of converting bagasse into fuel. Direct burning is one option. However, production of charcoal produces a saleable product; and the technology is simple and can be carried out under farm conditions (Ffoulkes *et al.* 1980).

A pilot scale gasifier adapted to burn bagasse was developed in the Dominican Republic (A. Lindgren, unpublished data), but the process has yet to be commercialized.

PERSPECTIVES

Many areas remain to be investigated. For example, the bagasse produced at farm level is potentially of higher nutritive value than factory bagasse, because extraction is less complete and there is more residual sugar (20 vs 2%). However, it also has more moisture (60–65 vs 50%) and the sugars ferment rapidly, also high pressure steam is not an on-farm commodity.

Selective consumption of the sugar-rich pith by goats or sheep, with the residual fibre being used for fuel, is the simplest of the systems being investigated. For cattle, some degree of prehydrolysis, following grinding or chopping, appears to be necessary to produce an edible product. Sodium hydroxide treatment gave encouraging results in one on-farm trial (CIPAV 1986, unpublished data) but cheaper and less-polluting hydrolysing agents such as acetic and phosphoric acids are now being investigated.

Another promising development is the use of earthworms to scavenge microbes active in the aerobic decomposition of bagasse supplemented with biodigester effluent. The worms will finally be harvested by 'grazing' poultry directly on the exhausted substrate and will be a valuable supplement to the carbohydrate-rich cane juice (Fig. 16.1).

The feeding of cane juice to other monogastric species, such as ducks, geese and turkeys, has still to be researched. However, no difficulties are anticipated since these species readily adapted to liquid molasses feeding (R. Perez 1987, personal communication).

The use of sugar cane and forage trees in an integrated farming system permits high animal and fuel productivity per unit area. Thus with 3 ha in sugar cane and 1 ha in legume trees, it is possible to fatten 150 pigs (30–100 kg), and feed a herd of 75 goats or sheep and 4 cows, fattening the progeny to mature weight of 25 and 450 kg, respectively. Feed supplement purchases are 7·5 tonnes of a 40% protein supplement for the pigs and 7·5 tonnes of poultry litter and 7·5 tonnes of multi-nutrient blocks (urea/minerals) for the cattle and goats. There is a surplus of fuel in the form of biogas from the livestock excreta, fibre from the sugar cane bagasse and wood from the legume trees. The technology is labour-intensive, and provides activities for the extended family.

Investment needs, excluding land purchase, are approximately US$20 000 and the annual income, less direct costs, of US$10 000 supports an internal rate of return of the order of 30–40 per cent over a 10 year period.

CONCLUSIONS

Much more work remains to be done before the full potential contained in mixed sugar cane/forage tree associations can be exploited. The perspectives for an integrated system based on these crops are huge. In the Cauca valley in Colombia, sugar cane averages 80 tonnes of millable stalk/ha/year over an area of 100 000 ha. Individual estates regularly report up to 150 tonnes/ha/year. The food and fuel that can theoretically be produced from these yields surpass severalfold the best that can be produced by cereal grain cropping, in whatever latitude.

There is therefore sound justification for believing that, in tropical regions, sugar cane grown for biomass will one day be assigned the niche presently occupied in temperate latitudes by the cereal grains. The technology, whose gestation began in Cuba in the late 1960s with the use of molasses as a grain substitute for livestock (Preston *et al.* 1967 & 1968), has now matured into the logical first option as the basis for 'self-reliance' in feed and fuel production in the tropics.

REFERENCES

CIPAV (1986) *Resultados preliminares del programa de validacion de tecnologias agropecuarias en base a recursos disponibles.* CIPAV, Cali, Colombia.

CIPAV (1987) *Nuevos sistemas agropecuarios en base a los recursos disponibles.* CIPAV, Cali, Colombia.

D'Mello, J. P. F. (1982) Toxic factors in some tropical legumes. *World Rev. Animal Prod.* 18: 41–46.

Esnaola, M. A. and C. Rios (1986) Hojas de 'Poro' (*Erythrina peoppigiana*) como suplemento proteico para cabras lactantes. *Trop. Animal Prod.* (in press).

FAO (in press) *FAO Expert Consultation on Sugar cane as Feed.* R. Sansoucy, G. Aarts and T. R. Preston (Eds). FAO, Rome.

Ffoulkes, D., R. Elliott and T. R. Preston (1980) Feasibility of using pressed sugar cane stalk for the production of charcoal. *Trop. Animal Prod.* 5: 125–130.

IFS (in press) *Molasses as Livestock Feed.* International Foundation for Science, Stockholm.

Lewis, C. and M. Slessor (1982) *Bio-energy Resources.* Chapman and Hall, London.

Ly (1987) Metabolism of molasses by pigs. *Paper presented at Seminar held 13–18 July 1987 at Universidad de Camaguey, Cuba on Molasses as Livestock Feed.* International Foundation for Science, Stockholm.

Mena, A. (1986) Sugar cane juice for all types of livestock. In: *FAO Expert Consultation on Sugarcane as Feed.* R. Sansoucy, G. Aarts and T. R. Preston (Eds). FAO, Rome.

Naidoo, G., C. Delaitre, and T. R. Preston (1977) Effect of maize and fish meal supplements on the performance of steers fed steam-cooked bagasse and urea. *Trop. Animal Prod.* 2: 117.

Posso, L. (1987) Sugar cane juice as a feed for poultry. *Paper presented at Seminar held 13–18 July 1987 at Universidad de Camaguey, Cuba on Molasses as Livestock Feed*. International Foundation for Science, Stockholm.

Preston, T. R. (1980) A model for converting biomass (sugar cane) in animal feed and fuel. In: *Animal Production Systems for the Tropics,* pp. 158–171. International Foundation for Science Publication No. 8, Stockholm.

Preston, T. R. and R. A. Leng (1987) *Matching Ruminant Production Systems with Available Resources in the Tropics and Subtropics*. PENAMBAL Books Ltd., Armidale, NSW.

Preston, T. R., A. Elias, M. B. Willis and T. M. Sutherland (1967) Intensive beef production from molasses and urea. *Nature* 219: 727–728.

Preston, T. R., N. A. MacLeod, L. Lassato, M. B. Willis ans M. Velasquez (1968) Sugar cane products as energy sources for pigs. *Nature* 219: 727–728.

Solarte, A. (1987) Sugar cane juice as a feed for rabbits. *Paper presented at Seminar held 13–18 July 1987 at Universidad de Camaguey, Cuba on Molasses as Livestock Feed*. International Foundation for Science, Stockholm.

Wong You Cheong, Y., J. T. d'Espaignet, P. J. Deville, R. Sansoucy and T. R. Preston (1974) The effect of steam treatment on cane bagasse in relation to its digestibility and furfural production. In: *Proceedings of the 15th Golden Jubilee Congress of the International Society for Sugar Cane Technology (South Africa)*. Hayne and Gibson, Durban.

17

Alternative Crops for Europe

R. S. Tayler

INTRODUCTION

In the UK, and no doubt in other parts of Europe too, interest in alternative crops has been stimulated over the last few years as a result of surpluses in the production of our major crop and livestock enterprises. These surpluses have arisen because of increased productivity resulting from technical developments in breeding and in crop and livestock management, and because of the encouragement to production provided by the EEC's pricing policy. At the same time, the excess production is difficult to market outside Europe, partly because of increasing production in other parts of the world and partly because of the widening gap between EEC prices and world prices.

Wheat can be taken as an example of the situation. World production is rising by about 3 per cent per year,the rate being greatest in the more developed parts of the world (Table 17.1). The EEC consumes about 80 per cent of its present production and seeks to export most of the remainder, but whilst EEC farmers can sell into community stores for about £105 per tonne, the world price is now about £35, only half what it was as recently as 1984. Across all farm products, the result has been that EEC financial support for farming has doubled since 1981, to more than £22 billion ECU per year (about £14 billion) and could double again by the early 1990s if present trends are allowed to continue. These are the fruits of allowing expansion to be led by production rather than by demand. It is necessary that changes occur and the key to change is that production should increasingly be determined by demand. Thus any alternative enterprises must both be suited to the environmental conditions of some part of Europe and also be expected to find a market.

Table 17.1: *Wheat Production (Mt per year)*

	USA	EEC-12	Canada	Australia
1970–73	46	48	15	9
1985–86	66	71	28	16
Increase	43%	48%	87%	78%

Source: Adapted from Murphy (1987).

Helpful developments may come in one of two ways. Firstly, it may be new markets or new uses for existing crops, or secondly we can seek to develop alternative crops. In examining these, this account gives no attention to the detailed economics of production. This is not only because the economics are difficult to assess at a preliminary stage of a new development, but also because the economic climate of EEC farming is bound to change. For example, a farmer paying high rent for his land and carrying substantial interest charges for borrowed capital must produce output of high value if his business is to remain viable. But if changing economic circumstances force him out of production, his eventual successor may be less heavily encumbered and capable of gaining a livelihood from a lower value gross output.

NEW MARKETS AND USES FOR EXISTING CROPS

The development of new markets outside Europe for price-supported crops, such as cereals, cannot occur on any substantial scale unless and until EEC prices move closer to world prices. But this does not apply to crops whose price is unsupported by the Common Agricultural Policy. None of these is a major crop in terms of the area grown, but many may provide useful export opportunities for a few growers. For example, some Dutch flower growers have developed an appreciable market for their products in parts of the US. We must increasingly be on the lookout for the creation and development of such opportunities. The same process is happening within and between the countries inside Europe. For example, UK producers of winter brassicas like Brussels sprouts are expanding their sales on the continent; the more open winter weather conditions of Britain's maritime climate gives them an advantage over continental producers. Each country or district must seek to exploit any special circumstances it may enjoy. Growers of pot plants and container plants in some European countries have benefited in the past from access to low and fixed interest loans, and to subsidized fuel costs. Perhaps also because of superior marketing skills, they supply about 40 per cent of the UK demand, valued at over £30 million. British growers now seek to recapture a greater share of this market.

It will be particularly valuable for European farmers if new uses can be developed for the major crops they currently grow, because they have many important attributes. These crops are already well adapted to the areas in which they are grown, their management is well understood, their performance is proven, and in many cases their output is greater than in other parts of the world. This particularly applies to the cereals. Alternative uses on a large scale can probably only be found in industrial applications, and these are outside the scope of this paper except for brief comment on one of them. The possible use of cereals or beet to produce ethanol as a petrol additive has been mentioned so often that one hesitates to raise it again. On present prices and costs it is not economic, although this partly depends on what saleable by-products are also produced. One Swedish plant is handling 25 000 t of wheat per year, and each tonne produces 200 litres of ethanol, 250 kg of starch, 150 kg of carbon dioxide and 400 kg of cattle feed containing 30 per cent protein (Dunn 1987).

Ethanol as an additive has an energy value comparable to petrol, and its use would both reduce a pollution problem and lessen the intervention cost of buying and storage. The use of a 10 per cent ethanol additive in community petrol could use 30 million tonnes cereals/year. It is therefore a development with many political implications and any decision about it is likely to take into account political factors as well as economic considerations.

ALTERNATIVE CROPS

The second approach to the problem of surpluses, and the one that occupies the bulk of this paper, is the possibility of developing new crops in Europe. They may be entirely new crop species, but more likely they will be existing minor crops, perhaps from other parts of the world. Potatoes, turnips, rapeseed and sugar beet are the major crops which have been introduced into northern Europe in the last three centuries, but none was truly new (Williams 1978). Many plants introduced into cultivation were successful fodder crops before genetic selection and processing technology developed their potential for human food. Both rape and beet were grown for fodder before being used for oil and sugar, and soya bean was a minor fodder crop in the USA for almost a century before being used for grain production.

Success as a weed can be an indication of suitability for cultivation. It suggests adaptation to soil and climate and a synchrony of development with existing crops which eliminates the impediment of late ripening. Weed status as an annual is also a clear indicator of effective seed production and of ready establishment from seed. On the other hand, a new crop plant may exhibit a spread of seed maturing which is inappropriate for modern once-over harvesting, may shatter and shed seeds too easily, or exhibit too much dormancy, or alternatively can display an absence of seed dormancy which in a wet season can result in germination before harvest has been completed. Carruthers (1986) considers a classification of alternative plants in terms of their characteristics as a crop, their novelty, the novelty of their use, and requirements for post-harvest management and development.

The need for climatic adaptation requires some comment because it is the factor which prevents cultivation, or widespread cultivation, in Europe of many crops which it would be useful to grow. For example, rubber from *Hevea brasiliensis* is clearly inappropriate for European production. Less obviously, it is also true that the EEC has relatively few areas suited to soya bean production, and there are several crops, for example sunflower and navy bean, whose climatic requirement prevent cultivation in the UK. The critical need for a seed crop is that the photoperiodic conditions must be appropriate for flowering, and that the crop must germinate, develop a leaf canopy, flower and then fill and mature its seeds, within the available growing season. For example, some crops which cannot easily be grown in the UK are cultivated in other parts of Europe where the length of the growing season is little different. The significant difference is that mean temperatures in the UK are lower, and it is temperature which mainly governs the length of the different stages of growth. This can be an advantage: the longer duration of the grain-fill period in cereals and oilseed rape is one reason why UK yields are greater than in other warmer areas, but it is a disadvantage if maturation cannot be completed.

The breeding of cool-adapted varieties for Europe has made slow but steady progress. Maize and sunflower cultivation has moved steadily northward, but in these and other crops we may be approaching the limit of what is possible by conventional methods. A more detailed knowledge is being accumulated of the plant biochemistry which allows adaptation to cool conditions, and this coupled with the development of genetic engineering techniques may allow more rapid progress in the next 10–15 years. But complex changes are required for adaptation and may still be difficult to achieve. Instead, what might appear to be more dramatic modifications may genetically be simpler to obtain; for example, an oil-rich pea, or a beet plant that stores protein rather than sugar. This implies that it may be easier to adjust the physiology of an existing crop so that it yields the required product than to adapt an exotic crop to European conditions. However, these intriguing possibilities are at least a decade away; what others may be more immediately available?

Flax

Flax (*Linum usitatissimum*) is a crop which is well adapted to Europe's cooler climates. The useful yield is the fibres contained in the stem which are loosened on the farm by a partial rotting process called retting. The process is completed in a scutching factory, usually run by a producers' cooperative, where the fibres are beaten free of the stem debris.

Linen fibre is a major constituent of a few textiles and a minor constituent of a wider range. It can be used for both furnishing fabrics and clothing and lends itself well to the currently fashionable 'crumpled look'. At present it forms 2 per cent of world textile production but it is expected to be able to take a gradually increasing, but still modest, share of the market even though it is more expensive than cotton and many of the synthetic fibres.

Western Europe currently produces about 10 per cent of world flax output, some 50–60 000 t per year from 60–65 000 ha. This is mostly grown in the Normandy and Flanders area of France and Belgium. An organization called International Linen Promotion is the UK branch of a grouping of European growers and manufacturers and it expects the market to be able to absorb without difficulty an expansion of European production of at least 5 per cent (or about 3000 ha) per year. Though small in relation to the cereal area, this could be a useful addition to Europe's cropping, but there is difficulty in finding a method of retting suited to modern conditions.

There are two traditional methods of retting. In one case the cut sheaves of flax stems are soaked in tanks of water to induce the required partial rotting. This is a labour-intensive process and also a potential cause of watercourse pollution when the tank contents are discharged. Furthermore, the scutching capacity will not be able to handle all the crop at once and so the retted crop must be dried to prevent deterioration. The alternative traditional method is 'dew' retting: the crop is left cut in the field, and provided it gets the right combination of morning dew and daytime sun, retting will occur with very limited handling and often without any post-retting drying cost. It is thus a highly weather-dependent method. For this reason, recent research has attempted to develop a more reliable chemical technique.

Retting can be induced by an application of the herbicide glyphosate to the standing crop at flowering, and for a time this seemed likely to be a very useful technique. However, it has proved extremely difficult to get appropriate distribution of glyphosate on the stems, with the result that variable rates of retting occur, producing a sample of fibre which is too variable in quality and which contains too many green stems to be usable. Thus dew retting is the only suitable system available at present and this means that flax production will need to be confined to areas where suitable weather conditions can be expected with reasonable reliability. These conditions have not been precisely defined but might include some parts of the south-eastern counties of England. The development of chemical retting would greatly reduce the climatic limitations on production.

Flavour and Medicinal Plants

This is a diverse group of species; some of them contain vegetable oils of specialized composition, others contain essential oils, oleo-resins, alkaloids or glucosides. The early pharmaceutical industry relied almost entirely on plant materials, but today the main sources in decreasing order of importance are: chemical synthesis, fermentation, animal extracts and plant extracts. Despite the declining importance of plant production, a number of species are still used in conventional, ethnic and alternative medicine; Svoboda (1984) lists over forty which are currently used in medical preparations. Many materials are produced in Eastern Europe, some by gathering rather than cultivation, but increasing problems of contamination by heavy metals may encourage brokers and processors to welcome local and more controlled sources of supply. An increasing interest in alternative medicine may increase the demand for these products, but an increasing capacity to produce some of them by tissue culture and fermentation will eventually reduce the need for field production.

Notable amongst recent developments is the increasing demand for gamma linolenic oil. This can be extracted from the seed of both evening primrose (*Oenothera biennis*) and borage (*Borago officinalis*). Evening primrose does not fit easily into rotation because it needs to be sown well before the autumn or be raised in soil blocks for subsequent planting out. Borage has the greater oil yield, is spring-sown and quick growing, but exhibits staggered ripening making choice of harvesting date difficult. The future of both crops is uncertain because recent reports have suggested that the black current residues left after juice extraction may be a useful source of the same product.

One of the most widely used essential oils is peppermint, derived from *Mentha piperita*. It is a constituent of many food, confectionery, drink and health preparations; the UK uses some 800 t per year valued at over £10 million. The crop is grown in small quantities in the UK, and in other European countries, especially Italy, but virtually all the peppermint consumed in the UK is imported from the USA. Increased local production would seem feasible though quality can vary considerably between both varieties and environments so that users will be cautious about changing their source of supply. It is unlikely that any grower would want to produce one essential oil alone; prices fluctuate markedly both within and between seasons. A range of crops would confer more stability and could make it economic to install distillation equipment on the farm which might increase the profitability of the business.

Salad, Vegetable and Fruit Crops

A consequence of the increasing interest in a healthy diet is some increase in the consumption of fruit and vegetables, and a demand for greater variety in these items. Some growers may be able to exploit opportunities of this type; for example the root crop salsify (*Tragopogon porrifolius*), or leaf crops like miner's lettuce (*Montia perfoliata*, syn. *Claytonia perfoliata*) mentioned by Guenault (1985), or ice plant (*Mesembryanthemum crystallinum*). There are many possibilities but all will need to be preceded by market research and promotion.

Amongst the fruits, various raspberry (*Rubus idaeus*) and blackberry (*R. fruticosus*) crosses could be produced to supply a created market. Cloudberry (*R. chamaemorus*) and blueberry (*Vaccinium corymbosum*) are suited to organic soils in high rainfall areas and may also be suitable for promotion. There has been a significant increase in fruit juice consumption which could bring benefits to some growers both by expansion in the area of some existing crops and the introduction of new ones. For example, sea-buckthorn (*Hippophae rhamnoides*) has a yellow-fruited hybrid with a high vitamin and carotenoid content in a juice of attractive aroma. Chokeberry (*Aronia melanocarpa*) has an anthocyanin content which can be used as a natural colourant for other fruit juices.

Oilseeds and Grain Legumes

In world terms, soya bean is the annual crop which is dominant in both the protein and vegetable oil markets. A wide range of types are available with different growth habits and for different lengths of growing season and different latitudes and climates, but mostly restricted to a range from the warmer temperate zones to the cooler or more elevated tropics. Small areas of soya bean are grown in the countries of southern Europe, but there are no cultivars clearly suited to other parts of the continent. Oilseed rape (*Brassica oleracea*) has become the major local oilseed crop, but there is also an increasing area of sunflower (*Helianthus annuus*). Cool-adapted types are being developed and have the added advantage that the content of unsaturated fatty acids is greater with slow ripening in cool conditions. But the same conditions also greatly increase the incidence of infection with the fungus *Botrytis cinerea* and there seems little prospect of improving resistance genetically.

Pearl lupin (*Lupinus mutabilis*) contains an oil of good composition, in a concentration similar to soya bean. It may be useful as a source of both protein and oil, but, because it has only so far been used as a Third-World smallholder crop, varieties suited to mechanized farming will have to be developed. Small areas of white lupin (*L. albus*) are grown in Europe, but its fleshy pod and delayed ripening make it unsuited to the cooler areas, and it has a low oil content. Chickpeas (*Cicer arietinum*) and lentils (*Lens culinaris*) are temperate crops whose cultivation within the EEC could reduce the need for imports. There can however be major problems of seed quality and of disease if maturity in both these crops is not induced by the onset of dry conditions.

Cereals

There may be some scope for the gradual development of some speciality cereals. Buckwheat (*Fagopyrum esculentum*) could have a place as a coarse flour for biscuit-making or perhaps as a breakfast cereal, and for livestock feed. It does well on poor soils and some types can produce a yield in 100 days from sowing. Quinoa (*Chenopodium quinoa*), an Andean grain crop, has a 14% protein content and is well-adapted to cultivation on light soils. Amaranth (*Amaranthus leucocarpus*)[2] contains 16% protein with a high lysine content and no gluten. It may have a valuable place in specialized diets, and it can also be used as 'popcorn'. Finally, gold of pleasure (*Camelina sativa*) produces a seed with 33% protein and 40% oil. It is already used in the birdseed trade, and because the oil is high in long-chain fatty acids it could also be a replacement for sperm whale oil.

---

2. Regarded by some authorities as a synonym of *A. hypochondriachus*. The genus urgently needs revising (G. E. W.)

LIVESTOCK ENTERPRISES

Any consideration of Europe's cropping must take into account the needs of its livestock industry. The slow trend towards vegetarianism reduces land requirements for food production because a smaller land area is required for a crop-based diet than if the crops are first fed to livestock. Nevertheless, there are some opportunities amongst animal enterprises. There has already been some increase in the production of milk and milk products from sheep and goats and this is likely to continue, meeting specialist dietary and health needs. A more substantial impact may be made by fine wool production. At present most of Europe's requirements are imported; UK imports alone amount to £70 million per year. Merino or Merino-type sheep could be kept in the UK and import replacement could utilize several hundred thousand hectares of grassland.

CONCLUSION

A number of possible alternative crops have been mentioned in this account and there are probably others which could have been included. But all of them together will not solve Europe's problem of surpluses, although their contribution will be very important to individual farmers and groups of farmers. In seeking to maintain their living standards, farmers will want to 'add value' to their output in whatever way they can. The on-farm washing, grading and packing of vegetables is already common. Production of flower and flower-bulb packs, alkali treatment of straw, and farm production of stone-ground wheat flour from organically grown wheat are other examples. At least one farmer produces dog biscuits from his own wheat crop and even buys in from his neighbours as well.

The current surpluses are caused by the past success of farmers and agricultural scientists in raising output. The solution to these problems lies in further technical innovation and in market research and promotion; the greatest opportunity almost certainly lies in the development of industrial uses for crop products, exploiting their value as continually renewable resources. The agricultural industry must seek to be as successful in solving its problems as it was in creating them, but it will need to be done in cooperation with other industries.

REFERENCES

Carruthers, S. P. (Ed.) (1986) *Alternative enterprises for agriculture in the UK.* University of Reading, Centre for Agricultural Strategy. Report No. 11.
Dunn, N. (1987) More grain from surplus grain. *The Furrow* 92, 2: 21.
Guenault, B. (1985) New plants and products as food. *Proc. Nutr. Soc.* 44: 31–35.

Murphy, M. C. (1987) Future economic strategy for combinable crops. In: *Combinable crops prospects for the future*. Proceedings of the 19th NIAB Crop Conference, 1986, pp. 84–110. National Institute of Agricultural Botany, Cambridge.

Svoboda, K. P. (1984). *Culinary and medicinal herbs*. Technical Note No. 237, West of Scotland Agricultural College, Auchincruive, Ayr.

Williams, W. 1978. New crops and agricultural systems. In: *Proceedings of the British Crop Protection Conference: Weeds*, pp. 1005–1012. British Crop Protection Council, Thornton Heath, Surrey.

SPECIFIC CROPS

18

Vernonia galamensis: a Promising New Industrial Crop for the Semi-arid Tropics and Subtropics

R. E. Perdue, Jr., E. Jones and C. T. Nyati

INTRODUCTION

An account of our knowledge of *Vernonia galamensis* as of early 1984 was provided by Perdue *et al.* (1986). Therefore, we will cite only two additional references and one then in press. We will summarize highlights of the 1986 paper, provide a general account of the more significant unpublished agronomic research in Zimbabwe during 1984–86, and then explain what must be accomplished to establish *V. galamensis* as a new industrial crop. Although several additional germplasm collections have been acquired and are now being increased in Zimbabwe, all agronomic research to date has been with the unimproved and very uniform germplasm collected in Ethiopia.

During the 1950s, an Agricultural Research Service (ARS) screening programme identified *Vernonia anthelmintica* as a good source of seed oil rich in vernolic (epoxy) acid (Fig. 18.1). Seed, from India, contained 27% oil, of which 67% was vernolic acid. Epoxy oils were then and still are used to manufacture plastic formulations, protective coatings and other products.

ARS chemists investigated the oil to identify potential markets, and showed that the oil or products derived from it could be used as a plasticizer-stabilizer in the manufacture of polyvinyl chloride. ARS agronomists developed improved lines with up to 31% oil containing up to 75% vernolic acid but yields were limited by poor seed retention. Seed from the first-formed flower heads were shed before those from the later flower heads matured.

V. anthelmintica, an annual, belongs to *Vernonia* section Stengelia, all other species of which are African. When it became evident little success was likely in developing a variety of *V. anthelmintica* as a new crop for the United States, exploration in eastern and southern Africa was undertaken in 1966–67 to acquire seed of related species. The best was *V. lasiopus* seed from western Uganda with 20·5–22·2% oil containing 75·2–81·3% vernolic acid. None had good seed retention and all were perennials. Both agronomic and utilization research on *Vernonia* in the US ended.

197

Figure 18.1: *Vernolic acid (A) and trivernolin (B)*.

It is the epoxy groups of such triglyceride oils that make these materials useful in plastics and coatings products. They serve as plasticizers (for flexibility), stabilizers (to inactivate agents in plastics that otherwise cause them to degrade) and as highly reactive sites where one triglyceride molecule can become attached to adjacent molecules, and these to others, to form interlocking polymer networks. The double bond (C=C) can be epoxidized to create a triglyceride even more valuable than this natural product.

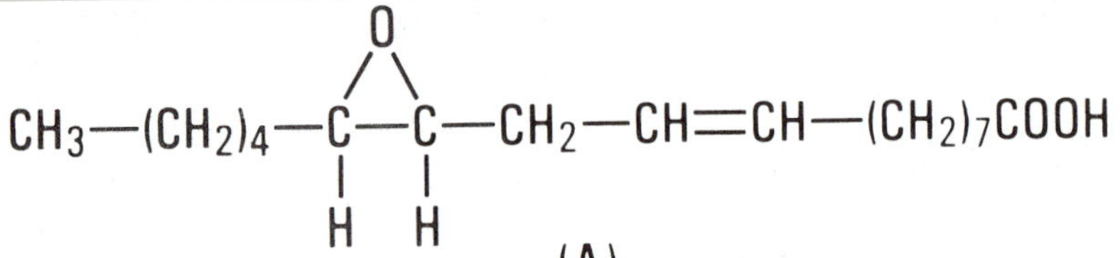

$$CH_3-(CH_2)_4-\overset{O}{\overset{/\backslash}{C}}-\underset{H}{\overset{}{C}}-CH_2-CH=CH-(CH_2)_7COOH$$

$$\overset{H}{\underset{}{}}\ \overset{H}{\underset{}{}}$$

(A)

$$CH_2-O-\overset{O}{\overset{||}{C}}-(CH_2)_7CH=CH-CH_2-\overset{O}{\overset{/\backslash}{CH}}-CH(CH_2)_4CH_3$$

$$CH-O-\overset{O}{\overset{||}{C}}-(CH_2)_7CH=CH-CH_2-\overset{O}{\overset{/\backslash}{CH}}-CH(CH_2)_4CH_3$$

$$CH_2-O-\overset{O}{\overset{||}{C}}-(CH_2)_7CH=CH-CH_2-\overset{O}{\overset{/\backslash}{CH}}-CH(CH_2)_4CH_3$$

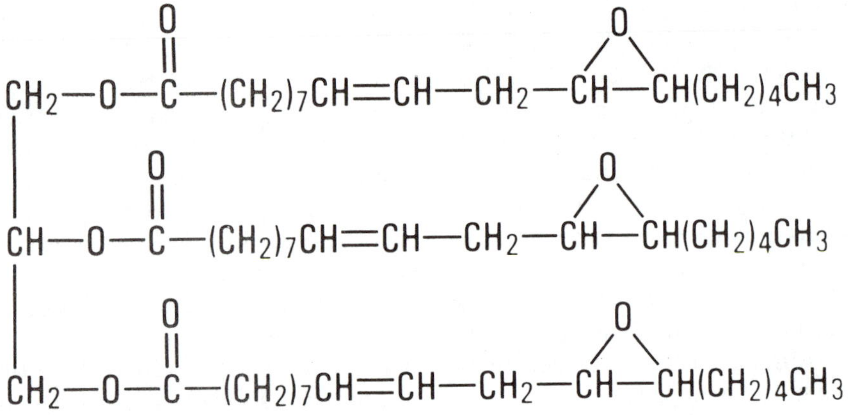

(B)

VERNONIA GALAMENSIS

Prior to the *Vernonia* exploration in Africa, R. E. Perdue, Jr. visited Ethiopia on another mission and observed an interesting non-stengeleoid annual *Vernonia* in an arid area (annual rainfall ca. 60 cm) near Harare. Plants were fully mature, and most stems were dry, brown and leafless. Seed retention was complete (Fig. 18.2). This plant was initially identified as *V. pauciflora*; it is now known as *V. galamensis*. Seed contained 41·9% oil with 72·6% vernolic acid. This unimproved germplasm contained about 30% more vernolic acid than the best improved varieties of *V. anthelmintica*.

Figure 18.2: *Vernonia galamensis ssp. galamensis var. ethiopica.*

Voucher specimen for the initial seed collection in Ethiopia (× 4/5). The 3 fully mature seeds (upper left) were removed from the uppermost flower head.

Vernonia galamensis

Seed of *V. galamensis* were planted at Experiment, Georgia. Plants were 1·2–1·5 m tall and flowered, but produced no seed. Seed were subsequently increased in a greenhouse at Glenn Dale, Maryland. Plants flowered in November and seed matured in December, long after the first frost. Subsequently, seed from this increase were planted at Kericho, Kenya by Leonard Bates, African Highlands Produce Company. Seed harvested in March 1976 contained 40% oil with 80% vernolic acid. A much larger planting at Kericho was made in early July 1977, and the crop was harvested in late February. Success here prompted new interest and provided a supply of seed as a source of oil for further utilization research.

Bates' success prompted Gilbert (1986) to undertake an in-depth taxonomic study of *V. galamensis*. He concluded that within the complex formerly called *V. galamensis*, a locally endemic population in northern Tanzania should be segregated as a new species, *V. filisquama*, and plants earlier recognized as *V. afromontana* are better considered as a subspecies of *V. galamensis*. According to Gilbert's concept of *V. galamensis*, this widely distributed species includes 6 subspecies, one of which includes four varieties (Figs. 18.3 & 4).

Figure 18.3: *Distribution of Vernonia filisquama and V. galamensis subsp. galamensis.*

V. *filisquama* is recorded from several locations in northern Tanzania (arrow, box, lower left). V. *galamensis* subsp. *galamensis* includes 4 varieties distributed from West Africa east to Sudan and Ethiopia, then south to Zimbabwe and Mozambique, a. var. *galamensis*, b. var. *petitiana* (arrows show 2 isolated records in Somalia). c. var. *ethiopica* d. var. *australis*.

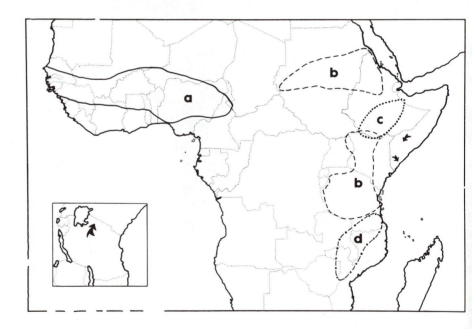

Vernonia galamensis

Figure 18.4: Distribution of 5 subspecies of Vernonia galamensis found only in East Africa (Uganda, Kenya and Tanzania), the centre of diversity of this species.

Subsp. *nairobensis* (solid line), subsp. *afromontana* (stippled), subsp. *gibbosa* (dotted line), subsp. *mutomoensis* (dashed line), subsp. *lushotoensis* (black; arrows show isolated collections in Uganda).

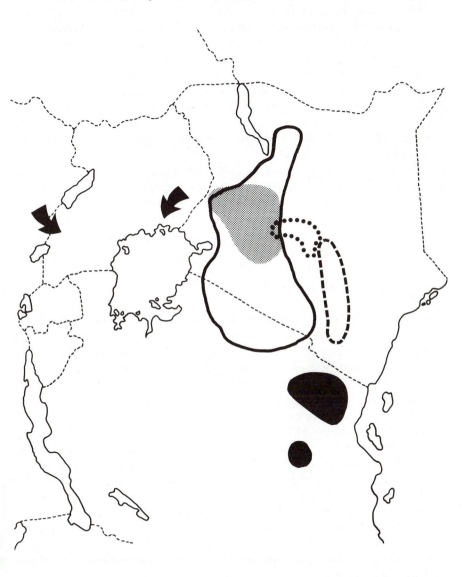

Gilbert's research showed that this widely distributed species is highly diverse and its centre of diversity is in East Africa. It shows us where we must explore to capture maximum genetic diversity (Perdue, in press). This research also showed that *V. galamensis* characteristically occurs on porous, well-drained soils, a critical observation that would later permit us to understand why some trial plantings failed.

V. galamensis oil was evaluated as a raw material for epoxy coatings. The work also evaluated processing conditions necessary to handle the seed, extract the oil, and then to refine the oil for further use. Film-forming characteristics were evaluated by spreading the oil on steel panels, which were then baked in an oven. Coated panels were evaluated for hardness, elongation, and deformation and resistance of coatings to alkali, acid and solvent. The oil proved suitable for baked films or coatings. Physical properties of the films were outstanding. They had good flexibility and resistance to chipping and excellent adhesion. Panels could be readily cut, drilled, and trimmed without loss of adhesion and without chipping at the cut edge. There was good resistance to alkali, acid, and solvents. From this research it was apparent the *V. galamensis* oil had potential as a raw material for the coatings industry.

With this encouraging research, many small samples were distributed. They were planted at other sites in Kenya, and in Australia, Jamaica, Pakistan, Puerto Rico, Taiwan, Tanzania and Zambia. Responses at most of these sites were not impressive and, at some, plants appeared diseased. But all these trials were made before we gained an understanding of the environmental requirements of *V. galamensis*. Most of these sites did not meet those requirements.

AGRONOMIC RESEARCH

Agronomic research on *V. galamensis* was initiated in Zimbabwe in 1983. Small trial plantings were made at the Botanical Garden in Harare and at four field stations of the Ministry of Agriculture's Department of Research and Specialist Services. The Harare trial and another at the Lowveld Research Station, Chiredzi, confirmed the good seed retention initially observed in Ethiopia. Seed yield at Chiredzi was impressive.

With this initial success, a 0·7 ha planting was established at the Coffee Research Station, Chipinge in November 1984. The objective was to obtain a large supply of seed as a source of oil for further utilization research. Plants grew to 3 m tall before they flowered. Seed were harvested prematurely for fear of lodging and total loss of the crop. Most were immature. Even so, they yielded 38·5% oil with 75·5% vernolic acid.

In another 1984 trial at Chipinge, plants were 'topped' (upper part of plants removed) to determine how they would respond to removal of apical dominance. Plants produced lateral branches which tended to produce mature flower heads at the same time. This enhanced uniformity of seed maturity and reduced time from planting to maturity.

A date-of-planting trial was initiated at Chiredzi in 1984–85 because several observations suggested *V. galamensis* is photosensitive. Seed were sown at mid-month, December through April. All plants flowered and produced mature seed, the plot seeded in December flowered when plants were almost 3 m tall, the plot seeded in April flowered when plants were 1 m tall. Clearly, later planting is desirable.

Effect of topping was investigated further at Chiredzi in 1986. The most encouraging response was in a plot where plants were topped at 15 cm above ground. Many branches formed from the base and each branch produced 3–4 flower heads, all of which matured at the same time. Indeed, individual branches of these plants were identical to the herbarium specimen that documented the original collection in Ethiopia (Fig. 18.2). Plants along the edge of the plot were especially impressive. Seed matured much earlier than on plants elsewhere in the plot and on much shorter branches.

Other research on *V. galamensis* was conducted in Zimbabwe in 1985 and 1986 to determine response to fertilizer and spacing, but more information is needed before we can draw reliable conclusions. Agronomic research continued in 1987. It is now centred at Chiredzi where it is the responsibility of C. T. Nyati.

While much remains to be accomplished to gain a full understanding of *Vernonia* agronomy, we have expanded our knowledge substantially since the first trial planting in Zimbabwe in 1983. We know that a well- drained soil is an absolute requirement. On well-drained soil plants grow erect with a single stem until the first flower head develops after which branches are formed. There is a characteristic response on poorly-drained soils. Terminal growth ceases before flowering and the upper part of the plant withers and dies. Branches then develop from the base but subsequently wither and die without flowering. When drainage is intermediate, plants will remain green and produce a few flower heads but seed yields are very low.

Seed germinate quickly but seedling vigour is poor and early weed control is a problem. According to an observer long resident in Ghana, the West African variety is so common in some areas of northern Ghana as a weed that seed could be profitably harvested from natural stands. This germplasm may provide better seedling vigour.

Judging from its natural distribution, *V. galamensis* is adaptable as a crop within 20° of the Equator. It was grown with moderate success, however, at Lahore, Pakistan (ca. 31°30' N) (Aziz *et al.* 1984) so must have much broader adaptability.

Herbarium specimens of *V. galamensis* have been collected in areas with annual rainfall from 38 to 185 cm (Gilbert 1986) but crop development should be focused on semi-arid areas where it is better adapted and where it will not compete with food crops. We are confident that the plant shown in Fig. 18.5 will prove characteristic of plants grown in excessively wet areas. This drawing illustrates a specimen collected from cultivation at Kericho, Kenya where annual rainfall is 185 cm. With such high rainfall, after the first terminal flower head, a plant will produce long secondary branches, each with a new terminal flower head and subsequent tertiary branches resulting in a diffuse branching system

Figure 18.5: *Vernonia galamensis ssp. galamensis var. ethiopica.*

From specimen collected by L. Bates at Kericho, Kenya, January 19, 1978. Habit including root × 2/5, single flower head (bottom) × 9/10, seed and flowers × 4. The Kericho planting was with seed collected in Ethiopia. The Bates specimen was collected ca. 1 month before seeds were harvested and shows the undesirable long stems, diffuse branching system, and lack of uniformity in seed maturity, characteristics of excessive rainfall. Compare with Fig. 18.2.

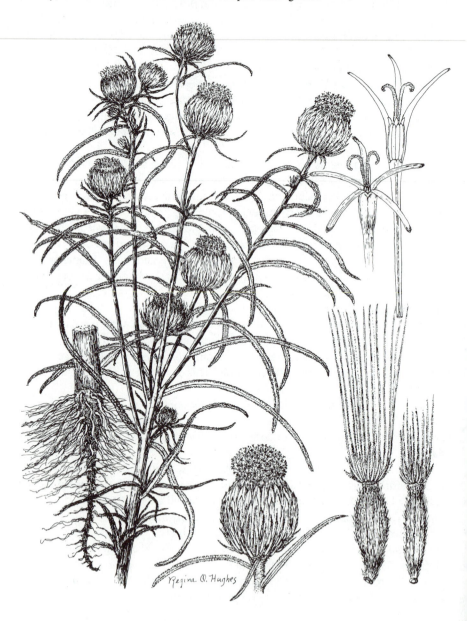

and lack of uniformity in seed maturity. Fig. 18.2, in contrast, illustrates development we think will prove characteristic of semi-arid environments, few flower heads on short stems and uniform maturity.

Seed yields of 1·5 t/ha (1627 lb/acre) were recorded in Zimbabwe. We believe yield can be doubled or possibly tripled by better management and breeding of better varieties when more germplasm is available. (Yield of soybeans has more than tripled in the United States since this crop was first grown for its seed during the 1920s.)

V. galamensis is photosensitive; it flowers in response to very short days. In the US it has flowered in a greenhouse in early spring as day length increased, and in late fall as day length decreased.

In Zimbabwe, time from planting to harvest has been 5 to 7 months; in Zambia, however, seed were mature 4 months after planting. During his taxonomic study of the species, Gilbert observed specimens of small ephemerals only 20 cm tall with a single flower head, representing germplasm that must have matured seed substantially less than 4 months after germination. The long growing period now being experienced in Zimbabwe should be overcome by improved management and by breeding when more germplasm is available.

V. galamensis appears to have good resistance to disease and insects, though such problems may arise when it is grown in extensive monoculture.

FUTURE WORK

Even though substantial progress has been made, much remains to be learned before *V. galamensis* can be established as a new crop. Research must be focused in two directions – agronomics and oil utilization.

Plant exploration must be undertaken to acquire more germplasm of *V. galamensis*, and other *Vernonia* species. Substantial effort was devoted to developing *V. anthelmintica* as a new crop for the US. Later we learned that *V. galamensis* is far superior, not only in seed retention, but also in yield of vernolic acid. While exploration should be focused on *V. galamensis*, collectors should obtain seed of all available *Vernonia* species for analysis of oil and vernolic acid content to determine if another species is superior. Areas like India, Africa, and subtropical and more temperate areas of South America should be explored. Extensive exploration for *V. galamensis* should yield germplasm with a higher yield of seed oil, and/or a higher yield of vernolic acid in the oil. Exploration must also seek germplasm with pest resistance should pest problems appear in extensive monoculture, even better drought resistance, reduced sensitivity or different sensitivity to day length, and more seedling vigour to overcome weeds.

After germplasm has been collected and evaluated, breeding must be undertaken to incorporate desirable characteristics into the available Ethiopian germplasm that already has so many desirable characteristics.

205

There must be more utilization research. In 1985, the senior author contacted 25 scientists, some in industry laboratories, and others in universities who are knowledgeable about the needs and interest of industry. There was much enthusiasm about vernonia oil because it is so unique, and 11 agreed to evaluate samples of oil in a variety of applications. But industry will not take this seriously until a product is in commerce and agriculture proves it can provide a reliable supply at reasonable cost.

Utilization research has shown there is a potential market for vernonia oil in at least three areas: (a) plasticizer-stabilizer for polyvinyl chloride, (b) chemical coatings, and (c) interpenetrating polymer-networks with polystyrene to form unique plastics. The first is an existing market where vernonia oil, or fully epoxidized vernonia oil (EVO) would have to compete primarily with epoxidized soybean oil (ESO) based on properties as well as economics. Because the composition of EVO is significantly different from ESO, there may be a niche for EVO in products where properties are significantly improved by the use of *Vernonia* oil or EVO.

The most promising area for a new market in the near future is chemical coatings, an area where ESO is now used. Coatings experts advise there are certain applications ESO cannot meet because it does not have the required properties. This is the area where further research on vernonia oil should first be focused. The most pressing need now is a market study, especially in the coatings field, to identify potential markets. This should be followed by additional laboratory research on the oil to supplement information now available, and resolve unanswered questions pertinent to potential markets identified by the market survey. The third area, interpenetrating polymer networks, holds promise for the future and must be explored.

Currently, the need is for a small market even though it is small. With oil or seed in international commerce, even in small amounts, agriculture can prove it can provide a reliable supply of seed; other industries will take *Vernonia* oil seriously; and the market will expand.

ACKNOWLEDGEMENTS

We are plant scientists without expertise on chemistry and utilization of *Vernonia* oil. We are indebted to chemists in industrial and other laboratories, and especially to Dr. K. D. Carlson, who have offered encouragement and guidance.

REFERENCES

Aziz, P., S. A. Khan and A. W. Sabir (1984) Experimental cultivation of *Vernonia pauciflora* – a rich source of vernolic acid. *Pakistan J. Sci. Ind. Res.* 24 4: 215–219.

Gilbert, M. G. (1986) Notes on East African Vernonieae (Compositae). A revision of the *Vernonia galamensis* complex. *Kew Bull.* 41, 1: 19–35.

Perdue, R. E. Jr., K. D. Carlson and M. G. Gilbert (1986) *Vernonia galamensis*, potential new crop source of epoxy acid. *Econ. Bot.* 40, 1: 54–68.

Perdue, R. E. Jr. (in press) Systematic botany in the development of *Vernonia galamensis* as a new industrial oilseed crop for the semi-arid tropics. Paper presented at Systematic Botany – a Key Science for Tropical Research and Documentation, "Natur och Kultur" Symposium, the Royal Swedish Academy of Sciences, Stockholm, Sweden, September 14–17, 1987.

19

Santalum acuminatum Fruit: a Prospect for Horticultural Development

D. E. Rivett, G. P. Jones and D. J. Tucker

INTRODUCTION

Santalum acuminatum belongs to the family Santalaceae, a family of about 30 genera and 400 species in tropical and temperate regions of the world. Six of the 25 species of *Santalum* are found in Australia, and of these five are confined to the continent and one, *S. album*, found outside, especially in India where it is cultivated for sandalwood (Hegnauer 1973; George 1986).

Santalum acuminatum, commonly called quandong or native peach, is widespread in southern Australia except for coastal regions of the south-east. It is a small tree or shrub up to 10 m high. The grey-green leaves are short-stalked, opposite, tapered at both ends with a short hooked tip when young. The small cream or dull orange flowers occur in spring in bunches at the end of branchlets. The fruit has a fleshy outer layer, that turns vivid shiny red (sometimes yellow) when ripe, enveloping a large wrinkled 'stone'.

GERMINATION AND PROPAGATION

Germination will occur within two months of removal of the seeds from the ripe fruit but the yield improves with more mature seeds (Grant & Buttrose 1978). The seeds are prone to fungal attack when removed from the stone and must first be treated with a fungicide to assure successful germination. The optimum temperature for germination is between 16 °C and 20 °C. A thick white root appears after about three weeks and the seedling can then be transferred to a pot with a host plant. The host is necessary because *Santalum*, in common with other members of the Santalaceae, are semiparasitic, particularly in the early stages of growth. Successful work on grafting and tissue culture suggest potential alternative means of propagation (Barlass *et al.* 1980).

In the presence of cytokinin, shoots were produced in quantity from all cultured aerial parts of seedlings. Shoots were also regenerated from explants of a 5 year old mature tree. These results offer the promise of clonal propagation as a useful adjunct to breeding programmes.

TREE GROWTH AND FRUIT YIELD

Experimental orchards have been established by the CSIRO Division of Horticultural Research in hot semi-arid areas of South Australia and Victoria. Smaller private orchards have also been established on farm land from Western Australia to the east coast (Mills 1977). The South Australian orchard, approximately 200 trees, is established in an area of red-brown earth of neutral to slightly acidic pH. The natural rainfall is supplemented by highly saline (1·3% chloride) bore water. After seven years the trees were reported to be averaging around two metres high (Sedgley 1982). Trees of up to eight metres high have been reported in Western Australia by Mills (1977). The trees flower in January and February (midsummer) and the fruit ripens in September to October. A trend towards biennial bearing is evident in some trees but not in others. The first trees began fruiting in the third year with a maximum yield of 10 kg per tree in the seventh year. There have been reports of trees in private gardens producing up to 23 kg of fruit (Saggers 1977), however some trees had not borne any fruit by this time. An indication of fruit characteristics can be gained from Table 19.1.

Table 19.1: Measurements of Fruit Characters from Ten Fruit Samples of Fruiting Trees in the Quorn (South Australia) Orchard in 1980.

Fruit Character	Mean measurement	Coef. var. between trees	Coef. var. within trees
Fruit diameter	22·7	10·1	7·6
Fruit length	23·9	12·9	9·0
Fruit weight	5·9	26·0	17·5
Flesh thickness	2·7	18·0	15·8
Fresh wt. of flesh	3·6	28·3	18·8
Dry wt. of flesh	0·9	28·9	18·9
Stone diameter	15·6	9·1	6·4
Stone weight	2·2	27·2	17·4
Shell thickness	2·2	16·5	10·1
Kernel diameter	9·2	10·2	7·0
Kernel weight	0·4	28·6	16·7
Shell weight	1·8	29·8	

The outer flesh of the fruit is red and pulpy, it can be eaten fresh but generally lacks appeal because it is usually quite acidic. However it can be cooked in pies, made into jam or chutney or served stewed. It is also readily dried to give a product which must be rehydrated before eating.

COMPOSITION OF FRUIT

The flesh of the quandong is nutritionally valuable partly due to a high content of vitamin C, reputably double that of oranges for equal flesh weights (Grant & Buttrose 1978). When analysed for other nutritionally important variables it gives values comparable to those of western foods from the same food category (Brand *et al.* 1983). The results are given in Table 19.2.

Table 19.2: *Composition of the Flesh of the Fruit of Santalum acuminatum per 100 g, compared with that of Orange and Peach.*

Analysis	*S. acuminatum*	Orange[1]	Peach[1]
Edible portion (%)	75	75	87
Water (g)	77	86	86
Protein (g)	1·7	0·81	0·63
Fat (g)	0·2	Tr.	Tr.
Carbohydrate (g)	19·3	8·5	9·1
Ash (g)	2·1	—	—
Vitamin C (mg)	16·4[2]	50	8
Na (mg)	51	3	3
K (mg)	659	200	260
Mg (mg)	40	13	8
Ca (mg)	42	41	5
Zn (mg)	0·2	0·2	0·1
Cu (mg)	0·2	0·07	0·05

1. From Paul & Southgate (1978). 2. Value for *S. lanceolatum* (*S. acuminatum* value not available).

The Kernel or Nut

The stone of *Santalum acuminatum* represents approximately 40% of the total weight of the fruit and therefore the economics of a fruit crop would be greatly enhanced if a use for the kernel could be found. Researchers at the CSIRO, Division of Horticultural Research have assessed the consumer acceptance of the kernel as an edible nut which was reputably eaten to some extent by the indigenous population (Cribb & Cribb 1974).

Kernels were tested, both raw and salted, after roasting with coconut oil. Neither product was well accepted, although the roasted product was preferred (Sedgley 1982). The main objection was a strong aroma which many people found unpleasant and in some cases nauseating. A distinctive sweet odour is usually apparent on cracking the nuts or slicing the kernels. Gas chromatographic analysis (Loveys *et al.* 1984) has identified the major aromatic volatile as methyl benzoate. Although several compounds were detected by the flame ionization detector, the major peak was associated with the characteristic aroma of methyl benzoate when the column effluent was sniffed.

The methyl benzoate content of the kernels varied considerably in the range of 32–1294 µg/g of fresh kernel. Wide variation was found not only between trees, which included a large standard deviation, but also between annual harvest from the same tree, however there is obviously potential for selection (Possingham 1986). Some loss of volatiles occur on storage and can be accelerated by placing the kernels in a vacuum oven.

Quantitative differences were readily distinguishable by a testing panel, particularly with the samples from the very low methyl benzoate specimens. The aroma of methyl benzoate was barely detectable from these kernels and although they were judged to be more acceptable than the methyl benzoate sample, consumer acceptance was still low.

Analysis of Kernel

The kernels of *Santalum acuminatum* have been found to have high levels of protein and oil (see Table 19.3).

Table 19.3: *Proximate Analysis of Santalum acuminatum Kernels compared with that of Brazil Nuts*

Analysis (%)	S. acuminatum[1]	Brazil Nuts[2]
Water	1·6	8·5
Oil	67·6	61·5
Protein	15·3	12·1
Starch	Tr.	2·4
Free sugars	3·1	1·7
Ash	1·3	–

1. Average mass of endocarp 2·8 g, kernel 0·48 g. Collected from York Peninsula, South Australia in 1979. Source: Jones *et al.* (1985). 2. Source: Paul & Southgate (1978).

If these values are typical of all *S. acuminatum* then their gross nutrient composition is similar to that of many nuts used as food, with high oil and protein and low starch and sugar content.

The potential food value of the kernels as sources of amino acids can be assessed by comparison with the idealized pattern of dietary essential amino acids for adults (WHO 1973). The kernels contain adequate quantities of most of the essential amino acids (see Table 19.4). However, samples collected from other locations in Australia had amino acid profiles deficient in sulphur amino acids when this comparison was made.

Whilst no individual non-protein amino acid was identified, up to 3·7% residue of unidentified ninhydrin-positive components were detected in some samples. These were spread over four compounds and are in sufficient quantity to warrant further study. Hydroxyproline and homospermidin have been detected in the leaves, fruits and seeds of some santalums, particularly *S. album* (for review see Hegnauer 1973). However neither of these compounds was shown conclusively to correspond to any of the four unknown ninhydrin positive products obtained from the acid hydrolysis of the kernel protein.

The major fatty acid in the oil-rich kernel has been shown by a number of authors to be trans-11-octadecen-9-ynoic acid. A seed oil constituent which is characteristic of two plant families, the Santalaceae and the Olacaceae (Gunstone & Russell 1955; Morris & Marshall 1966; Bu'Lock & Smith 1963).

Table 19.4*: Amino Acid Content of Santalum acuminatum*

Amino Acid (mol %)	S. acuminatum	WHO reference[1]
Lysine	3·9	1·9
Histidine	1·8	
Arginine	9·3	
Tryptophan	0·6	0·4
CySO₃H	0·04	
Aspartic acid	10·5	
Threonine	3·9	1·4
Serine	7·5	
Glutamic acid	14·5	
Proline	6·4	
Glycine	9·6	
Alanine	8·1	
1/2 Cystine	1·5	2·2
Methionine	1·4	
Valine	4·5	1·9
Isoleucine	2·9	1·7
Leucine	8·4	2·4
Tyrosine	3·2	1·8
Phenylalanine	2·6	
Methionine sulphoxide	0·2	
B[2]	0·3	

1. Reference dietary amino acid pattern (WHO 1973). 2. Total of four unidentified amino acids based on average colour factors. Source: Jones *et al.* (1985). The broad analysis is detailed in Table 19.3.

This fatty acid, common name santabic acid or ximenic acid, is present in relatively large amounts in santalums and is accompanied by small amounts of a second acetylenic fatty acid, stearolic acid (9-octadecynoic acid).

The gas chromatographic analysis of the methyl esters produced from the saponified oil of *S. acuminatum* are shown in Fig. 19.1 and Table 19.5. Twelve fatty acids have been detected, eight have been identified.

Similar analyses have been performed by Morris & Marshall (1966), Gunstone & Subbarao (1966) and by Jones *et al.* (1985) for *S. acuminatum* oil and for other Santalaceae seed oils by Hopkins *et al.* (1969).

According to Bu'Lock (1966), the acetylene fatty acids of the family are presumably formed by way of oleic acid and stearolic acid; the introduction of further double and triple bonds continues conjugately in the direction of the terminal methyl group. However an alternative possibility has been suggested by Morris & Marshall (1966). These authors put forward the hypothesis that santalbic acid may be produced by conjugative rearrangement of 12-octadecen-9-ynoic acid, and that the stearolic acid might be the precursor of the 12-octadecen-9-ynoic acid. Alternatively, this latter acid might be derived from linoleic acid by further dehydrogenation at the 9,10-position and stearolic acid

Figure 19.1: *Chromatographic Separation of the Methyl Esters of the Fatty Acids from S. acuminatum Seed Oil*

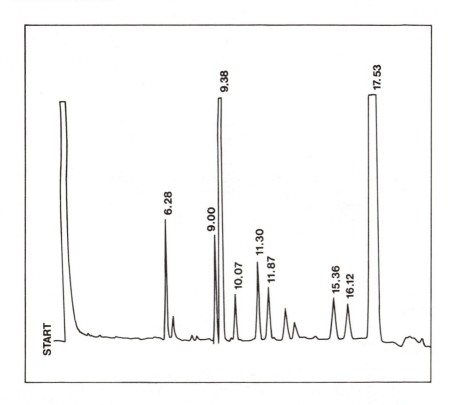

might then be an end product from oleic acid rather than a precursor for further desaturation.

An examination of the oil of *S. acuminatum* (Jones *et al.* 1985) by HPLC showed that the oil consisted predominantly of three triglycerides. Gas chromatographic analysis, after saponification and methylation, confirmed that these triglycerides contained oleic to santalbic acid ratios of 1:2, 1:2 and 2:1 respectively. It was inferred that controlled triglyceride biosynthesis in *S. acuminatum* is indicated.

Several long-chain acetylenic fatty acids have been shown to be physiologically active (Blain & Shearer 1965; Ahern & Downing 1970) inhibiting enzymes involved in some important biological pathways. The acetylenic fatty acid, crepenynic acid (octadec-cis-9-en-12-ynoic acid) has been studied with respect to mortalities of sheep grazing on mature seeded *Ixiolaena brevicompta* (Ford & Whitfield 1983), but evidence for its toxicity is inconclusive. The suggestion has been made by Downing *et al.* (1972) that toxicity problems may arise when an acetylenic acid has a structure analogous to a polyunsaturated fatty acid substrate for essential enzymes.

Table 19.5: *Gas Chromatographic Analysis[1] of the Methyl Esters of the Fatty Acids from S. acuminatum Seed Oil.*

Retention Time	Fatty Acid	Weight %
6·28	16:0	1·20
6·58	16:1	Tr
9·01	18:0	1·15
9·32	18:1	23·47
10·08	18:2	0·67
11·31	18:3	1·14
11·87	stearolic	0·87
15·36	unidentified	0·97
16·13	unidentified	0·86
17·53	santalbic	69·41

1. Chromatographic conditions: Column, 25 m bonded carbowax, 0·53 mm ID helium carrier gas. Initial temperature 170°C for 5 min then 10°/min. to 200°C maintained for 10 min then 20°C/min to 225°C.

The fact that *S. acuminatum* seeds have such a high level of santalbic acid, together with the doubts cast as to the safety of acetylenic fatty acids, would suggest that considerable caution be exercised before the consumption of the kernels as an edible nut could be recommended. Our current work on the metabolism of santalbic acid should resolve this doubt.

REFERENCES

Ahern, D. G. and D. T. Downing (1970) Inhibition of prostaglandin biosynthesis by eicosa-5,8,11,14-tetraynoic acid. *Biochem. Biophys. Acta* 216: 456–461.

Barlass, M., W. J. R. Grant and K. G. M. Skene (1980) Shoot regeneration *in vitro* from native Australian fruit-bearing trees – quandong and plum bush. *Aust. J. Bot.* 28: 405–409.

Blain, J. A. and G. Shearer (1965) Inhibition of soya lipoxidase. *J. Sci. Food Agric.* 16: 373–378.

Brand, J. C., C. Rae, J. McDonnel, A. Lee, V. Cherikoff and A. S. Truswell (1983) The nutritional composition of Australian Aboriginal bushfoods. *Food Tech. Aust.* 35: 293–298.

Bu'Lock, J. D. (1966) Biogenesis of natural acetylenes. In: *Phytochemical Group Symposium on Comparative Phytochemistry, March, 1965*, T. Swain (Ed.). Academic Press, London.

Bu'Lock, J. D. and G. N. Smith (1963) Acetylenic fatty acids in seed and seedlings of sweet quandong. *Phytochem.* 2: 289–296.

Cribb, A. B. and J. W. Cribb (1974) *Wild Food in Australia.* Collins,Sydney.

Downing, D. T., J. A. Barve, F. D. Gunstone, F. R. Jacobsberg and M. Lie Ken Jie (1972) Structural requirements of acetylenic fatty acids for inhibition of soybean lipoxygenase and prostaglandin synthetase. *Biochem. Biophys. Acta* 280: 343–347.

Ford, G. L. and F. B. Whitfield (1983) Fatty acid composition of *Ixiolaena brevicompta. Lipids* 18: 103–105.

George, A. S. (1986) Sandalwoods and Quandongs of Australia. *Australian Plants* 13: 318–319.

Grant, W. J. R. and M. S. Buttrose (1978) Domestication of the Quandong, *Santalum acuminatum. Australian Plants* 9: 316–318.

Gunstone, F. D. and W. C. Russell (1955) The constitution and properties of santalbic acid. *J. Chem. Soc.* 3782–3784.

Gunstone, F. D. and R. Subbarao (1966) The occurrence of linolenic and stearolic acid in *Santalum acuminatum* seed oil. *Chem. Ind.* (London): 461–462.

Hegnauer, R. (1973) *Chemotaxonomie die Pflanzen,* Vol. 6. Birkhauser Verlag, Basel and Stuttgart.

Hopkins, V. Y., M. J. Chisholm and W. J. Cody (1969) Fatty acid components of some Santalaceae seed oils. *Phytochemistry* 8: 161–165.

Jones, G. P., D. J. Tucker, D. E. Rivett and M. Sedgley (1985) The nutritional potential of the quandong (*Santalum acuminatum*) kernel. *J. Plant Foods* 6: 239–246.

Loveys, B. R., M. Sedgley and R. F. Simpson (1984) Identification and quantitative analysis of methyl benzoate in quardon (*Santalum acuminatum*) kernels. *Food Technology in Australia* 36: 280–289.

Mills, M. B. (1977) Observations on quandong trees. *West. Aust. Naturalist* 14: 15–17.

Morris, L. J. and M. O. Marshall (1966) Occurrence of stearolic acid in Santalaceae seed oils. *Chem. Ind.* (London): 460–461.

Paul, A. A. and D. A. T. Southgate (1978) McCance and Widdowsome. *The Composition of Foods,* 4th Ed. Revised. HMSO. London.

Possingham, J. (1986) Selection for a better quandong. *Aust. Hort.* 84: 55–56.

Saggers, J. (1977) Quandong or Wolgol? *West. Aust. Nutgrowing Society Yearbook* 3: 31.

Sedgley, M. (1982) Preliminary assessment of an orchard of quandong seedling trees. *J. Aust. Inst. Agric. Sci.* 1: 52–56.

WHO (1973) *Energy and Protein Requirements.* World Health Organization Tech. Rep. Ser. No. 522, Geneva.

20

Commercial Exploitation of Alternative Crops, with Special Reference to Evening Primrose

P. Lapinskas

INTRODUCTION

This paper will examine some aspects of the commercialization of novel crops, using the evening primrose (*Oenothera* spp.), as an example. The evening primrose is a biennial plant, normally sown in July in the Northern Hemisphere. As it establishes it forms a rosette of leaves which grows to 10–15 cm in diameter by the onset of winter. The plant overwinters in the rosette form and in the following spring the stem elongates to form the flowering plant, which may be 1–2 metres in height. Flowering usually starts in early June and during the flowering period fresh bright yellow flowers open every evening, to fade by the following day, hence the name of the plant: the evening primrose. The plant comes to maturity in late September – early October and this is the time at which it is harvested. The seeds are very small, ca. 0.5 g/1000 seeds, with an oil content of ca. 22%. This is a highly unsaturated oil which contains 8–10% of a fatty acid known as gamma linolenic acid, or GLA. This is the component of interest.

GLA is important because it is also an intermediate in human metabolism in the production of prostaglandin hormones. There is a sound theoretical basis and a considerable amount of clinical data which suggest that supplementation of the diet with evening primrose oil can have a beneficial effect on certain human diseases.

With this in mind we can now examine some of the factors which affect the commercial viability of potential new crops.

TECHNICAL FEASIBILITY

The factor which generally gets most attention is that of technical feasibility – that is, is it possible to grow the crop on a field scale? This is natural, as it is at this first hurdle that most candidate species fail. It is easy however mistakenly to think that if a new crop is technically feasible, then commercial exploitation will surely follow. For anything other than subsistence agriculture this is of course by no means the case, and there are a whole range of factors which come into play, of which economic feasibility is the most important.

ECONOMIC FEASIBILITY

The economic feasibility is determined by the balance of the price which can be obtained for a product as opposed to the cost of producing it. The price of the product will be initially determined by the intrinsic value to the consumer, which is going to be dependent upon his perceived need. Clearly however, the price will also be determined by the price of competing products. To apply this to evening primrose oil, the product has a high intrinsic value to the consumer, in that it concerns his health, and at the time it was introduced to the market there were no competing products, so there were no immediate limitations on pricing.

On the other side of the equation the cost of production is perhaps a little more complicated. Clearly it is going to be determined by the yield which can be obtained; the higher yield per hectare the lower the unit cost should be. But it will also be influenced by the costs incurred by the farmer in producing the crop and by the processing and distribution costs.

The farmers of course are trying to maximize their profit and therefore tend to grow the most profitable crops they can on their land and this can have an effect on production costs. For example, evening primrose grows best on the most fertile soils. However, the most fertile soils usually produce yields of, for instance, wheat, in excess of 10 tonnes/hectare in the UK. Therefore to minimize the unit cost, it is better to place evening primrose in less fertile soils which it is better able to exploit than wheat. In this way the economic cost of the production is optimized when the plant is grown under conditions which are less than optimal from the point of view of the plant's own performance. In other words, the best place to grow the crop may not be the place where the crop grows best.

Another factor affecting the cost of production is government intervention. In the EEC there are extensive crop subsidies for existing major crops and these subsidies can stifle the development of new and alternative crops. As an example of this, in 1979 there was an attempt to introduce agricultural lupins into the UK as a protein and oil crop. The lupins were directly competing with peas and beans which were receiving government subsidy. Lupins were not eligible for a subsidy because the production area was too small. However, a larger area was difficult to achieve in the face of such unfair competition. It was a 'Catch-22' situation. If governments wish to achieve diversification of agriculture they could consider a subsidy for novel crops in addition to the current mainstream crops. This could be administered to crops with an area of less than say 5000 hectares per annum (in the UK) – and be made payable on a hectare basis.

In summary, if the cost of production is too high, or the price obtainable too low, it does not matter how technically feasible a crop may be, it will not make any impact in commercial agriculture.

MARKET POTENTIAL

Given economic feasibility, there is still of course a limit to the size of market which the product can achieve and this will also depend on a number of factors. The need for the product is probably the most important, for if there is only a small market, then development costs are likely to be prohibitive, even though the crop is potentially able to make a running profit.

Similarly the absolute cost of the product will be a limiting factor if it is too expensive; the relative cost of the product is also limiting if the competing products are a similar or lower price. The level of quality is also important, as perceived by the buyer, and the market size could be limited by production capacity if there is an overriding supply constraint.

A further factor is the question of supply stability. If the supply is unstable this will limit the crop's acceptance in the market place. While fluctuations are common in established crops (such as for instance coffee) they tend to be more severe in new crops because yield is uncertain. In the case of evening primrose this is exacerbated by the length of the life cycle, and the small size of the demand relative to production capacity. The impact of such instabilities can be reduced for evening primrose by making good use of contracted production, where the price is related to the real costs of production rather than market forces; by growing in both Northern and Southern Hemispheres, to stagger sowing dates; and by investigating ways of shortening the life cycle.

RATE OF MARKET PENETRATION

There are therefore a number of limits to the size for the potential market achievable and this is perhaps obvious. Possibly less obvious are the limits to the rate at which a market potential can be achieved by a new crop and these limits can affect both the supply and the demand side of the equation.

On the supply side one of the constraints is the rate of germplasm multiplication. For a new crop which is in great demand it may take a while for the germplasm to be multiplied sufficiently to achieve the production required. Many commercial crops only achieve a five or ten-fold multiplication in each generation, which means that it can take six or seven years to achieve sufficient seed stocks for large-scale commercial planting. In evening primrose, this is not a problem, as a single plant can easily produce 150 000 seeds.

Another problem is that of grower resistance to a new crop. Growers are naturally reluctant to risk an untried crop because if the crop fails then they lose their income from that area of land. This is particularly true if any additional investment in equipment is required. The climate in this respect is, however, rapidly changing. Ten years ago UK growers were extremely reluctant to try anything new and their gross margins had to be guaranteed before they would even consider a new crop. Now, however, it is possible to put out new crops without too much difficulty and pick and choose between growers.

Similarly up until now contracts have usually been placed on an area basis, where the buyer contracts to purchase whatever is produced on a given area of land. This protected the grower from having surpluses which he would have to sell on his own behalf. This is beginning to change and it is likely that in future, for evening primrose, more and more contracts will be placed on a tonnage basis with a grower or grower-organization deciding how many hectares need to be planted to achieve the contracted seed quantity.

Incidentally it should perhaps be pointed out that growers do not always behave as rational economic animals and as an example I would quote the case of borage versus evening primrose. Borage is a short season annual crop which also produces an oil containing GLA. It is planted in April, whereon it grows very vigorously to a harvest in late July or early August. It is a very attractive plant to farmers, as it smothers weed competition and is very easy to grow. However, as the plant matures – and it matures over a period of time – the seed is dropped to the ground, it not being contained in any kind of pod or similar structure. Therefore there is a need for swathing or desiccation of the crop and if the weather should be bad then it is quite possible to lose a crop completely, with consequent total loss of income from that field. By comparison, however, most evening primrose crops fail at the establishment phase, so that a grower knows within a month to six weeks whether he will have an effective crop. Since evening primrose is planted in July, this means that if the crop fails he will still have time to follow with a replacement crop, such as winter wheat, and so will not lose income from the land in the following season. Logically therefore, if a farmer has had a failure with either of the crops it would be expected that he would be more likely to take a second crop after an evening primrose failure than after a borage failure, since his economic loss is much less. However the reverse is true, since the farmer will probably take the attitude that he was so close to success with the borage crop, that given a change in management and better luck next year he could have a good harvest, whereas with evening primrose he would consider that since he never got an establishment it is not a crop that he could work with.

Another source of resistance to new crops is from processors and this can occur even with established crops. An example of this is malting barley, where there can be considerable resistance from the maltsters to the introduction of any new variety. The reason is that they have already established the process criteria for the currently grown variety and any change requires them to repeat their development work. Examples of novel crops which have not been taken up by manufacturers are manifold, such as for instance fenugreek, which contains a source of diosgenin for use by pharmaceutical companies in the manufacture of the female contraceptive pill.

Another limiting factor to the rate of market growth is the financial uncertainty involved with the development of any new crop. For a commercial company developing a new crop this can cause considerable problems. This is particularly so in the case of evening primrose. Since it takes three years to take a crop from the first contract negotiations through to having the product derived from that crop on the shelves of the stores, it means that one has to be able to predict sales three to four years in advance. This is no easy task. Having determined how much raw material will be needed for the sales it is then necessary to determine how many hectares will need to be contracted in order to achieve that production. If the estimates are on the generous side then there is the risk of building up a substantial store of surplus seed and if the estimates are on the low side then there is the risk of running out of stock and hence having a catastrophic impact on the company's finance. Bearing in mind that each crop is planted before the results of the previous crop are known, it can be seen that there is a considerable uncertainty built into the planning of production of a crop like evening primrose and this must inevitably reduce the rate at which the crop can expand.

Most novel crops suffer from supply uncertainties, given that the yield from a contracted crop cannot be predicted with any great certainty. Evening primrose certainly is prone to crop failure and yet a good crop can yield up to 1700 kg/ha. This further contributes to the instability of supply which in turn limits the rate at which market penetration can be achieved.

On the other side of the coin there are also demand limits to the rate of penetration achievable by a novel crop. These are largely covered by the marketing function and would include consumer awareness of the product – that is to say, if the consumer does not know about it he certainly will not buy it – and also consumer resistance – if the consumer does not believe what is said about it he will not buy it either. Other factors also have to be taken into account, such as the resistance from distributors, who may take some convincing that a new product is going to achieve the sales predicted. Furthermore, if a new product is being introduced into an area which is already served by competing products, then the livelihood of these competitors is being attacked and they can be expected to take countermeasures.

In summary the achievement of the potential market size will certainly not be immediate and in fact may take even decades.

ROLE OF RESEARCH

The above discussion implies a static picture of supply and demand factors affecting the market penetration and potential market size of a new crop, but of course there are many additional factors which can influence the outcome of this balance and change it with time, of which probably the most important is research.

Novel crops have usually been little researched and so little is known of their agronomic requirements and generally only unimproved genotypes are available. In the market place they will have to compete with a range of established and highly researched crops and will hence be at a competitive disadvantage. An example of this situation is the agricultural lupin which as a source of protein and oil is competing directly with plants such as oil seed rape and soya bean. Since it has no particular advantage over these commodities it is clear that for success it must be competitive with them. There is therefore a threshold which the crop has to reach in terms of performance before it can be commercially grown. This means that a considerable investment has to be made over a long period of time without any significant commercial return being received and this is very difficult for commercial companies to do. To take an alternative situation, the evening primrose was introduced as a novel oil seed containing GLA and, with no competitors in this field, the prices could be organized so as to cover the cost of production at the start, almost irrespective of how high those costs were. Thus there was income generated from the very beginning, which made it possible to invest in research on agronomic aspects of the crop to reduce the production costs. This in turn increased the income and hence increased the incentive for doing further research.

In the absence of government assistance therefore, it is likely that only new crops which exploit new uses will be developed, and not better crops for existing uses. It would seem sensible therefore for governments to fund extensive research on new crops up to the point where it is possible for them to make money, or least likely to be possible, and at that point allow the private sector to take over. It is only by this means that a range of alternatives to existing crops are likely to be produced.

CONCLUSIONS

There are a range of factors which will determine the economic feasibility and market potential of a new crop and how fast that potential may be achieved. These factors have to be assessed together with the likely impact of modifying factors, such as research and governmental intervention, when judging whether or not a species is worth pursuing as a potential future crop.

21

Chenopodium Grains of the Andes: a Crop for Temperate Latitudes

J. Risi C. and N. W. Galwey

INTRODUCTION

The grain crop quinoa (*Chenopodium quinoa*), like maize, potatoes and *Phaseolus* beans, was one of the staple foods of the Inca empire. Its agricultural and culinary role was similar to that of barley in the Old World. However its protein content is a few percentage points higher than that of most cereal species (Table 21.1), and the protein has a better balanced amino acid composition, having a higher proportion of lysine and of essential sulphur-bearing amino acids cystine and methionine (Table 21.2). This high protein content does not result in a low energy value because the fat content is also higher than that of

Table 21.1: *Nutritional Value of Quinoa Grain*

Component	Range (%)		Average (%)
Moisture	6·8	– 20·7	12·65
Protein	7·47	– 22·08	13·81
Carbohydrate	38·72	– 71·3	59·74
Fat	1·8	– 9·3	5·01
Cellulose	1·5	– 12·2	4·38
Fibre	1·1	– 16·32	4·14
Ash	2·22	– 9·8	3·36

Source: Cardozo & Tapia (1979).

Table 21.2: *Protein Compositions of Quinoa and other Foodstuffs*

Amino Acid	Quinoa[1]	Wheat[2]	Barley[1]	Soya[2]	Milk[3]
Isoleucine	6·4	3·8	3·8	4·9	5·6
Leucine	7·1	6·8	7·0	7·6	9·8
Lysine	6·6	2·9	3·6	6·4	8·2
Methionine	2·4	1·7	1·7	1·4	2·6
Cystine	2·4	2·3	2·3	1·5	0·9
Phenylalanine	3·5	4·5	5·2	4·9	4·8
Tyrosine	2·8	3·1	3·4	3·5	5·0
Threonine	4·8	3·1	3·5	4·2	4·6
Tryptophan	1·1	1·1	1·3	1·3	1·3
Valine	4·0	4·7	5·5	5·0	6·9

Sources: 1. Data from Van Etten *et al*. (1963). 2. Data from Janssen *et al*. (1979). 3. Data from Johnson & Aguilera (1979) reported by Cusack (1984).

wheat or barley, though not high enough for oil extraction. The seeds are borne in a panicle at the top of the plant (Fig. 21.1) and are not shed spontaneously, so the crop can be combine harvested. Even today, quinoa is grown in a wide range of environments in the Andes, at latitudes from 2° N in Colombia to 40° S in Chile, and from sea level to an altitude of 3800 m. It is mainly grown in cool highland regions, where during the growing season the climate is fairly similar to that of temperate regions. The areas where it survives are mostly agriculturally marginal, are prone to drought and have soils of low fertility.

Unlike the other staple crops of the Andes, quinoa did not attain global importance following the Spanish conquest of the Incas: throughout the Colonial and Republican eras (since about 1530 AD) its cultivation steadily declined. Smartt (1985) has argued that before effort is invested in developing a minor crop outside its indigenous environment its suitability for other environmental conditions and its acceptability to consumers elsewhere should be questioned. Therefore the possible reasons for this decline of quinoa need to be considered. During colonial times the cultivation of quinoa was discouraged, possibly because of its honoured position in the Inca society and religion (Cusack 1984), and the wheat and barley introduced by the Europeans were adapted to many of the regions where it had been cultivated. The panicle of the plant, which resembles that of sorghum, shares with sorghum the disadvantage of exposing the seed to attack by birds. Like sorghum, quinoa has evolved a chemical defence: however, unlike the tannins which protect sorghum grains, the bitter saponins present in the grain of most varieties of quinoa can be washed out with cold water. Nevertheless this unfamiliar processing requirement may have discouraged the adoption of the crop, particularly in areas with cloudy climates where drying is a problem.

There are crops which have become successful outside their indigenous environments after a long delay: sorghum and the soya bean rose to prominence only in the past few decades as their nutritional and agricultural value became widely recognized. Since about 1975 the decline of quinoa has been arrested, as economic pressures have encouraged Peruvians and Bolivians to decrease their dependence on imported foodstuffs and the nutritional value of the crop has become more widely recognized. Concurrently, the crop has attracted an increasing amount of attention from agricultural researchers in the Andes. This work has been reviewed by Risi C. & Galwey (1984).

Quinoa is not the only chenopod of importance to mankind. A very similar plant, huauzontle (*Chenopodium berlandieri* subsp. *nuttalliae*), is cultivated in Mexico as a vegetable and for grain, and a low-growing species, canihua (*Chenopodium pallidicaule*), is grown for forage and grain on the Altiplano around Lake Titicaca. In the Himalayas, plants classified as *Chenopodium album* are cultivated for grain (Partap & Kapoor 1984). A form of *C. album* morphologically very different from the Himalayan cultivar, known as fat hen in Britain and lamb's quarters in North America, is distributed world-wide as an annual weed on arable land. The dividing line between a weed and a crop is

223

Figure 21.1: *Morphology of the Quinoa Plant showing the Variation in Shape of the Apical and Basal Leaves of the Three Cultivars.*

a) terminal inflorescence; b) axillary inflorescence; c) hermaphodite flower; d) female flower; e) seed.

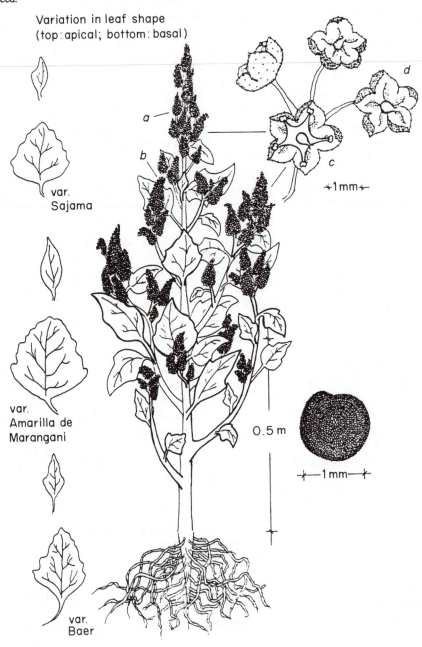

Variation in leaf shape
(top: apical; bottom: basal)

var.
Sajama

var.
Amarilla de
Marangani

var.
Baer

often thin: the seeds of *Chenopodium album* were used as food by the former inhabitants of Russia, Denmark, Greece and northern Italy, and considerable quantities were found in the stomachs of corpses preserved in the Danish peat bogs (Renfrew 1973).

One of the areas of the world where quinoa may have potential is eastern Britain, where continuous cultivation of cereals gives rise to problems of weeds, diseases and surpluses. Quinoa is better adapted to the climate here than some alternative break crops which have been tried, lupins, navy beans and sunflowers for example. Unlike these crops,it does not supply an obvious market. The grain could be used as an animal feed, provided that it can be produced at a competitive price and that the saponins can be eliminated, either by selection of low-saponin lines or (less probably) by processing. The price constraint will be less severe if it can be marketed for direct human consumption, a use which should no doubt initially be confined to the more experimental consumers in the 'health food' sector. Quinoa grain is already being sold on this basis in the United States, to be cooked like rice (D. Johnson and J. F. McCamant, pers. comm.).

In 1981 we initiated a programme to explore the potential of quinoa for cultivation in Britain, studying the effects both of crop genotype and of agronomic variables. We therefore assembled a germplasm collection of nearly 300 accessions from the entire geographical range of the crop; we studied the variation present, made selections and began to develop varieties by mass selection and by crossing and pedigree selection. We also identified those aspects of husbandry which could initially be determined by inference from standard practices or research results obtained in the Andes, and those to which priority should be given for investigation. It was clear that the crop is particularly suited to light, sandy soils which are relatively infertile and drought-prone (Narrea 1976; Junge *et al.* 1975; Canahua 1977), that fertilizer applications similar to those for oilseed rape would be suitable, and that the crop should be sown at depths between 1 and 2 cm (Bornas 1977; Etchevers & Avila 1979). On the other hand the appropriate sowing date in a cool temperate climate and the appropriate row widths and seed densities for mechanized agriculture were unknown, and no herbicide recommendations were available.

GERMPLASM EVALUATION

The germplasm collection, comprising 294 accessions, was grown out in two replications, and 19 characteristics were noted or measured on each accession. Some of the characteristics were discrete (e.g. plant colour), others were continuous (e.g. plant height). All the continuous characteristics varied significantly between accessions, and the discrete characteristics were clearly also under genetic control. The associations between the characteristics were explored by calculating chi-squared statistics, t statistics and correlation

Figure 21.2: *Groups of Accessions obtained by Hierarchical Cluster Analysis.*

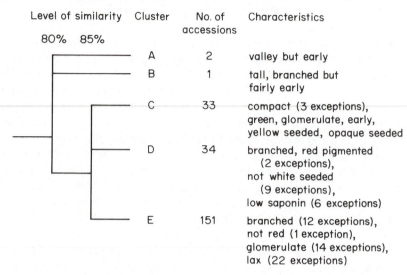

Level of similarity	Cluster	No. of accessions	Characteristics
80% 85%	A	2	valley but early
	B	1	tall, branched but fairly early
	C	33	compact (3 exceptions), green, glomerulate, early, yellow seeded, opaque seeded
	D	34	branched, red pigmented (2 exceptions), not white seeded (9 exceptions), low saponin (6 exceptions)
	E	151	branched (12 exceptions), not red (1 exception), glomerulate (14 exceptions), lax (22 exceptions)

coefficients: the latter are displayed in Table 21.3. The plant and inflorescence dimensions were all fairly strongly correlated, but the associations between the durations of developmental phases were surprisingly weak, suggesting that there is great scope for manipulation of the pattern of development through breeding. On the basis of the associations found, and of the 'passport data' available for the accessions, seven groups were defined (Table 21.4). In an attempt to group the accessions on a more objective basis, a hierarchical cluster analysis was performed on the data, using the single linkage method (which defines the proximity of two clusters as the proximity of their closest points) to form the clusters. The dendrogram produced by this analysis (Fig. 21.2) shows that the demarcations between the clusters were not very clear; moreover an attempt to define the characteristics common to each cluster produced numerous exceptions. Many accessions could not be included in a cluster at all. However one distinctive cluster was formed by two accessions, both identified as samples of the variety Amarilla de Marangani, which came from the Andean valleys north of the Altiplano but were earlier maturing than most such accessions. Another distinctive cluster was formed by the rather homogeneous group of accessions originating from southern Chile. The accessions in this cluster have early maturity, short stature and unbranched growth habit, and their day-length insensitivity enables them to flower and set seed during the long summer days of Chile or Britain, hence they are the accessions best adapted for cultivation here. Although all the plant characteristics required are found in these accessions, they lack the seed characteristics which we believe to be necessary, namely low saponin content to avoid processing costs, large size to minimize the proportion of fibrous pericarp, and a white pericarp to produce a colourless, versatile

Table 21.3: *Correlations between the Continuous Characteristics.*

	No.of plants	Plant height	Stem diam.	No. of prot'ns	Inflor. length	Inflor. diam.
Plant height	−0·005					
Stem diam.	−0·32	0·473				
No.of leaf margin protrusions	0·092	0·295	0·138			
Infloresc. length	−0·039	0·959	0·447	0·350		
Infloresc. diam.	0·023	0·081	0·409	0·306	0·827	
Relative saponin content	−0·158	0·230	0·206	0·178	0·284	0·258
Subperiod 1[1]	0·289	0·282	0·312	0·332	0·362	0·242
Subperiod 2[1]	0·065	−0·081	−0·028	−0·214	−0·183	−0·098
Subperiod 3[1]	−0·049	0·013	0·168	−0·024	−0·109	−0·010
Subperiod 4[1]	−0·059	0·086	0·140	0·036	−0·013	0·044
Subperiod 5[1]	−0·181	0·005	0·160	−0·232	−0·047	0·007

	Rel.sap.	Subp.1	Subp.2	Subp.3	Subp.4
Subperiod 1	0·281				
Subperiod 2	−0·028	0·586			
Subperiod 3	0·010	0·104	0·039		
Subperiod 4	0·080	0·186	0·090	0·520	
Subperiod 5	−0·015	0·121	0·368	0·328	0·085

Critical Values

D.F[2]	0·05	0·01	0·001
279	0·117	0·154	0·197
270	0·119	0·157	0·120
250	0·124	0·163	0·208

1. The subperiods are the lengths of time between sowing, germination, the two true leaf stage, flower initiation, anthesis and maturity. 2. The correlations for inflorescence characteristics have 270 DF and those for the fifth subperiod have 250. All others have 279.

product for the food processing industry. These characteristics are found widely scattered among accessions from further north in the Andes.

Various other attempts have been made to classify the diversity of quinoa genotypes: probably the most widely accepted is the definition of fi‎ by Tapia *et al.* (1980). These are the sea-level type from southern Salar type from the salt flats of southern Bolivia, the Altiplano typ‎ type from Peru and Ecuador, and the subtropical type, represented originally by a single plant found in the Yungas region of Bolivia. In order to determine whether this classification was borne out by the evidence from the germplasm collection, each accession was assigned to one of these ecotypes on the basis of the information available about its origin, and a canonical variate analysis was conducted on the continuous characteristics in order to define new axes which would separate the ecotypes as completely as possible. There was a good deal of overlap between accessions in different ecotypes (Fig. 21.3), except for the subtropical type which was too far away from the rest to be presented in the same figure.

Table 21.4: Groups of Accessions based on Associations between Pairs of Characters and Passport Data

Group	Characteristics
Chilean sea level accessions	Early maturity, short, unbranched plants, compact inflorescences, yellow seed, medium saponin content. A particularly homogeneous group.
Bolivian Salar accessions	Mostly having red-stemmed plants, amaranth-iform inflorescences, dark seed, high saponin content.
Accessions from Ilave market, Puno	Branched, red-stemmed plants, small inflores-cences, small seed. Low saponin content, sur-prising in heavily-pigmented landrace material.
Accessions from the Altiplano	Generally similar to those from Ilave.
Accessions from the Universidad Nacional Agraria, Lima.	Green plants, long, compact inflorescences, large seed, high saponin content. These charact-eristics reflect the activities of plant breeders.
Accessions from farmers' stores in the Mantaro Valley, Junin.	Green plants, lax inflorescences, large white seed, low saponin content.
Andean Valley accessions from Peru to Colombia	Late maturity, branched plants, lax inflores-cences, large seed. A diverse group.

SOWING DATE AND DENSITY

The Chilean sea-level variety Baer and the Peruvian valley variety Blanca de Junin were sown on 25 March, 14 April and 7 May 1982 at row widths of 0·8 and 0·4 m and within-row densities of 0·2, 0·4 and 0·6 g/m. In a second experiment sown on 15 March 1984 Blanca de Junin was replaced by Amarilla

Table 21.5: Number of Days from Sowing to Maturity

		Row width (m)						
		0·8			0·4			
		Within-row density (g/m)						
Variety	Sowing Date	0·2	0·4	0·6	0·2	0·4	0·6	SED
Baer	March	169	167	166	166	163	161	
Baer	April	158	156	154	157	153	152	
B. de J.	March	234	228	227	219	218	218	1·63

		Row width (m)					
		0·8			0·4		
		Sowing density (kg/ha)					
Variety		15	20		15	20	30	SED
Baer		150	148		149	148	147	
A. de M.		240	241		239	237	237	1·16

Figure 21.3: *Canonical Variate Analysis.*

The one-standard-deviation ellipses show the distribution of accessions of the Chilean sea-level ©, Salar (S), Altiplano (A) and Valley (V) types.

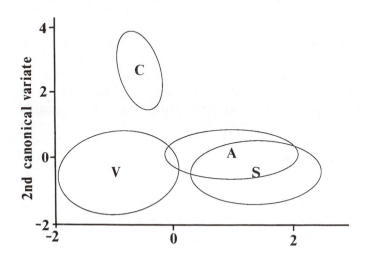

1st canonical variate

de Maranganì, row widths of 0·4 and 0·2 m were used, and instead of adjusting the within-row seed density, the overall sowing density was adjusted to give values of 15, 20 and 30 kg/ha. Weed competition was more intense following the later sowings, and indeed the plots sown in May had to be abandoned.

Table 21.6: *Grain Yield (g/m²)*

Variety	Sowing Date	Row width (m) 0·8 Within-row density (g/m) 0·2	0·4	0·6	Row width (m) 0·4 Within-row density (g/m) 0·2	0·4	0·6	SED
Baer	March	303	351	362	518	568	633	
Baer	April	149	223	258	242	390	354	
B. de J.	March	257	277	328	329	384	415	71·1

Variety	Row width (m) 0·8 Sowing density (kg/ha) 15	20	Row width (m) 0·4 Sowing density (kg/ha) 15	20	30	SED
Baer	494	500	616	696	636	
A. de M.	451	458	515	623	646	34·5

The valley varieties were consistently later maturing than Baer (Table 21.5). Increased sowing density caused slightly, but significantly, earlier maturity, but had profound effects on plant architecture and yield, effects which differed between the varieties. For instance the percentage of branched plants decreased at higher sowing densities, and this trend was more pronounced in the highly branched valley varieties than in Baer (Fig. 21.4). As the percentage of branched plants decreased, the percentage of stunted plants producing little grain increased. It might therefore be expected that Baer would show the greatest increase in grain yield as sowing density was increased. Baer was indeed more responsive than Blanca de Junin, which was low yielding in all cases (Table 21.6). However the yield of Amarilla de Maranganì continued to increase up to a sowing density of 30 kg/ha and exceeded that of Baer at this density. Higher yields were obtained from March than April sowings, but subsequent experience has shown that crop establishment is often poor following cold weather in March. The factors which influence establishment are not clear, but in general the problem seems to be most severe on heavy clay soils which are slow to warm up and prone to waterlogging. The sensitivity of quinoa to these stresses may be related to its small seed size (1–2 mm in diameter). Further investigation of these effects, and selection of genotypes which are tolerant of early spring conditions, is needed in order to develop a crop which makes maximum use of the growing season.

WEED CONTROL

In the subsistence agricultural systems of the Andes quinoa is generally weeded by hand, but this is not feasible for commercial production. A range of herbicides was therefore tested, choosing those used on sugar beet since this is the British crop most closely related to quinoa. Two herbicide regimes which gave effective weed control are suggested (Table 21.7), but many herbicides proved highly toxic to quinoa. This is not surprising since the control of fat hen is an important objective in all dicotyledonous arable crops.

Regime 1 will control a broader spectrum of broad-leaved weeds than regime 2, but may scorch the crop in wet conditions. For regime 2 the chemicals should be 'creamed' separately, the propyzamide diluted to half-quantity with water, the propachlor added, and the mixture diluted fully. Fusilade should be applied

Table 21.7: Herbicide Regimes

Weeds controlled	Commercial products	Rate	Growth stage
Broad-leaved: regime 1	Goltix	2·5 kg/200 l water/ha	pre-emergence
	Kerb 50 W	1·0 kg/200 l water/ha	4 true leaves
Broad-leaved: regime 2	Kerb 50 W	1·0 kg/200 l water/ha	pre-emergence
	Ramrod	3·0 kg/200 l water/ha	pre-emergence
Grasses	Fusilade 5	1·5 l/200 l water/ha	pre-flowering

Figure 21.4: *Effects of Variety and Sowing Density on the Percentage of Branched Plants.*

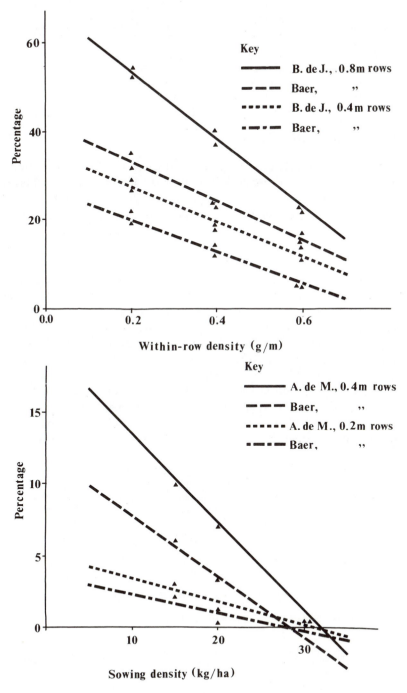

with a wetting agent added at a rate of 0·5 ml/l. The active ingredients of the products mentioned are given below.

Commercial name	Manufacturer	Chemical name	% chemical
Goltix	Baer	metamitron	100
Kerb 50 W	Rohm and Hass	propyzamide	50
Ramrod	Monsanto	propachlor	65
Fusilade 5	Plant Protection	fluarifop-P-butyl	12·5
Agral	Plant Protection	wetting agent	–

GENETICS AND SELECTION

Mass selection of plants with compact inflorescences was conducted for two generations in the Chilean sea-level variety Baer. In the first generation they were selected by eye and in the second generation by calculating the density of the inflorescence. Plants of all three generations were grown in the same season, from which it was clear that the selection was effective in increasing the inflorescence density, particularly in the second generation (Table 21.8). There was also some evidence of a correlated increase in yield. In order to combine the desired plant characteristics of the Chilean varieties with seed characteristics from elsewhere, and to study the genetic control of these characteristics, a diallel cross was made between accessions from a wide range of origins. The results from the F_1 generation of this cross showed that for most quantitative characteristics additive genetic effects were larger than dominance effects. Late flower initiation and high saponin content were generally genetically dominant, but there was evidence that more complex genetic effects influenced these characteristics in Amarilla de Marangànì. This variety is thus unusual in several respects. The segregants from the crosses involving a non-Chilean parent were too late-maturing to be immediately useful, but fortunately a substantial number of low-saponin segregants have been found in Chilean × Chilean crosses.

Table 21.8: Effects of Mass Selection in the Variety Baer

Selection cycle	Inflorescence density[1]		Yield	
	Mean	Standard deviation	Mean	Standard deviation
0	0·156	0·095	24·6	11·8
1	0·142	0·108	25·1	14·4
2	0·248	0·102	29·9	11·0
SED	0·0089	0·0089	2·83	1·22

1. Defined as grain yield/volume.

FUTURE PROSPECTS

In this project a great diversity of valuable genotypes of quinoa and much valuable information about the crop have been obtained. In view of the small amount of attention which it has received in the past and the progress which has been made, further exploration of its potential is more than justified. The most pressing need is for purification, multiplication and further evaluation of promising lines and for the initiation of further cycles of crossing and selection in order to bring the many desirable characteristics available in the germplasm into the most advanced lines. These activities should take priority over work of a more academic kind, though there are still several important questions unanswered. The problems of seedling establishment have already been mentioned. The degree of outcrossing in the field is not known reliably, though the frequency of types F_3 and F_4 lines suggests that it may be in the range of 10–15 per cent, rather than our original estimate of 0–5 per cent. Controlled hybridization is however difficult to achieve due to the small size of the flowers and the abundant production of pollen. A simple, reliable method of emasculation would be useful.

Developing quinoa in Britain is only one attempt to exploit its potential outside the Andes, but the considerations to be taken into account in other local circumstances would be analogous. The regions in which quinoa may have potential include other cool-temperate regions, drought-prone temperate regions such as Spain and the high plains of North America, and, perhaps especially, highland tropical regions such as Ethiopia and the Himalayas. Quinoa is being grown in Colorado to supply the American 'health food' market (D. Johnson and J. F. McCamant, pers. comm.), and is also being evaluated in Denmark (A. R. Denis-Ramirez and S. E. Jacobsen, pers. comm.), West Germany (E. Ritter, pers. comm.) and, for leaf protein production, in Sweden (Carlsson 1980). Such interest from developed countries may help to increase the prestige of quinoa in the Andes, reinforcing the growing interest of agricultural researchers there. This would lead to improved cultivation on an enlarged area in the crop's region of origin, where it has remained on the sidelines for so long.

REFERENCES

Bornas C., E. A. (1977) *Requesta de la quinua (Chenopodium quinoa* Willd.) *variedades Sajama y Kancolla a la profundidad de siembra en cuatro clases texturales de suelo.* Ingeniero Agronomo thesis, Universidad Nacional Tecnica del Altiplano, Puno, Peru.

Canahua M., A. (1977) Observaciones del comportamiento de quinua a la sequia. In: *Primer congreso internacional sobre cultivos andinos*, pp. 390–392. Universidad Nacional San Cristobal de Huamanga, Instituto Interamerican de Ciencias Agricolas, Ayacucho, Peru.

Cardozo, A. and M. E. Tapia (1979) Valor nutritivo. In: *Quinua y Kaniwa. Cultivos andinos. Serie Libros y Materiales Educativos* No. 49, M. E. Tapia (Ed), pp. 149–192. Instituto Interamericano de Ciencias, Agricolas, Bogota, Colombia.

Carlsson, R. (1980) Quantity and quality of leaf protein concentrates from *Atriplex hortensis* L., *Chenopodium quinoa* Willd. and *Amaranthus caudatus* L., grown in southern Sweden. *Acta Agric. Scand.* 30: 418–426.

Cusack, D. (1984) Quinua; grain of the Incas. *The Ecologist* 14: 21–31.

Etchevers B., J. and P. Avila T. (1979) Factores que afectan el crecimiento de quinua (*Chenopodium quinoa*) en el centro-sur de Chile. In: *10th Latin American meeting of Agricultural Sciences.*

Janssen, W. M. M., K. Terpestra, F. F. E. Beeking and A. J. B. Baisalsky (1979) *Feeding Values for Poultry.* 2nd edition. Spelderhold Institute for Poultry Research, Beekbergen, Netherlands.

Junge, I., P. Cerda and K. Alid (1975) *Lupino y quinoa, estado actual de los conocimientos y de las investigaciones sobre su empleo en alimentacion humana.* Departamento de Ingeniera Quimica, Universidad de Concepcion, Concepcion, Chile.

Narrea R., A. (1976) *Cultivo de la quinua.* Bulletin No. 5, Direccion General de Produccion, *Ministerio de Alimentacion,* Lima, Peru.

Partap, T. and P. Kapoor (1984) Investigation of the food value of chenopods. In: *Progress in leaf protein research. Current trends in life sciences,* Narrendra Singh (Ed.), pp. 99-101. New Delhi, India.

Renfrew, J. M. (1973) *Palaeoethnobotany.* Methuen, London.

Risi C., J. and N. W. Galwey (1984) The Chenopodium grains of the Andes: Inca crops for modern agriculture. In: *Advances in Applied Biology* Vol. 10, T. H. Coaker (Ed.), pp. 145–216. Academic Press, London.

Smartt, J. (1985) Evolution of grain legumes. II. Old and New World pulses of lesser economic importance. *Exp. Agric.* 21: 1–18

Tapia, M. E., A. Mujica S. and A. Canahua (1980) Origen distribucion geografica y sistemas de produccion en quinua. In: *Primera reunion sobre genetica y fitomejoramiento de la quinua,* A1-A18. Universidad Nacional Tecnica del Altiplano, Instituto Boliviano de Tecnologia Agropecuaria, Instituto Interamericano de Ciencias Agricolas, Centro de Investigacion Internacional para el Desarrollo, Puno, Peru.

Van Etten, C. H., R. W. Miller, I. A. Wolff and Q. Jones (1963) Amino acid composition of seeds from 200 angiosperm plants. *J. Agric. Food Chem.* 11: 399–410.

22

Ethiopian T'ef: a Cereal Confined to its Centre of Variability

M. R. Cheverton and G. P. Chapman

INTRODUCTION

Eragrostis tef is a staple cereal crop in the Ethiopian highlands where it is used to make enjara, a type of bread resembling a pancake. T'ef grain is also used for making other types of bread, for porridges, soup and alcoholic beverages. Its straw is used for fodder and to reinforce mud plastering. As a cereal crop t'ef is presently confined to Ethiopia, but it has been introduced to several other countries where it is grown for forage.

NUTRITIONAL VALUE

T'ef grain is nutritionally very important in the Ethiopian highlands where many people eat it two or three times daily, preferring it to other cereals. According to the Ethiopian Nutrition Survey (Anon. 1959 and cited by Jansen *et al*. 1962) it provides two thirds of the protein consumption of many Ethiopian and presumably a large proportion of the total energy in their food as well. T'ef grain is nutritious, being rich in energy (353–367 kcal/100 g), protein (8·6–11·5%), and minerals, especially iron (0·011–0·033%) and calcium (0·1–0·15%). In common with other cereals it is deficient in the essential amino acid lysine. Synopses of several analyses are given by Tadessa Ebba (1969).

GROWTH HABIT

T'ef is a slender, more or less upright annual tufted C4 grass of the subtribe Eleusininae in the tribe Eragrostideae of the Gramineae subfamily Chloridoideae. Its inflorescence is a panicle which is usually loose with branches spreading radially in all directions, but it may be unilaterally ramified and semi-compact or compact with the branches adpressed to the rachis and obscuring it from sight. The minute free-threshing grains are 0·8–1·5 mm long by 0·4–0·8 mm wide and may be from very pale fawn to dark reddish brown in colour.

GRAIN TYPE

T'ef is described and sold according to the grain colour, 'white' grain fetching a considerable premium over 'red' grain. Total annual production is estimated at 1·22 Mt, an area of 1·34 Mha being devoted to the crop (Information from the Central Statistics Office for 1979–1984).

PROFITABILITY AND YIELD

At a price of about 80 to 95 Birr per quintal (Mojo market, February 1987), i.e. £250-£300/t, t'ef production would be worth in excess of £300 M annually on the open market. T'ef always commands a good price substantially above other cereals. At times of food shortage, the price would be much higher than those quoted above. T'ef is the single most important food crop in Ethiopia in terms of the area grown and total production, and its potential contribution to the rural economy is very great. Grain controls are such that farmers are forced to sell some of their grain below the open market price, but the effect of these is beyond the scope of this paper.

The national average yield is officially estimated at 0·9 t/ha. Using 'improved' varieties and improved cultural practices including fertilizers, yields of 1·7–2·2 t/ha may be obtained on farmers' own fields (Anon. 1979). Other sources give widely variant yields, Asrat Felleke (1965) reporting 1·65–2·53 t/ha for trials of nine varieties under experimental conditions, the Agriculture of Ethiopia (Anon. 1954) quoting 1·7–1·9 t/ha as a general yield, and Brown & Cochemé (1970) reporting 0·55 t/ha as an average yield under traditional practice and estimating ten times this under ideal conditions.

PESTS AND DISEASES

The principal t'ef growing regions are in Shoa, Gojam and Gondar provinces at altitudes from 1700 to 2400 m, but it is grown to a greater or lesser extent in all provinces and from sea level to over 2800 m. T'ef is relatively resistant to disease and pests both before and after harvest (Tadessa Ebba 1969). However, *Uromyces eragrostidis* (t'ef rust) is widely distributed and *Helminthosporium miyiakei* (head smudge) can cause considerable damage. *Spodoptera exampta* (an army worm) and *Schistocerca gregaris* (a locust) can devastate crops. *Antherigonia hyalipennis* (a muscid) causes wilting and drying of the panicle when the larvae eat the base of the inflorescence. Huffnagel (1961) reports some post harvest losses to rodents, insects and moulds.

AGRONOMY

T'ef is resistant to drought but most cultivars require at least three good rains during their early growth and a total of 200 to 300 mm of water. Some rapid maturing cultivars may obtain the 150 mm they need from water retentive soils when planted at the end of the rainy season and require little or no extra water. T'ef may be grown after maize and sorghum crops have failed due to drought and still provide a useful yield (Seyfu Ketema 1986).

T'ef is a successful crop on vertisols and vertic soils due to its ability to withstand waterlogging during its early growth stages. In wet years without drainage t'ef may yield more than wheat on such soils and subsequently more than faba beans even though in a better drier year t'ef might be outperformed by wheat and nearly matched by beans (S. Jutzi, pers. comm.).

The resistance of t'ef to disease, pests, drought and waterlogging make it a staple crop on which farmers can depend. This stability, together with its good storage properties and its preference as a food and consequent high price make t'ef an attractive crop for many farmers in Ethiopia. Even in regions such as near Sheshamane, Shoa Province, where other crops such as maize may offer substantially higher yields and reasonable reliability, some farmers still grow a field of t'ef as well, benefiting from its high price, the later sowing date for t'ef facilitating husbandry of the two crops.

IMPROVEMENT

The characteristics of t'ef make it well suited to Ethiopia and this has been recognized by farmers who have made it the country's most important food crop, yet elsewhere t'ef was until recently unknown and even where it has been introduced as a forage crop, the cooking skills to make enjara from the grain have not gone with it, and thus its potential as a cereal has been overlooked. T'ef is thus a crop which is relatively new to science and systematic improvement, yet is of vital importance to a large population, the insecurity of which has been tragically highlighted by the most recent of a series of famines. T'ef might also hold promise for other countries where its qualities could counter similar problems to those in Ethiopia. It is in this context that a breeding programme has been started with research at Wye College, University of London supporting work at the Institute of Agricultural Research and Debre Zeit Research Station in Ethiopia with funding from Britain's Overseas Development Administration, the German agency GTZ and OXFAM.

The primary aim of the project is to develop cultivars which can increase the reliable yield of t'ef, taking into account the current situation of Ethiopian farmers and their prospects for the future as well as the uses which will be made of the grain and straw and the qualities needed in them. One current constraint on t'ef production is the very limited availability of fertilizers and other artificial inputs. As there can be no guarantee that this will change in the near future, a breeding programme must take this into account or it will fail to achieve its objectives if inputs remain unavailable. This in turn raises an interesting question, for if the environment can be changed by artificial inputs, a crop can be aided by plant breeding to adapt advantageously to its new environment, whereas in a situation where the environment cannot be changed the crop is presumably already adapted to that environment. The options for the plant breeder are thus reduced. Two options which remain are (a) to exploit the differences between agricultural objectives (largely a matter of yield and quality) and the result of natural selection which is normally to maximize fitness and thus ensure the survival of the genes contributing to fitness, and (b) seeking goals which were unattainable by natural selection within the time it has had to operate due to lack of suitable available genes. The former might be achieved for instance by redirecting resources presently devoted to interplant competition towards grain yield, and the latter by bringing together geographically or otherwise isolated genotypes.

Ethiopia is a centre for diversity for many genera and is almost certainly the centre of origin for t'ef. The environment varies considerably both spatially and temporally and it would seem likely that t'ef has developed ecotypes adapted to the various conditions.

REFERENCES

Anon. (1954) *The Agriculture of Ethiopia 1*. College of Agriculture, Dire Dawa, Ethiopia.

Anon. (1959) *Ethiopia Nutrition Survey, Report*. Inter-departmental Committee on Nutrition for National Defense, Washington DC.

Anon. (1979) *Institute of Agricultural Research. Handbook on Crop Production in Ethiopia*. IAR, Addis Ababa, Ethiopia.

Asrat Felleke (1965) *Progress report on cereal and oil seed research 1955–1963*. Exp. Stn. Bull. 39, Imperial Ethiopian College of Agricultural and Mechanical Arts, Branch Experiment Station, Debre Zeit, Ethiopia.

Brown, L. H. and J. Cochemé (1970) Agrometeorology survey of the highlands of eastern Africa. *Nature and Resources, UNESCO* 6(3): 2–10.

Huffnagel, H. P. (1961) *Agriculture in Ethiopia*. FAO, Rome.

Jansen, G. R., L. R. Di Maio and N. L. Hause (1962) Amino Acid Composition and Lysine Supplementation of tef. *Agric. Food Chem.* 10: 62–64.

Seyfu Ketema (1986) Food self sufficiency and some roles of tef (*Eragrostis tef*) in the Ethiopian agriculture. Paper presented at National Workshop on Food Strategy for Ethiopia held at Aleymaya University of Agriculture, Hararge, Ethiopia. December 8–12, 1986. (mimeo.)

Tadessa Ebba (1969) *T'ef (Eragrostis tef) The cultivation, usage and some of the known diseases and insect pests. Part I*. Exp. Stn. Bull. 60, Haile Selassie University, College of Agriculture, Dire Dawa, Ethiopia.

23

New Small-grained Cereals which may have Value in Agriculture

C. N. Law

INTRODUCTION

Today cereal breeders are faced with the hard task of redefining objectives. In Europe in particular, breeders have achieved the earlier objective of producing high-yielding wheats and barleys and are well on the way to making the improvements in grain quality recently demanded of them. The cereal mountains are in some ways a monument to their achievements, although the size of the mountain is more related to political and economic influences rather than to breeding. In the less developed world, the green revolution of the 1960s has transformed the agricultures of many parts of the subtropics, extending from the Indian subcontinent to Mexico and South America. Perhaps Africa is the only major land-mass that has yet to benefit greatly from these achievements. For most cereal breeders, therefore, the search is on for new goals.

In the developed world, these new goals for cereal breeding will probably be concerned with improving efficiency through increasing yield per unit area of land whilst restricting or using more effectively inputs such as chemical fertilizers, herbicides and fungicides. A likely objective for many parts of the world will be the extension of cereals to areas where at present diseases and environmental stresses limit their growth. However, both these approaches are conceived as taking place by selection from within the existing cultivated crop. They cannot be regarded as leading to the creation of new crops, which is the subject of this paper.

THE DEVELOPMENT OF NEW CEREAL CROPS

There are four possible areas of activity from which a new cereal crop might emerge:

(a)   From the exploitation of existing species which have not been used directly or widely in agriculture.

(b)   Through the directed transfer of a few genes into an existing cereal crop to create a totally different product to that normally provided.

(c) From the exploitation of a wild species by the transfer into it of genes from cultivated forms.

(d) By producing new synthetic polyploids, often with one or more of their genomes derived from established crop plants.

The first of these areas already features in many of the papers presented in this Symposium and there is therefore no need to describe further examples.

The second area is the domain of genetic engineering proper, since the objective will be to create a totally different product to that normally exploited from a crop. This is most likely to result from transgenic transfers rather than from transfers within a species or even from related species. At present the means for making transfers of this nature in the cereals are either not fully perfected or are just not available. Undoubtedly, this will not be the case in the near future. However, because the technology of gene transfer is not fully developed in cereals, this second area of activity will also be omitted from this paper. In passing, though, it is relevant to point out that, even given the technology, there is still the very important question to be answered concerning which genes should be targeted. These will need to be sufficiently distinct, giving rise to novel products if a new crop is to be created. Indeed, the form of enquiry into new crops being promoted in the Symposium may be one of the best ways of identifying worthwhile targets for genetic engineering in the future.

This leaves the remaining two areas of activity, and these will be the main subjects of this paper.

The introduction of genes from cultivated cereals into wild species to create a new cereal crop

A frequent approach in plant breeding is to introgress genes from the wild species into the related cultivated crop by means of sexual hybridization. This has been very successful and there are many instances of new genes, particularly for resistance to diseases and pests, being introduced into cereals (Gale & Miller 1987). The reasons for this success, certainly in the case of genes for resistance, is the simplistic nature of their inheritance combined with the existence of efficient screening procedures allowing easy selection. For the genetically more complex characters, such as the many which relate to yield, successes are not so forthcoming. A recent example of an attempt in this area occurs in the wild tetraploid wheat, *Triticum dicoccoides* ($2n = 4x = 28$) which grows widely in the Middle East. *T. dicoccoides* is, of course, one of the progenitors of the cultivated durums and bread wheats, and until recently was mainly of interest to cytogeneticists studying wheat evolution. However, Avivi (1979) observed that forms of *T. dicoccoides* existed with levels of grain protein as high as 40%. Moreover, these high levels of protein were found in accessions with large, bold grains and having vigorous vegetative growth. The levels of protein were thus not a consequence of poorly developed grains or reduced growth.

Several different groups throughout the world have taken these lines with the aim of introducing the high protein character into cultivated forms with varying degrees of success. In no instance has it been possible to produce exactly the high levels of protein in the grain of the recipient cultivated crop. A typical example of the levels of achievement is described in Table 23.1 from work being undertaken by T. E. Miller and S. M. Reader at the Institute of Plant Science Research. This illustrates some improvement of protein levels amongst the durum selections, but nowhere near the high levels of the wild species.

Table 23.1: Mean Yields and Percentage Proteins of Several Lines Resulting from the Initial Cross of an Accession of Triticum dicoccoides with Two Durum Wheats, Cando and Mexicali, followed by Repeated Backcrossing to the Durum Parents combined with Selection for Non-Brittle Rachis and other Durum Characters

	Yield (g)	% Protein	Difference from durum parent
Cando	50·4	14·8	
C27C-46/3–2	46·20	17·0	2·2***
C27C-46/3–6	72·44	16·5	1·7***
C27C-46/3–7	46·49	16·6	1·8***
C27C-46/3–10	35·48	18·6	3·8***
Mexicali	86·47	15·8	–
M76M-6/2–5	96·17	16·1	0·3
M76M-6/2–6	79·65	17·7	1·9**
M76M-6/2–7	91·99	16·3	0·7
M76M-6/2–10	82·96	17·0	1·2**
M27M-8/3–1	73·73	18·0	2·2***
M27M-8/3–16	89·11	17·4	1·6***
M27M-8/3–17	95·48	19·4	3·6***
M27M-8/3–18	72·29	17·8	2·0***
Original *Triticum dicoccoides*		28·0	

** P 0·01–0·001   *** P <0·001

It is likely that one of the reasons for this failure is the large number of genes involved in controlling the character. If this is the case, then the task of accumulating all the high protein genes in one line, disassociated from linked deleterious genes, will be formidable, although recent developments in molecular marker techniques may make this less daunting (Burr *et al.* 1983).

However, the failure to achieve the 30 or more per cent protein by conventional selection may really be a failure of breeding strategy. For instance, rather than using *T. dicoccoides* as the non-recurrent parent, it might be better to consider the wild tetraploid as a new crop and try to introgress a few genes from the cultivated durums necessary to adapt *T. dicoccoides* for use in agriculture. Because of the fragile nature of *T. dicoccoides* spikes, an obvious gene for introduction from the durums would be for non-brittle rachis. Other genes that may need to be considered would be those for disease resistance and also for adaptation. The point, however, is that the numbers of genes to be transferred may be relatively few, particularly if the emphasis is on the creation of a new crop which is not constrained by the need to compete with the cultivated wheats with respect to yield and other performance characters. The objective would be to produce a new cereal with grain protein levels as high as those obtained in legumes even though yields may be a half to two thirds of that of wheat in the UK today. Yield levels of this order would incidentally be as good if not better than those obtained for peas and beans, so that protein yield from such a new crop would be higher. This new crop would also have value as a source of gluten for industrial usage, not to mention other uses that would almost certainly emerge once such high protein grain was available to the food industry at large.

The development of new synthetic polyploids

Turning now to the possible exploitation of synthesized polyploids as new crops. This was a very active area in the 1950s and 1960s, many cereal-like synthetic polyploids being produced (Bell 1950). Only one has, however, survived and that is triticale – the allohexaploid between diploid rye and tetraploid wheat. The reason for the success of triticale is probably due to the fact that it combines the genomes of two highly selected cultivated species whose genetic contributions complement each other with respect to a range of characters. Thus the high yields of wheat can be combined with the greater hardiness of rye and so on.

Triticale, however, has a phenotype which is basically comparable to that of existing cultivated cereals such as wheat. It is not like most of the synthetic amphiploids that were produced, where the deleterious effects of the added genome were so evident that the new polyploid could not be considered as remotely competitive with present day cereals grown under intensive agriculture. It was for these reasons that they were discarded.

However, in the context of creating a completely new crop rather than as a near-substitute for existing cereals, it could be argued that this rejection may have been short-sighted and ill-considered. Under certain growing conditions, or if certain new products were desirable, then synthetic amphiploids might have the beginnings of new cereal crops which could have advantages over presently cultivated cereals such as wheat.

A possible example of this is the development of a cereal which can grow in saline soils. Under high to moderate levels of salt in the soil, wheat fails to give any grain return at all. A new cereal to replace wheat under these circumstances might be worth serious consideration, even if its yield levels were low.

Amongst the relatives of wheat are several that show much greater degrees of tolerance to salt than the cultivated cereal. One of the most tolerant is *Elymus farctus* (syn. *Agropyron junceum*), a diploid species ($2n = 2x = 14$) which grows in littoral habitats in many parts of the world. When grown in culture solutions containing half-strength sea water or in saline soils, *E. farctus* survives and produces grain, whereas wheat soon dies. More importantly, however, the amphiploid between *E. farctus* and wheat exhibits the salt-tolerance of the wild diploid (see Fig. 23.1). The genes for salt tolerance are thus expressed on a wheat background and are presumably dominant, therefore, to the recessive genes for the susceptibility present in wheat. This is an important and desirable result because the accepted way of handling this type of material is to try and introgress the genes for salt-tolerance into wheat by further hybridization and chromosome engineering. This would not be possible unless the genes for tolerance were expressed in the amphiploid.

Figure 23.1: *Bread Wheat Diploid Elymus farctus and the Amphiploid between them after days growth in Salt Solution (concentration 250 mol/ m3 or approximately half-strength sea water).*

Clearly, the wheat plants have died, whilst the amphiploid (in the middle) and *E. farctus* have flowered and set seed (Forster *et al*. 1987).

However, the tolerance of the amphiploid could suggest an alternative strategy in which the amphiploid itself is the basis of a new cereal crop.

An inter-crossing programme in the manner used in the development of triticale could be expected to improve yield levels of the amphiploid, and it is even conceivable that, given a sufficiently wide genetic base in the wheat donors, yield levels comparable to wheat itself could be achieved. The possibility of creating such a new crop is being actively considered at IPSR (Cambridge Laboratory) in conjunction with CIMMYT, Mexico and the Centre for Arid Zone Studies at Bangor, UK.

It is probable that many other synthetic cereal polyploids could provide partial and even complete solutions to the problems of adapting cereals to growth in fringe areas and to new climates. As mentioned, a large number of amphiploids were produced many years ago. Seed of such amphiploids is still available and viable, so that an evaluation of their potential for creating new cereal crops could be a worthwhile venture.

CONCLUSION

Perhaps the major point that emerges from this discussion of the potential for the creation of new small-grained cereals is the need to reappraise breeding objectives, particularly in relation to the problems of the developing world where the achievements of high yields may not always be an essential requirement of a new crop. In the case of soil salinity and the exploitation of such areas for cereal growing, the comparison is not between different levels of yield, but between the loss of yield altogether and a reasonable level of grain return. Similar choices are likely to be the case when other severe environmental constraints are considered. Even in the developed world, a niche for a new cereal crop may be there and awaiting exploitation. Thus the introduction of a modified *T. dicoccoides* into agriculture is only likely to be attractive to industry if a new product or marked increase in the quantity of a product were to be made available.

The genetical variation for achieving many new objectives is unquestionably present today amongst the cultivated cereals and their relatives. Genetic engineering will further extend this variation. The creation of a new cereal crop is thus more a problem of defining the correct objectives, and this in turn is dependent upon understanding the limitations imposed by environments and the agricultural industry itself. Above all, though, it is the imagination and ingenuity of the breeder that will be the decisive element in producing any new cereal crop in the future.

REFERENCES

Avivi, L. (1979) High grain protein content in wild tetraploid wheat *Triticum dicoccoides* Korn. *Proc. 5th Int. Wheat Gen. Symp.* publ. Indian Soc. Gen. Plant Breeding, IARI, New Delhi, 372–381.

Bell, G. D. H. (1950) Investigations in the Triticinae. I. Colchicine techniques for chromosome doubling in interspecific and intergeneric hybridisation. *J. Agric. Sci.* 40: 9–18

Burr, B., S. V. Evola, F. A. Burr and J. S. Beckmann (1983) The application of restriction fragment length polymorphism to plant breeding. In: *Genetic Engineering Principles and Methods*, Vol. 4, J. K. Setlow and A. Hollaender (Eds). Plenum Press, New York.

Forster, B. P., J. Gorham and T. E. Miller (1987) Salt tolerance of an amphiploid between *Triticum aestivum* and *Agropyron junceum. J. Pl. Breeding* 98: 1–8

Gale, M. D. and T. E. Miller (1987) The introduction of alien genetic variation into wheat. In: *Wheat Breeding – its scientific basis*, F. G. H. Lupton (Ed), pp. 173–210. Chapman and Hall, London and New York.

24

Crop Plants: Potential for Food and Industry

N. Haq

INTRODUCTION

Man's need plays a central role in sustaining agricultural productivity. The accelerating population growth and the ecological hazards in many developing countries make it necessary that scientists maintain a constant search for improved crop varieties and endeavour to develop locally grown under-utilized crop plants, which are already used as a subsistence crop by farmers. This would raise farmers' income and make more effective use of marginal land. A vast genetic wealth of these crop plants remains to be exploited. However, the adaptation, cultivation and future of 'new crops' depend, in my opinion, on three factors: crop development, socio-economic conditions and acceptability of the crop, production and the market development. I am endeavouring here to highlight two groups of crops, which I believe have received little attention for research and commercial exploitation although their potential as food and as rural income earner only needs to be emphasized. The crops discussed here are small millets belonging to the cereal group and dhaincha (*Sesbania bispinosa*) belonging to the legume group.

SMALL MILLETS

The small millets, *Eleusine coracana, Setaria italica* and *Panicum miliaceum* constitute a major source of energy and protein for millions of people in Africa and Asia. Table 24.1 shows the world production and yield per hectare of millets. They are well adapted in adverse agro-ecological conditions and consequently they form the most important food crops of the arid and semi-arid tropics. The millets are grown mainly for feed grain in the western hemisphere but these crops play an important role in the economy of many developing countries as they can be used for food, fodder, feed, brewing and for cottage

Table 24.1: *Production of Millets*

	1979–81			1985		
	World	Africa	Asia	World	Africa	Asia
Area (1000 ha)	42221	15914	23266	42621	16968	22695
Yield (kg/ha)	664	629	692	740	685	773
Production (Mt)	28037	10002	16093	31559	11615	17541

industry. This paper reports the characters and prospects of the first two millets but concentrates on the potential of common millet (*Panicum miliaceum*) as little has been done on improving this crop.

Finger millet (*Eleusine coracana*)

The finger millet, also known as African millet, koracan, ragi etc. is an annual, robust cereal with many tillers and branches (Table 24.2). It grows up to 1·2 m and has narrow grass-like leaves. The inflorescence has about 6 spikes, each ca. 10 cm long and arranged digitally, thereby giving the crop its name. The crop has been grown predominantly in Africa and in Asia but Uganda, Zambia and India are the main users. The botany and cultural practices have been described by Acland (1971) and Purseglove (1972). Finger millet is a tetraploid (2n = 4x = 36) and is closely related to a wild species *E. indica* (diploid) which is found throughout the world. It is the African subspecies *E. indica* subsp. *africana* (syn. *E. africana*) which is believed to be the wild progenitor of *E. coracana* (Clayton *et al.* 1974; Cobley & Steele 1976) and Purseglove (1972) considered that Central Africa could be the centre of origin.

Table 24.2: *Distribution and Plant Characteristics of Small Millets*

Species	Common Name	Distribution	Life Form	Plant Type (in m)
Eleusine coracana	Finger millet	Asia, Africa (both high and low land areas), Europe	Robust annual	0·7–1·3
Panicum miliaceum	Proso millet	Central & eastern Asia (Japan, Russia) Middle East, eastern Europe.	Erect annual	0·3–1·1
Setaria italica	Foxtail millet	Asia (China, Japan, Mongolia, India), North Africa, eastern Europe	Erect annual	1·0–1·8

The crop can be grown in a wide range of soils but free-draining sandy loams are preferred. The crop grows profusely in areas with an average rainfall of 900–1250 mm. However, types with a degree of drought resistance and varieties with a wide range of maturity periods exist in the germplasm resources (Tables 24.3 & 4).

The seed yield varies considerably from area to area (Table 24.4), perhaps because of difference in cost input, crop management and suitability of the genotypes. Table 24.5 shows the chemical composition of the seed. Purseglove (1972) has reported a wide variation in protein content with a good content of cystine, tyrosine, tryptophan and methionine.

Table 24.3: Cultivation Conditions

Species	Soil	Temp (°C)	Daylength	Rainfall (mm)	Altitude (m)
E. coracana	variety of soils but reasonably fertile, free-draining sandy loam	18–30	short	800–1250	up to 2400
P. miliaceum	poor soils	hot	variable	little	up to 2100
S. italica	wide range of soils, light sand to heavy clays		variable	750–1200 susceptible to long period drought, cannot tolerate waterlogging	up to 2000

Table 24.4: Some Quantitative Characters

Species	Crop Dum. (days)	Seeding Rate (kg/ha)	Grain Yield (kg/ha)	1^{ary} Prodn. Area
E. coracana	70–80	India 20 Uganda 5–10	India: 600–800 Uganda: 1800 (average), 5000 (irrigated)	India, Uganda, Zambia
P. miliaceum	60–90	8–11	India: 450–650, 1000–2000 (irrigated)	India
S. italica	70–120	8–10	800–900	Japan, China, India, SE Europe

Table 24.5: Percentage Chemical Composition of Small Millets

Species	Protein*	Fat	Carbohydrate	Fibre	Ash
E. coracana	14·0	7·7	76·1	7·8	5·0
P. miliaceum	18·0	1·1	68·9	2·2	3·4
S. italica	9·7	3·5	72·4	1·0	1·5

* Highest value considered

Relatively few diseases and pests affect finger millet but most serious is the blast (*Piricularia* spp.) which damages the leaves, and causes them to dry out prematurely. The caterpillars and grasshoppers also cause damage to the crop. The grain is almost immune from insect attack and can be stored for long periods without the use of insecticides.

Table 24.6: Utilization of Small Millets

Plant Product	E. coracana	P. miliaceum	S. italica
Grain	Flour, porridge, beer, malting	Flour, porridge, bread by mixing with wheat	Flour, porridge, pudding, beer
Straw	Fodder, fuel, cottage industries	Forage, fodder, fuel, cottage industries	Hay and silage, cottage industries

The utilization of finger millets has been illustrated in Table 24.6. The small grains, only 1–2 mm in diameter, are usually reddish-brown or white in colour, were investigated for their use in food, feed and in industry and their results are promising.

Foxtail millet (*Setaria italica*)

The foxtail millet, also known as Italian or Siberian millet, is a native of north and eastern Asia, north Asia and eastern Europe. The crop is adapted to low rainfall and can be grown in a wide range of soils, from light sands to heavy clays. The plant is an annual, up to 1·9 m high and with a variable daylength threshold.

The growth, maturity and yield (Table 24.4) of foxtail millet varies depending on cultivars, climatic constraints and the time of sowing. Prasad *et al.* (1985) recorded genetic variability in time to maturity, grain yield, number of tillers. Similar variability in height of plant, length, breadth and compactness of inflorescence, number, colour and length of bristles and colour of grain has also been reported by Purseglove (1972). Cobley & Steele (1977) reported approximately 12 variable groups of cultivars on the basis of size and shape of panicle and colour of grain.

Foxtail millet contains about 10% protein and its chemical composition is reported in Table 24.5.

The crop species is remarkably free from many pests and diseases and this make it a potential source of resistance for other species in the genus.

Table 24.6 shows the utilization of foxtail millet. One of the important uses of this crop is as a catch crop when paddy fails. In the western hemisphere its use is restricted to bird feed, and as an important fodder, particularly in the US.

Common millet (*Panicum miliaceum*)

Panicum miliaceum, commonly known as proso millet, hog millet or common millet is not known in the wild state and is now mainly cultivated in eastern and southern Asia. The origin of this crop has not yet been ascertained, although it is believed to be from India or the eastern Mediterranean. The crop is also grown in Africa.

Common millet displays perhaps the lowest water requirement of any cereal studied. It generally matures between 60–70 days and can be grown in poor soil and in hot and dry weather. The crop grows well in plateau conditions and at high altitudes.

The plant is an erect, mostly free-tillering annual, up to 1 m high. Variation has been observed in inflorescence size and shape, in stem and seed colour. The growth and maturity time of the crop depends on the cultivar and climatic conditions. At Southampton, under greenhouse conditions (11 hrs daylength,

25 °C day and 21 °C night temperature), maturation period varied from 45 to 60 days from a trial involving seven lines.

The yield varies from 450 kg to 2000 kg/ha, depending on the amount of irrigation, but little work has been done on improving the crop by breeding. Furthermore, there has been little effort in germplasm collection, despite this crop's immense potential in hot, dry land areas of Africa for its early maturing character.

The proso millet suffers from few diseases, such as downy mildew, ergot and smut.

The husked grain is considered nutritious and is eaten whole, boiled or cooked like rice. It is fermented as a beverage in Ethiopia. In the US it is particularly valued as hog food. The grain contains 10–18% protein (Table 24.5) with a biological value of 56%. Feeding trials with rats indicate that it has a lower supplementing value than the protein of wheat or rice. The grain also contains starch as the major carbohydrate with minor accounts of sugars and dextrin.

The crop can be grown as a catch crop when main cereal crops fail and also as an intercrop with other cereals and legumes.

DHAINCHA

Sesbania bispinosa (syn. *S. aculeata, S. cannabina*), commonly known as 'dhaincha' is a quick growing annual herb or shrub belonging to the family Fabacea. As a legume it fixes atmospheric nitrogen with its root nodules. It is reported to grow in most tropical and subtropical regions (Table 24.7). It grows well in loamy, clayey, black and sandy soils and is highly resistant to drought; it also withstands waterlogging and salinity (Table 24.8).

Table 24.7: Distribution and Plant Characteristics of Sesbania bispinosa

Common Name	Dhaincha, Jantar
Distribution	Indian subcontinent, SE Asia, China, tropical Africa, West Indies, S Italy, S Africa, Australia
Life form	Shrub or annual
Plant type	up to 4 m

The plant grows to 4 m during its 6 month growing season. The stem is usually 2–3 cm in diameter but can reach up to 5 cm, green and sparingly prickly. The branches and undersides of leaves are beset with small hooked prickles, hence the plant is also known as the prickly sesban.

The seed contains 27% protein and 33% gum, the stem contains 15·6%, the leaves 30% and the whole plant 22·9% protein (Table 24.9). The seeds are easily harvested and the yield may be up to 1·5 t/ha when plants are topped twice during the growing season to encourage branching for greater flower and seed

Table 24.8: Cultivation conditions for Sesbania bispinosa

Soil	Saline, alkaline wastelands, loam, clay, black and sandy soils. Can withstand waterlogging.
Temp °C	20–27 °C (warm temperate to tropical)
Daylength	short day
Rainfall	550–1100 mm
Altitude	up to 120 m
Some quantitative characters:	
Crop duration	150–180 days
Seeding rate	50–60 kg/ha
Grain yield	600–1500 kg/ha
1ary prodn. area	India, S.E. Asia, China, S. Africa

production. At present, the crop is mainly used as a green manure, fodder and as a rotation crop (Table 24.10). There are several reports on its multi-purpose uses and its possible commercial exploitation but, to my knowledge, there has been no report on systematic germplasm collection or on genetic and agronomic improvement of this crop species.

Table 24.9: Percentage Chemical Characteristics of Sesbania bispinosa

(% dry wt.)	Crude protein	Crude fibre	Ash	Calcium	Phosphorus
Stem	15·6	39·4	7·6	0·9	0·4
Leaf	30·3	18·5	10·7	1·2	0·3
Whole plant	22·9	28·9	9·1	1·1	0·4
Fodder	25·1	23·6	9·3	1·2	0·3

Table 24.10: Present Uses of Sesbania bispinosa

Use	Production
Green manure crop	15–22·5 t/ha of green manure can return to soil (adding 150 kg N/ha).
Fodder	16·2 kg/day fodder can increase the body weight of bullocks in the order of 8·5 kg/day.
Animal feed	Can yield 57·8% protein (d.w.) after extracting gum from seed.
Rotation crop	Plant nodulates vigorously and improves soil fertility. Mostly used as rotation crop with rice.
Firewood	The stems make useful firewood.

The plant has four commercially attractive characteristics (Table 24.11) and these have not yet been exploited. These are: gum, which could be used industrially as a food additive and for soil stabilization; meal which could be valuable for an animal feed and texturized vegetable protein (TVP); stems can be processed to yield a fibre suitable for papermaking; the species also fixes

Table 24.11: Comparative Study Between Sesbania and Two Guar Gums

Materials	Apparent viscosity	Plastic viscosity	Yield point	Gel strength (lb/100 ft^2)
Guar gum (I)	50	15	71	19
Guar gum (II)	115	30	170	86
Sesbania gum	143	52	182	51

nitrogen in the soil and thus has potential for intercropping and for rotation and could be valuable for low input agriculture.

Gum

The sesbania gum can be used in similar ways to guar gum and these include: oil well drilling fluids, papermaking, textile sizing and printing paste, explosives, coal slurries and effluent flocculents, soil conditioners, food additives, cosmetics and pharmaceuticals. Here I will only discuss two uses of sesbania gum, as our experiments showed promising results for its exploitation. The seeds used for the experiments were obtained from Bangladesh and India.

USE IN DRILLING MUDS

Table 24.11 shows the results of a series of tests carried out to investigate the potential of sesbania gum as a drilling fluid, particularly for use in oil drilling mud compositions. The tests were carried out at room temperature (18 °C) for a concentration of 1·5 per cent by weight according to specifications of API RP.

The main significance of these results is that the viscosity characteristics of sesbania gum composition are better than those for guar gum. Particularly the gel strength of sesbania gum is significantly lower than guar gum, which makes it more attractive than guar gum as a viscosifying ingredient for drilling fluids. Furthermore, we have found that sesbania gum does not degrade at temperatures even up to 85 °C; guar gum degrades at 70 °C and is therefore limited to use in drilling relatively shallow wells. In contrast with sesbania gum, a bottom hole temperature of at least 85 °C will typically be reached at depths of 2700–3000 m, which would allow sesbania gum drilling mud to be used in the vast majority of wells drilled in the world.

SOIL STABILIZATION

An assessment of *Sesbania* grown for soil stabilization was carried out following the method of Hedrick (1954). An important improvement called for in the behaviour of certain soils is for them to be able to withstand heavy rainfall without excessive amounts losing structure and washing into streams and gullies.

In these experiment *Sesbania* was compared with commercial xanthan gum (Table 24.12) and it is clear from the results that the gum additions have a dramatic effect on diminishing the washing away of soil by water, and perhaps surprisingly sesbania gum seems to be more effective than xanthan.

Table 24.12: *Soil stabilization*

Treatment	Control	*Sesbania*	Xanthan
Dry weight treated (g)	40·0	40·0	40·0
Sieved gently 1 mm, dried (g)	7·4	28·5	23·0
Retained stabilized soil (%)	18·75	71·25	57·5

Sesbania meal

Table 24.13 relates analytical results of gum free meal. The protein, oil and fibre levels indicate that the meal could be a useful protein source in livestock feeds. The lysine and methionine levels are low compared to soya and winged bean but nevertheless the meal could still make a significant contribution to the overall protein content of a mixed feed. Meal has already been used as poultry feed in South Africa.

Table 24.13: *Chemical Composition of Sesbania Meal*

	Gum-free Meal (%)
Moisture	9·80
Oil	4·85
Protein (N × 6·25)	37·80
Fibre	10·90
Ash	4·25
Lysine	1·76
Methionine	0·25

Potential fibre for pulp and paper

When young, the plant stem resembles a reed stem, with a thin fibrous outer stem, filled with soft pith. However, as the plant matures, lignification of the stem takes place and the pith in the centre is replaced by woody tissue. The stem in the mature plant thus essentially consists of solid though light wood with a density of approximately 300 kg/m^3 bone-dry weight/green volume.

It appears from Table 24.14 that the length of the fibre is quite short compared to typical paper-making hardwoods. The strength properties of bleached pulp are comparable to birch and eucalyptus (Table 24.15) and the results indicate that sesbania fibre can be used for printing papers and corrugating medium. The high brightness levels in bleaching are encouraging. These features indicate that a low cost pulp could be produced.

Table 24.14: *Comparison of Physical and Morphological Characters Between the Fibres of Sesbania bispinosa, Betula pendula (syn. B. verrucosa), (Birch), Pinus silvestris and Bambusa arundinacea*

Characters	S. bispinosa	B. pendula	P. sylvestris	B. arundinacea
Basic density (g/cm³)	0·3	0·51	0·41	0·48
Fibre length (mm)	0·96	1·10	2·90	2·80
Fibre width (mm)	0·02	0·02	0·028	0·016
Fibre wall thickness (µm)	1·4	1·8	3·2	3·1
Fibre slenderness (length/width)	48	55	104	178
Wall fraction (linear %)	14	18	23	38
Wall fraction (area %)	26	34	40	62
Vessel fraction %	10	21	–	15

Table 24.15: *Percentage Chemical Composition of Sesbania bispinosa, Betula pendula, Pinus silvestris and Bambusa arundinacea*

	S. bispinosa	B. verrucosa	P. silvestris	B. arundinacea
Resin	2·6	2·7	2·0	3·5
Lignin	22·2	19·5	26·6	24·5
Acetyl	4·8	4·2	1·4	1·9
Uronic acid	5·0	3·5	2·1	2·8
Ashes	0·53	0·3	0·2	5·4
Silica	0·23	–	–	3·8
Carbohydrate composition				
Galactose	0·9	1·1	4·1	1·2
Glucose	69·8	62·8	66·0	68·0
Mannose	2·7	2·7	17·3	0·7
Arabinose	0·6	0·5	3·3	1·8
Xylose	26·0	32·9	9·4	28·3
Wood composition (extracted ash-free substances)				
Lignin	22	20	27	27
Cellulose	46	40	41	45
Glucomannan	4	4	17	1
Glucuronxylan acetate	27	34	11	24
Galactan, araban, etc.	1	1	3	2

Fibres are in short supply in developing countries and a local source like *Sesbania* fibres would be invaluable.

It was estimated in an experiment with *Sesbania bispinosa* (Hussain & Khan 1962) that one hectare of land in Pakistan can produce 1500 kg of *Sesbania* seed and this can yield 450 kg gum, 1000 kg meal, 15 tonne (bone-dry) of fibre and about 150 kg nitrogen fixed by nodules.

CONCLUSIONS

The crops discussed above merit further investigation and commercial exploitation. These crops have been grown in the tropics for centuries but play a limited role in agricultural systems in these countries. In their present form they are minor crops and tend to be lost in more sophisticated farming systems. Large- scale cultivation of these crops is restricted by the lack of improved genotypes and economic production technology back-up and market outlets. The genetic diversity present in these crops and their broad range of possible adaptation make them suitable for wide cultivation and indicate that their use could be extended beyond subsistence agriculture in the tropics, particularly in Africa. The short duration of diverse ecotypes of *Panicum* spp. may be extremely useful in Africa where natural hazards necessitate the use of these potential crops in the existing cropping systems as rotation crops, intercrops, and as catch-crops in various cultivations of food legumes, cereals and industrial crops, including plantation crops.

The growing demand for food, a variety of food products and industrial raw materials calls for interest and investment in developing these crops. Germplasm collections from the recognized ecotypes, their evaluation and confirmation of ecotype groupings need to be carried out. The development of variety/varieties for wide adaptation, good quality and including by-products production and multi-purpose usage, needs to be encouraged. The development of production technology for large-scale cultivation and their inclusion in existing farming systems will make these crops more versatile. The development of a variety of food products from small millets, e.g. health foods, high fibre food, flakes, weaning foods, etc., using appropriate milling machinery, also requires investigation. The development of suitable varieties with straw for forage and fodder and the use of straw for other purposes, such as various cottage industries, needs to be encouraged. Further development on processing of the various potential products and their utilization needs to be carried out.

Lastly, the multi-disciplinary approach to development of these crops will not only enable us to establish them more widely in agricultural systems, but will also help to provide subsistence food to the farmers and develop the rural economy.

Acknowledgements

I owe my thanks to Kins Plants Ltd., Epsom, Surrey for allowing me to use their data on *Sesbania* spp.

REFERENCES

Acland, J. D. (1971) *East African Crops*. Longman, London.

Clayton, W. D., S. M. Phillips and S. A. Ronvoize (1974) *Flora of Tropical East Africa. Gramineae (Part 2).* Crown Agents, London.

Cobley, L. S. and W. M. Steele (1976) *An Introduction to the Botany of Tropical Crops.* Longman, London

Hedrick, R. M. (1954) Laboratory Evaluation of Polyelectrolytes as Soil Conditioners. *J. Agric. Food. Chem.* 2: 182–185.

Hussain, A. and D. M. Khan (1962) Nutritive value and Galactomannan content of Jantar – *Sesbania aculeata* and *Sesbania aegyptica. Pakistan J. Agric. Res.* 1, 1: 31–35.

Prasad, D. K., A. Ahmed, A. R. Rahman and S. H. Khan (1985) Genetic variability and correlation study in Foxtail millet. *Bangladesh J. Agric. Res.* 10 1:59–64.

Purseglove, J. W. (1972) *Tropical Crops: Monocotyledons.* Longman, London.

25

Cocona (*Solanum sessiliflorum*) Production and Breeding Potentials of the Peach-tomato

J. Salick

INTRODUCTION

Of all domesticated species of plants, Solanaceae is one of the most common families, with such important crops as the potato, tomato, tobacco, eggplant and peppers.

Still there are many under-exploited species within the family that have great potential. Of these, cocona or the peach-tomato (*Solanum sessiliflorum*) seems most likely to add significantly to lowland tropical subsistence and possibly commercial production.

The tropical fruit has an unusual and appealing flavour, as well as favourable characteristics for nutrition, agronomy, processing, marketing and cooking. Cocona is exceptionally high in iron and contains vitamins A, C and niacin. These vitamins and minerals are particularly important for women and children in tropical environments. In Peru, cocona is valued for its medicinal properties, especially beneficial for the kidneys, liver, and skin. Agronomically, cocona is capable of producing on the acid, infertile soils that are so common in the tropics, and yet will produce profusely on better soils. It travels, stores, and processes extraordinarily well. The culinary versatility of cocona is indicated by its English name. The peach-tomato is a fruit and used in sweet recipes as a peach; it also has a tangy flavour and is used in savory recipes similar to a tomato. The failure to exploit cocona in the past may very well have a geo-political bias, with tropical crops receiving much less attention from temperate economies. Previous studies on cocona include taxonomy (Whalen *et al.* 1981), economic botany (Schultes 1958; Patiño 1962; Schultes & Romero-Casteñeda 1962; Heiser 1968, 1971, 1972), biology (Fernandez 1985), and agronomy (Rodriguez & Garayar 1969; Benza & Rodriguez 1977; Pahlen 1977). Cocona could add significantly to development, commerce, subsistence farming and backyard gardening. To these diverse ends an overview of basic cocona agronomy is presented here, summarizing the most recent studies (Salick 1986, 1988a&b) on genetics, production, fertilization, density and time-of-planting.

PRODUCTION AND VARIETAL DIFFERENCES

Cocona was collected from various regions of the upper Amazon in the Peruvian tropical lowlands. More than 25 varieties were grown on alluvial soils at the experimental station of the International Potato Centre (CIP), San Ramon, Peru (850 m altitude, 1929 mm rainfall). Seeds were treated in 2% gibberellic acid for 24 hours and planted in trays containing a mixture of sand, organic matter and fertilizer for two to four weeks. Seedlings were then transplanted into 'Jiffy Pots'. When these plants were at least 15 cm tall (six to eight weeks) they were planted in the field at a spacing of 1 m^2 and fertilized at the rate of 50 g/plant of 160N-160P-200K-60Mg or 500 kg/ha. Varieties were planted in pots of 36 plants (5 m × 5 m) with three replicates. For field production where exact replication is unnecessary, less labor-intensive planting schemes can be used as with tomatoes.

During the initial growth phase of two months the plants were irrigated during periods of drought. Harvesting began after six months of growth in the field and plants were removed from the field after eight months because of crop rotations. Production had dropped but not terminated completely, thus the figures presented are not maximum production, but do represent a convenient annual production cycle. Data were taken on growth, leaf area, and flowering, as well as fruit production and is presented here. Extensive notes on diseases and insects are available, but generally pests and pathogens are the same as those on other Solanaceous crops. Pest management for tomatoes and potatoes is directly applicable to cocona.

Cocona production ranged from a mean of 29 ± 15 tons/ha for a small fruited variety to a mean of 55 ± 22 tons/ha for a large fruited variety. Table 25.1 presents a representative range of cocona varying in fruit size and shape. On these relatively fertile alluvial soils large fruited varieties produce better. Use varies, however, so that the large fruits grown for pulp and small fruits for juice do not replace one another.

Table 25.1: *Summary Statistics on Seven Varieties of Cocona*
Production levels range from 29 to 55 tons per hectare which is considerably greater than tomato production in the tropics.

Fruit			Production Data			
Size	Shape	Origin	Varietal Number	Weight (g)	Number of fruits/plant	Total prodn. (tons/ha)
small	round	Iquitos	6188	24	119 ± 58	29 ± 15
small	round	Iquitos	6196	42	87 ± 36	37 ± 20
medium	round	Iquitos	7010	56	83 ± 37	46 ± 24
medium	long	Palcazu	7021	37	95 ± 28	35 ± 12
large	flat	Yurimaguas	6174	194	22 ± 10	43 ± 22
large	long	Pacalpa	61163	141	39 ± 14	55 ± 22
large	long	Palcazu	7011	215	24 ± 17	52 ± 28

Overall, the results show consistently high production levels, greater than tomatoes in the tropics (Villareal 1980). Even without cocona's other advantages, based on production figures alone, it would seem beneficial to promote the well-adapted peach-tomato in tropical areas where tomato production is difficult.

BREEDING FOR FRUIT SIZE AND SHAPE

The seven varieties reported in the production statistics were used to make crosses for a study of inheritance of fruit characteristics. Crosses were made by emasculating flowers before opening, and upon opening, pollinating these flowers manually with pollen extracted using a tuning fork. Other flowers were bagged before opening and still others emasculated and left unpollinated to assure that neither autogamy, pseudogamy nor natural pollination after emasculation were taking place. A complete diallel cross (Fig. 25.1), involving seven varieties, was carried out in the first year and in the second year 56 varieties were planted in a randomized block design with three replicates.

Figure 25.1: Diallel cross.

The surprising outcome is the predominance of maternal inheritance for fruit characteristics. The large females gave rise to large offspring regardless of the male with which it was crossed. Vegetative characteristics varied negligibly

Table 25.2: *Cocona Production at Four Planting Densities*
There are significant differences in production (tons/ha) with planting density of both a) the large fruited variety, and b) the small fruited variety. Production is greatest at highest densities.

		Planting densities			
		0·5 m²	1·0 m²	1·5 m²	2·0 m²
Large variety					
	I	72	47	30	14
Repetitions:	II	100	30	30	15
	III	81	14	18	9
Small variety					
	I	44	29	14	9
Repetitions:	II	60	26	20	11
	III	62	22	18	7

among varieties. The male line did have a limited effect in that there was some quantitative component to fruit character inheritance. The crosses with large females and large males gave rise to some very large offspring with high production which were immediately taken by the farmers with whom I worked.

The maternal inheritance continued through the next generation of selfed lines with no indication of segregation. At this stage it cannot be determined conclusively whether the maternal inheritance is due to cytoplasmic inheritance or to some more complex mechanism (e.g. pseudogamy triggered by pollen). Maternal inheritance can be advantageous in breeding, especially by small farmers or indigenous people because it reduces the need for selection among a segregating population, backcrossing, and multiplication. There is still potential for improvement of cocona production by selecting among preferred varieties and breeding the quantitative portion of inheritance.

Density Trials

Presently, cocona production in Peru is limited to gardens or fields with few plants at very low densities. Even the small commercial farmers plant cocona at maximum densities of 2 m × 1 m, while the mature cocona plant covers about 1 m². If commercial production were attempted, how could cocona be planted? Research on tomatoes shows that high densities increase production.

Cocona planting densities at 2·0 m², 1·5 m², 1·0 m² and 0·5 m² were compared with both a large and a small fruited variety. The results (Table 25.2) were highly significant. At a planting density of 1 m², results from the production experiment were confirmed with the large fruited variety yielding 47, 33, and 30 tons/ha and the small fruited variety 29, 26 and 22 tons/ha in the three replicates. The superior production of the large fruited variety was confirmed at all densities. The highest planting density (0·5 m² or 20 000 plants/ ha) yielded the most, up to 100 tons/ha under the optimal conditions of alluvial soils, irrigation, fertilization and intense management. For cocona, the problem

260

with high planting densities is moving between plants during manual cultivation and harvest. Mechanization might facilitate high density planting. Cocona has a resistant peel and bruises little, which would lend it to mechanization. It would need to be bred for determinant growth.

FERTILIZER TRIALS

Fertilizer levels in the previous experiments were standard, 50 g of 160N-160P-100K-60Mg per plant, but production varies with varying inputs. To maximize differences in fertility levels this work was done on the acid, infertile soils with high aluminium contents on the weathered terraces of the upper Amazon in Iscozacin, Palcazu Valley, Peru. Five levels of fertilizers were tested; fertilizer levels tested were none, 300 kg/ha and 600 kg/ha with and without calcium. Fertilizer used was 1 urea: 3 superphosphate : 2 KCl, or 23 kg N, 69 kg P_2O_5, 60 kg K_2O in 300 kg of fertilizer.

There are two general trends with cocona production and soil fertility. Plant mortality decreases and fruit production increases with increasing soil fertility (Fig. 25.2). Comparing fruit production between large and small fruited varieties indicates that the small fruited variety does not respond to fertilizer whereas the large fruited variety does. These results may explain why some small farmers on poor lands with minimum inputs prefer small fruited varieties. Production of juice (small fruited varieties) and pulp (large fruited varieties) should vary with soil fertility and inputs. With comparatively slow production on poor tropical soils it is also important that systems of production be appropriate. Studies on underplanting plantains and cassava with cocona (Salick 1988b) provide one such option.

Time of Planting

Large and small varieties of cocona were planted three times during the year (Fig. 25.3): at the beginning of the rainy season, late in the rainy season, and during the dry season. Highest production and lowest mortality resulted from planting at the beginning of the rainy season. Seedlings wilt and die quickly with drought, whereas mature plants withstand some water stress when sunshine is important for fruit maturation. Plants and fruits maturing during the rainy season are susceptible to disease, and fruit sunscorch was found in the dry season. Prolonged drought at any stage will cause death. Cocona planted at the beginning of the rainy season and harvested during the dry season in these time-of-planting trials produced more fruit than any of the previous experiments at equivalent densities.

Figure 25.2: Trends in Cocona production with soil fertility.

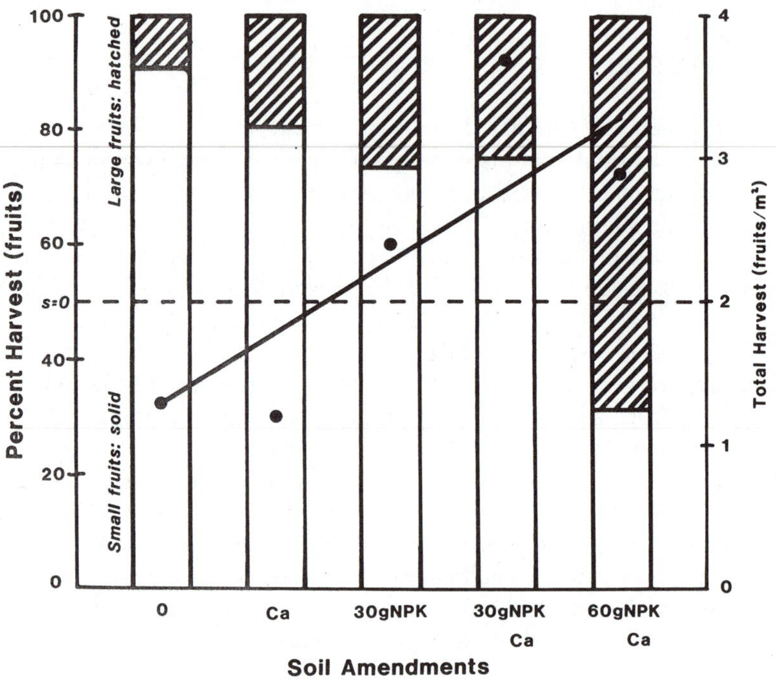

CONCLUSIONS

On reviewing the data, cocona production potentials in the humid tropics are outstanding. Large fruited varieties are available for pulp production as are small, juice varieties. Inheritance of fruit shape and size is largely maternal, facilitating propagation and multiplication, while the quantitative component allows breeding advances. Increasing planting density and inputs increases production up to 100 t/ha under optimal conditions.

Cocona is a versatile fruit appropriate for subsistence farming or industrial processing depending on the variety, land capability, management and capital inputs. Based on nutritional analyses and culinary experiments the fruit should be promoted for health and flavor. The potential for cocona is both under-exploited and under-explored. There is a great need for further work spanning many disciplines from plant breeding to agroindustry, from nutrition to food technology. To initiate programs and sustain progress beyond these preliminary investigations, it is important to give support and provide guidelines

Figure 25.3: *Planting times.*

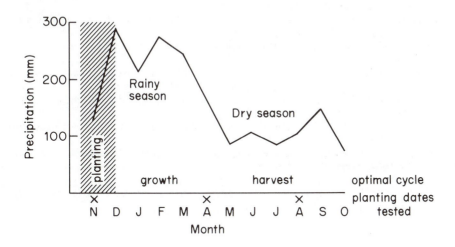

for research and development of *Solanum sessiliflorum*, the tropical peach-tomato.

ACKNOWLEDGEMENTS

For generous financial and logistical support, I thank the New York Botanical Garden and the Institute of Economic Botany with grants from the Mellon Foundation, the Peruvian Mission of the United States Aid to International Development, the International Potato Centre (CIP), and the Proyecto Especial Pichis-Palcazu of Peru. Personally, I would like to express my gratitude to the late Dr. Michael Whalen for his inspiration and encouragement of my work on cocona.

REFERENCES

Benza, J. C. and J. B. Rodriguez (1977) *El cultivo de la cocona*. Universidad Nacional Agraria, La Molina, Peru.

Fernandez, E. (1985) Biologia floral de *Solanum sessiliflorum* e *Solanum subinerme* (Solanaceae), na região de Manaus. Am. M. S. thesis, Universidade do Amazonas, Manaus, Amazonas.

Heiser, C. B., Jr. (1968) Some Ecuadorian and Colombian Solanums with edible fruits. *Ciencia y Naturaleza* 11: 1–9.

Heiser, C. B., Jr. (1971) Notes on some species of *Solanum* (Sect. *Leptostemonum*) in Latin America. *Baileya* 18: 59–65.

Heiser, C. B., Jr. (1972) The relationships of the naranjilla, *Solanum quitoense*. *Biotropica* 4: 77–84.

Pahlen, A. von der (1977) Cubiu *(Solanum topiro* (Humb. & Bonpl.)), uma fruiteira da Amazonia. *Acta Amazonica* 7: 301–307.

Patiño, V. M. (1962) Edible fruits of *Solanum* in South America historic and geographic references. *Bot. Mus. Leafl.* 19: 215–234.

Rodriguez F., R. and H. Garayar M. (1969) *Cultivo de cocona, maracuya, y naranjilla.* Ministerio de Agricultura, Lima, Peru.

Salick, J. (1986) *Cocona (Solanum sessiliflorum). Ethnobotany of the Amuesha.* USAID/ Peru, Lima.

Salick, J. (1988a) Population genetics of domestication, the case of cocona *(Solanum sessiliflorum)* along the upper Amazon. *Evol. Biol.* 22: (in press).

Salick, J. (1988b) Economic botany of cocona *(Solanum sessiliflorum). Adv. Econ. Bot. (in prep.)*

Schultes, R. E. (1958) A little-known cultivated plant from northern South America. *Bot. Mus. Leafl.* 18: 229–244.

Schultes, R. E. and R. Romero-Casteñeda (1962) Edible fruits of *Solanum* in Colombia. *Bot. Mus. Leafl.* 19: 235–286.

Villareal, R. L. (1980) *Tomatoes in the Tropics.* Westview Press.

Whalen, M. D., D. E. Costich and C. B. Hieser (1981) Taxonomy of *Solanum* Section Lasiocarpa. *Gentes Herb.* 12: 4–129.

26

Review of the African Plum Tree (*Dacroydes edulis*)

J. N. Ngatchou and J. Kengu

INTRODUCTION

For nearly fifty years, most cultivated plants in Cameroon, West Africa were grown either for food or for commercial and industrial purposes. Much research work has been done on these crops; interesting results have been obtained and better farming techniques have been adopted. Efficient chemical formulations have been recommended for the protection of plants against pests and diseases. Important improvements were obtained at the level of transformation and conservation of the harvest.

Despite all the efforts that were made to increase crop production, it is noticeable that a considerable number of potential crops having a certain food and commercial value in the traditional African background, have not yet been seriously investigated. The principal reason being that research was, until very recently, programmed and executed by foreign organizations, whose researchers have shown a certain indifference to traditional African cultures.

In order to fill this gap, the programme committee of the Institute of Agronomic Research (IRA) decided to include the study of *Dacroydes edulis* in their 'Various fruits' and 'Plant genetic resources' projects.

Although it is difficult to find statistics concerning national production and commercialization of *D. edulis*, there is good reason to believe, seeing the place it occupies in the market, that its cultivation will reasonably increase revenue to agriculturalists.

Besides, the oil from the *D. edulis* contains the following principal fatty acids: palmitic acid 36·5%, stearic acid 55·5%, oleic acid 33·9% and linoleic acid 24·0% (Ucciani & Busson 1963). The production of oil for food or for cosmetic industries could reach 7–8 t/ha compared with 3 t/ha of palm oil (Giacomo 1982).

To this economic consideration must be added agrobiological considerations. Whilst oil palm (*Elaeis guineensis*) and coconut (*Cocos nucifera*) grow in hot and humid areas of the littoral, South and South-West Provinces, *D. edulis* grows very well in nearly all the ecological zones of the country except the North and extreme North provinces, which have a very low rainfall. This is also the case with the groundnut, (*Arachis hypogea*), while other oil-bearing plants are often confined to narrow ecological areas.

SYSTEMATIC POSITION

Dacryodes edulis is a member of the large family Burseraceae. According to Chevalier (1916), this family is represented by four genera in equatorial Africa:

Aucoumea, with only one species, *A. klaineana*, the famous Okoume (Gabon, Zaire, Cameroon).

Canarium with 85 species spread throughout all tropical regions of oceanic and eastern Asia, Africa and Madagascar, and is represented in Cameroon by *C. schweinfurthii.*

Santiria has 20 species, essentially in Malaya, with at least one species in central Africa, well known in Cameroon as *S. trimera* (Ebap).

Dacryodes has 11 species in West and Central Africa. The genus according to the revision by Lam (1932) has a total of 34 species grouped in three sections; section Archidacryodes in tropical America with two species (Antilles and Peru), section *Pachylobus* in tropical Africa with 19 species and section *Curtisina* in Indo-Malaya with 13 species.

Gabon appears to be the centre of distribution for the genus because among the 19 species that occur in Africa, 11 species are found there (Aubreville 1962).

The family of Burseraceae has not yet been studied for the Flore du Cameroon but the following species are currently recognized: *D. buettneri, D. edulis, D. igaganga, D. klaineana* and *D. macrophylla*, all of which produce edible fruits.

The following synonyms have been recorded for *Dacryodes edulis: Pachylobus edulis, Canarium edule, C. mubafo, C. saphu, Pachylobus saphu, P. edulis* var. *mubafo* and *Soreindeia deliciosa*.

Okafor (1983) investigated the varietal deliminations within *D. edulis*. His study, based on morphological features, leads to the following conclusion:

> The species involves two taxa, which in the author's opinion should be regarded as varieties, since they differ from each other in having distinctive but imperfect, to some extent, overlapping combination of characters which are not clearly or exclusively correlated with geographical location. The distinction between the two varieties of *D. edulis* are briefly as follows
>
> var. *edulis* – fruit large, usually more than   5 cm long by 2·5 cm wide.
>
> var. *parvicarpa* – fruit small, more or less conical; usually less than 5 cm long by 2·5 cm wide.

But the reality seems to be more complex, seeing the great diversity of shapes and sizes that characterizes the fruits and leaves of the African plum or bush butter trees cultivated in Cameroon.

ECOLOGY AND ORIGIN

The African plum is an oil-bearing fruit tree in the humid intertropical regions originating from southern Nigeria, Congo and perhaps Cameroon. It is no doubt one of the rare cultivated species whose origin is truly African and cultivated by the natives of Central Africa, Gulf of Guinea and the interior basin of Congo. Its cultivation must have stretched from Cameroon, Zaire, Gabon across to Uganda and to central Angola (Aubreville 1962).

It is an extremely plastic plant, tolerating a wide range of day length, temperature, rainfall and edaphic factors; growing equally well at both high and low altitudes.

It occurs throughout Cameroon, except for the North and extreme North provinces, and presents phenological variations in keeping with local parameters of climate and soil.

DESCRIPTION

A dioecious evergreen tree up to 8–12 m high when grown under cultivation in the open, up to 45 m in the forest (Chevalier 1916). Trunk cylindrical and straight according to Philippe (1957) and Bourdeaut (1971), but this is considered by the present authors to be an over-simplification; up to 1·5 m in diameter. Leaves compound, imparipinnate with 4–12 pairs of leaflets. Inflorescence axillary or terminal, forming pyramidal panicles; male and hermaphrodite inflorescences 8–25 cm long, female inflorescences 5–15 cm long. Fruit an ellipsoid to globular or conical drupe, 4–12 (-15) cm long, 3–6 cm in diameter, with edible pulp. For more detailed descriptions see Troupin (1950 & 1958) and Okafor (1983).

FLOWERING AND FRUITING

Flowering

Flowering commences in early January in the Littoral and Centre provinces. For an individual tree it will last for about a month but since certain varieties are early and others late, flowering is spread over a three month period. February 15th is the optimum flowering date; by April 15th no more flowers can be found except in exceptionally late varieties.

The terminal bud of an inflorescence forms the flowering shoot for the following year. Should all the flowers fail to produce fruit, as is often the case with male plants, the terminal bud will, a month after the flowers have fallen, begin to produce a new flowering shoot.

Pollination and fertilization

The flowers are not showy but have a very strong perfume, an olfactory signal which attracts pollinating insects. The strongly agglutinate pollen is also indicative of entomophilous pollination. According to Giacomo (1982) ca. 82 per cent of the pollinating insects are bees (*Apis mellifera*).

The ovary is biloculary, each with two ovules. After fertilization of the four ovules, three fail to develop leaving only one to develop normally and produce seed (Aubreville 1962).

Fruits and Seeds

All the trees, irrespective of whether male, female or hermaphrodite, produce fruit. What varies is the degree of fructification. The female inflorescences are naturally very productive. The hermaphrodite inflorescences are average while the male have poor and very haphazard production. The dissection of many flowers from the same tree have not enabled us to establish any morphological differences. Furthermore we can give no reason for the fertility of certain atrophied ovaries in male flowers.

The drupes are rose coloured when young, changing to deep blue at maturity. The pericarp, which represents half the weight of the whole fruit, consists of a very thin, waxy and coloured epicarp and a pulpy, edible mesocarp ca. 0·5 cm thick, light rose, light green or whitish, and varied flavour. The endocarp is thin, smooth and plated externally with a barely visible escutcheon. The seed is light green, composed of two fleshy, five-lobed cotyledons.

The fruit has a very high nutritive value. The fresh pulp contains 33–65% oil depending on the origin of the plant and the degree of maturity. Besides fatty acids the fruit also contains certain indispensable amino acids, of which the more important are leucine, valine, isoleucine, tyrosine, arginine, cystine, threosine and lysine (Busson 1965).

Usually there is only one seed; the shape varies in the fruit, but with the ventral surface always slightly flattened. At maturity a radicle, ca. 2 cm long, is already developed. The seed sometimes germinates *in situ* while the fruit is still hanging on the tree.

The rate of germination based on 200 seeds collected from the Yaounde area and germinated at Nkolbisson in propagation pits using a substrate of more or less decomposed sawdust was compared with a similar collection from South-west province germinated at Barombi-Kang in polyethylene bags containing black soil. The seeds from the former site first germinated after 14 days, and after 11 days at Barombi-Kang. The results are shown in Fig. 26.1.

Fertility after storage was examined for batches of seed stored with and without pulp at air temperature (ca. 25 °C). Twenty seeds from each batch were sown and their percentage germination calculated. It was found that seeds

Figure 26.1: *Graph of Germination*

Seeds from: 1. South-west Province. 2. Centre Province.

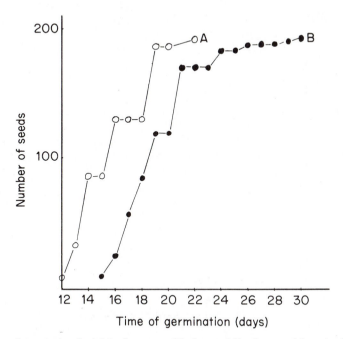

without pulp remained viable for up to 21 days while those with pulp for only 7 days (Fig. 26.2).

VEGETATIVE PROPAGATION

At the end of the trials of vegetative propagation (grafting on trunk, grafting following the Fokert modified method, slit grafting, propagation by slips, layering in tuft of shoot from pollarded stump, aerial layering) Philippe (1957) came to the following conclusion: '*Dacryodes edulis*, because of the weak vitality of its vegetative apparatus, and with its inability to produce adventitious roots easily, is a rebel fruit-tree with classic asexual reproduction.'

Only aerial layering produced some results. Our own propagation trials using slips and different quantities of hormones (Exuberone and Rootone) and fungicides (Captan, Tirane, Orthodifolatan and Ridomil), were a failure. But they need to be repeated. The formation of cicatricial callus as well as the production although slow and infrequent, of some roots, observed on some propagates, are encouraging signs.

Figure 26.2: *Viability of Seeds.*

A. Seeds with Pulp. B. Seeds without Pulp.

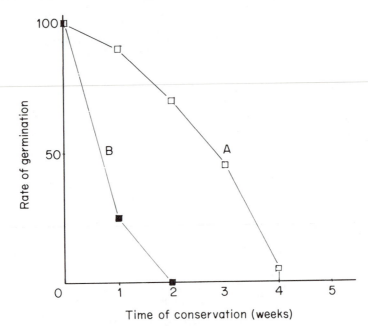

UTILIZATION

The most widespread usage is the consummation of fruits boiled in salted water, roasted in hot ashes or fried. The medicinal utilization of bush-butter-tree by all ethnic Congolese is widespread. Tribal divisions affect differences in the prescriptions which vary not only in function between different tribes but often within the group, from one individual to another. The juice of the leaves is used in the ear against otitis; a decoction of the leaves is used for fever, stiffness and headache. A decoction of the bark is used for various oral complaints (Bouquet 1969).

According to Chevalier (1916), *Dacryodes edulis* can reach a height of 45 m and a diameter of 1·5 m when it grows in the forest. The timber of *Dacryodes* could be substituted for 'acajou' (a mahogany). This great height plus interesting wood characters makes the bush-butter-tree suitable for veneers and for cabinet work.

CONCLUSIONS

The present review takes note of previous literature and preliminary observations both in the laboratory and the field. It raises certain problems and suggests areas for subsequent research. Nevertheless, without claiming to have completely investigated the floral biology and the mechanism of fertilization, we now know that there exist three types of flowers distributed on three types of plants: male, female and hermaphrodite plants.

D. edulis is a species for which no improvement work has been realized until now. Eventually vegetative multiplication should enable us to speed up the improvement of this plant and ultimately introduce clonal plantations of highly productive homogeneous material.

REFERENCES

Aubreville, A. (1962) *Flore du Gabon. No. 3, Irvingi   aceae, Simaroubaceae, Burseraceae.* Museum National   d'Histoire Naturelle, Paris.

Bouquet, A. (1969) *Féticheurs et medecine tradionnelle du Congo (Brazzaville).* Memoire ORSTOM, Paris.

Bourdeaut, J. (1971) Le safoutier (*Pachylobus edulis*). *Fruits* 26, 10: 663–665.

Busson, F. (1965) *Les plants alimentaires de l'ouest Africain: étude botanique, biologique et chimique.* Leconte, Marseilles.

Chevalier, A. (1916) *Les végétaux utiles de l'Afrique Tropicale Française – La forêt et les bois du Gabon.* A. Challamel, Paris.

Giacomo, R. (1982) *Etude de la biologie florale du safoutier (Dacryodes edulis) au Gabon.* (Unpublished).

Lam, H. J. (1932) Burseraceae of the Malay Archipelago and Peninsula. *Bull. Jard. Bot. Buitenz.* 111, 12 3-4: 281–561.

Okafor, J. D. (1983) Varietal delimitation in *Dacryodes edulis* (G. Don) H. J. Lam. (Burseraceae). *Int. Tree Crops J.* 2, 3–4: 255–265.

Philippe, J. (1957) Essais de reproduction végétative de "NSAFOU" *Dacryodes (Pachylobus) edulis. Bull d'inform. INEAC* 4(5):320–327.

Troupin, G. (1950) Les Burseraceae due Congo Belge et due Ruanda-Urundi. *Bull. Soc. Roy. Bot. Belg.* 83: 111–126.

Troupin, G. (1958) Burseraceae. In: *Flora du Congo Belge et du Ruanda-Urundi.* 7: 132–146. Institut National pour l'Etude Agronomique due Congo Belge, Brussels.

Ucciani, E. and F. Busson (1963) Contribution à l'étude des corps gras de *Pachylobus edulis* Don (Burseracées). *Oléagineaux* 18, 4: 253–255.

27

Study on the Processing of *Balanites aegyptiaca* Fruits for Drug, Food and Feed

I. M. Abu-Al-Futuh

DISTRIBUTION AND DESCRIPTION

Balanites aegyptiaca, family Balanitaceae, is a tree which in the Sudan is known as heglig and its fruit as lalobe. It is widespread in the drier areas of tropical Africa, from Mauritania to Nigeria eastwards to Ethiopia, Somalia and East Africa; also in Israel and Arabia.

Balanites (Suliman & Jackson 1959) is a savanna tree which attains a height of more than 6 m; it has a spherical crown and tangled mass of long thorny branches. Leaves are subsessile or shortly petiolate, grey green in colour, orbicular, rhomboid or obovate in shape, 3·5(-6) × 2(-5) cm, apex acute or rarely obtuse. Spines are simple or very rarely bifid, up to 5 cm long alternate in the leaf axils. Flowers are yellow-green in colour up to 1 cm in diameter. The ripe drupe is yellow-brown in colour, up to 4 cm long and 2·5 cm in diameter (Fig. 27.1). The fruit has a thin, brittle epicarp, a fleshy mesocarp and a woody endocarp or shell (Fig. 27.2) containing the oilseed or kernel, known in the Sudan as damlouge (Fig. 27.3).

Other species of *Balanites* indigenous to Africa include *B. wilsoniana* with fruits 8–11 cm long and 5·8–6·5 cm in diameter; *B. rotundifolia* with fruits up to 3 cm long and 2·5 in diameter and *B. pedicellaris* with fruits 3 cm long and 2·2 cm in diameter (Hardman 1969).

OCCURRENCE IN SUDAN AND USES

In Sudan the major occurrence of *Balanites aegyptiaca* is in Darfur, Kordofan and Kassala Provinces. The fruit is considered as an article of diet among the Acholi and Nilotic tribes; it is sucked by school children as a confectionery. The oil is extracted from the kernel by boiling; the oil is used in cooking, the exhausted kernel is used as a meal. In folk medicine an aqueous decoction of the fruit is used as a purgative, vermifuge and in the treatment of stomach ailments. A very strong infusion is used as a fish poison. In some African countries, an alcoholic beverage and drinks are prepared from the fruit.

A preliminary reconnaissance survey in the Sudan has indicated that the total wild resource of *Balanites* fruits exceeds 400 000 tons/year (Brown 1979). The recommended area of potential for development is the area of Hill Catenas in

Processing of Balanites aegyptica

Southern Kordofan and Southern Darfur: El Fula – Abu Zabad – El Khwei – Sugae Elgamel – Tawesha – Gadad Rasel – El Fil – south of Nayala.

Figure 27.1: Lalobe, the Fruits of Balanites aegyptiaca.

Figure 27.2: The Shells of Balanites aegyptiaca.

Figure 27.3: Damlouge, the Kernels of Balanites aegyptiaca.

CHEMICAL COMPOSITION OF THE FRUIT

From the industrial point of view, the mesocarp and kernel are the most important parts of the fruit. The weight of the fruit ranges between 10–15 g, of which 5–9% is epicarp, 28–33% is mesocarp, 49–54% is endocarp and 8–12% is the oily kernel.

The composition of the kernel is shown in Table 27.1. There are also steroidal saponins which, on hydrolysis, yield up to 2% diosgenin/yamogenin.

Table 27.1: Composition of Balanites aegyptiaca Kernel, compared with Peanut.

Composition (% net wt.)	Balanites kernel			Peanut
Moisture content	2·3	–	4·5	3·9
Crude protein	26·4	–	29·9	19·0
Fat	46·0	–	50·6	48·0
Crude fibre	2·4	–	3·3	2·8 – 3·0
Ash	2·8	–	3·3	2·5 – 3·0

The composition of the mesocarp is shown in Table 27.2; steroidal saponins are also present which, on hydrolysis, yield up to 4% diosgenin/yamogenin.

Some of the steroidal saponins from *Balanites* have been identified as Balanitin-1,2 and 3 (Hung-wen Liu & Koji Nakanishi 1981).

Table 27.2: *Composition of Balanites aegypiaca Mesocarp*

Constituent	% net wt.
Moisture	22·32
Ash (sulphated)	4·85
Protein (crude)	5·56
Fat (crude)	0·10
Carbohydrates (total)	65·51
Fibre	1·72
Vitamin C	0·02

PROCESSING OF THE FRUIT (Abu-Al-Futuh 1983)

The *Balanites* fruit was first softened in water, and the mesocarp removed as a slurry. The remaining shell enclosing the kernel, the 'nut', was dried and cracked (Fig. 27.4). The Tropical Development and Research Institute (TDRI) is now finalizing a special *Balanites* decorticating machine which includes a nut size grader, a feeder/elevator, a decorticator and a kernel separator/shell remover, with a working capacity of 250 kg/h nut input.

The mesocarp slurry was then fermented to produce ethanol and other products for industrial use. The ethanol was distilled out and the spent mash acid-hydrolysed, extracted and purified to produce steroidal sapogenins for the steroid industry. Alternatively the mesocarp slurry may be incorporated into animal feed (El-Khidir & Thomsen 1982).

The separated kernels were pressed in Rosedawn Maxoil Duplex Expellers. The whole kernels were then cooked and finally single pressed under high pressure. The crude oil was either refined to produce an edible oil or used for soap-making. The remaining *Balanites* seed cake was then either used for animal feed or for the production of steroidal saponins.

The crude oil produced was refined using the continuous caustic soda method. The physico-chemical properties of the oil were compared with the standard limits of some commercial edible oils (Table 27.3). The crude *Balanites* oil was used to prepare a laundry soap containing ca. 70% *Balanites* oil, the other constituents being tallow, caustic soda, silicate and other additives.

Alternatively, the kernels may be processed for human consumption; the whole kernel being leached in water at 60 °C to remove the bitter principle and then dried and roasted. The roasted kernels were salted using a 5% sodium chloride solution to produce a bright brown, edible nut with similar taste to peanuts.

The *Balanites* cake obtained after expressing the oil was experimentally processed for the production of steroidal saponins. The ground cake was per-colated with 90% ethanol. The ethanolic extract was then concentrated under vacuum and was further purified and spray-dried to yield a yellowish amorphous hygroscopic saponin powder.

Figure 27.4: *Flow Diagram for Processing the Fruit of Balanites aegyptiaca.*

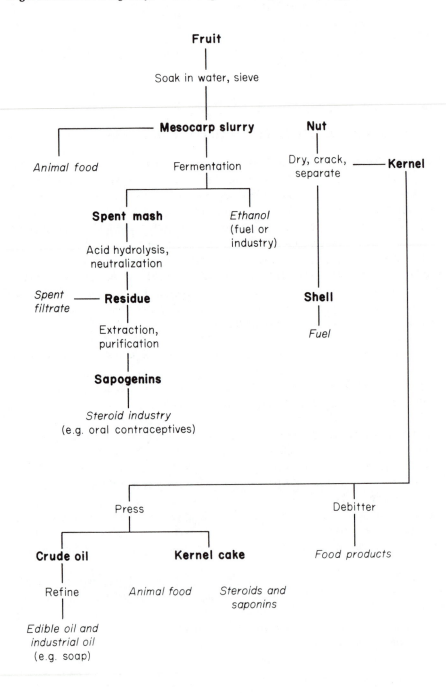

Table 27.3: *Recommended International Codex Standards for some Edible Oils as Compared with Analytical Data Obtained for Balanites Oil*

Characterization of oil	Balanites kernel	Soya bean	Sesame seed	Peanut	Cotton seed	Sun-flower
Specific gravity (25 °C)	0·918	0·917-0·921	0·914-0·919	0·910-0·915	0·916-0·918	0·91-0·915
Refractive index (25 °C)	1·471 1·471	1·470-1·476	1·470-1·474	1·467-1·470	1·468 1·472	1·47 1·472
Iodine number	98	120–141	104–120	80–106	99–119	110–143
Saponification no.	189	189–195	187–195	187–196	198–198	188–194
Unsaponified matter (wt %)	1·2	<1·5	<2·0	<1·0	<1·5	<1·5
Steroidal saponin	Negative	–	–	–	–	–
Free steroidal sapogenins	Negative	–	–	–	–	–

The main intermediate products obtained from *B. aegyptiaca* fruits were the following: the mesocarp, the kernel comprising crude oil and cake and the shell (Fig. 27.3).

UTILIZATION OF FRUIT PROCESSED PRODUCTS (Abu-Al-Futuh 1983)

Saponins

In view of their future utilization by the pharmaceutical industry, *Balanites* saponins were tested for some biological activities. These saponins were found to possess a pronounced cathartic effect and anti-fertility activity (Abu-Al-Futuh & Babikir, unpublished data 1987). They were also found to possess high molluscicidal activity (Hung-wen Liu & Nakanishi 1981). The effects of acute toxicity of these saponins on Nile fish were also examined (El-Rabaa & Abu-Al-Futuh, unpublished data, 1986). Recently, the saponins were applied topically as a cream and showed good response to the healing of lesions caused by cutaneous leishmaniasis (Abu-Al-Futuh, Khalifa & Ishag, unpublished data 1987).

Diosgenin

Diosgenin is a starting material for the production of steroidal drugs by partial synthesis. It is mainly supplied from species of *Dioscorea* (Hardman 1969). Although multinationals have partially shifted to plant sterols and microbial degradation for the synthesis of steroids, it is still expected that diosgenin from *Balanites* will be able to find a market at regional level within Africa and the Middle East.

Roasted *Balanites* Kernels

This product was found acceptable and tasteful to the consumer; it has a good aroma and high nutritive value.

Refined Oil

This has good frying properties, comparable with the oils of cotton seed and peanut. When market analysed, consumers rated the oil equal to cotton seed oil.

Balanites Kernel Cake

Feeding trials on ruminants indicated that 20 per cent *Balanites* kernel cake can replace 30 per cent cotton seed cake in their rations, and resulted in nearly the same or slightly higher performance. The palatability of the rations was acceptable to the experimental ruminants; no adverse effects were noticed. The *Balanites* cake diet was also found to be substantially cheaper. On the other hand, feeding *Balanites* cake has revealed some toxic effects.

Shell

This could certainly increase the profitability of the process by being used as fuel or for the production of charcoal and particle board.

ACKNOWLEDGEMENTS

I would like to thank UNIDO for initiating the *Balanites* project and the Federal Republic of Germany for financial support. Special consideration is given to Mr. Horst Koenig, Senior Industrial Development Officer, Agro-Industries Branch, UNIDO, for his persistent enthusiasm for this project. Thanks are extended to the Industrial Research and Consultancy Centre, Khartoum, where most of the pilot work was undertaken. I am grateful to the Technical Centre for Agricultural and Rural Co-operation, and the University of Khartoum for sponsoring my participation in the International Symposium on New Crops for Food and Industry.

REFERENCES

Abu-Al-Futuh, I. M. (1983) *Balanites aegyptiaca; an unutilized raw material potential ready for agro-industrial exploitation.* Report UNIDO/10.494, R-83–54325; Order No. PB 84–131085, UNIDO, Vienna.
El-Khidir, O. A. and K. Vestergaard Thomsen (1982) The effect of high levels of molasses in combinations. *Animal Feed Sci. Tech.* 7: 277–286.

Brown, G. D. (1979) *Balanites aegyptiaca*. Draft Final Report No. F/79/42, UNIDO, Vienna.

Hardman, R. (1969) Pharmaceutical products from plant steroids. *Trop. Sci.* 11(3): 196–228.

Hung-wen Liu and K. Nakanishi (1981) A micromethod for determining the branching points in oligosaccharides based on circular dichroism. *J. Amer. Chem. Soc.* 103: 7005–6.

Suliman, A. E. G. M. and J. K. Jackson (1959) The Heglig Tree (*Balanites aegyptiaca* (L.) Del.). *Sudan Silva* No. 9, Vol. I, Leaflet No. 6.

28

Variation of Fruit Production and Quality of Different Ecotypes of Chilean Algarrobo *Prosopis chilensis* (Mol.) Stunz

M. Pinto and E. Riveros

INTRODUCTION

The Chilean algarrobo tree (*Prosopis chilensis*, family Leguminosae is widely distributed in the dry regions of Chile, Peru, Argentina and Bolivia. In Chile, it is found between 18–34° S and 70–71° W.

Algarrobo fruits have been used by natives of these regions for centuries as a food source and also as animal fodder. Likewise, the leaves of this tree represent a very interesting alternative for forage during drought periods (Habit *et al.* 1981). At present, however, in Chile the algarrobo is used only as a firewood source, and rarely as forage. This fact, together with its elimination from places where underground water has been discovered and is being utilized for grapevine culture, is seriously endangering the existence of this species.

Unfortunately, the great variation shown by algarrobo both in fruit production and quality, does not allow for intensive cultivation nowadays. However, this variation can be a very important characteristic for its future genetic improvement.

The objective of this work is to present some results concerning the variation of both fruit production and quality in different ecotypes of the Chilean algarrobo.

Variation in Fruit Production

The self-incompatibility of Chilean algarrobo (Hunziker *et al.* 1975; Mooney *et al.* 1982) and its evolution in very different ecological conditions are perhaps the main causes for the great variation in fruit production, which is a characteristic of this species, and is undoubtedly genetic (Hunziker *et al.* 1975; Felker 1982). However, prior to the selection of highly productive algarrobo ecotypes, it seems important to consider other factors that may be related to fruit variation. Thus environmental and physiological factors could have a strong influence in growth (Meyer *et al.* 1973) and in production variations shown among different parts of the tree, between individual trees and for different seasons within the same individual (Mooney *et al.* 1982; Salvo 1986).

Production Variation among Ecotypes

Some examples of the production from 11 ecotypes during two seasons are presented in Table 28.1. In this case, the great variation in production during the same period is clearly observed. Thus, within the same population it is possible to find individuals with a high productivity while others may produce nothing at all.

Table 28.1: *Production of Different Algarrobo Ecotypes during 1984–85 and 1985–86 Seasons*

Tree No.	Location	Approximate age (years)	Production (kg/tree/yr)	
			1984–85	1985–86
Ch 1	Chacabuco	20	0·30	3·6
Ch 2	Chacabuco	20	1·1	5·6
Ch 3	Chacabuco	20	0·82	1·1
H 6	Hurtado	100–120	60·0	90·0
E 10	Elqui	50	10·0	0·0
E 2	Elqui	40	15·0	0·0
Hco 5	Huasco	30	–	130·0
Hco 6	Huasco	30	–	100·0
Co 16	Combarbalá	40	30·0	0·0
Co 17	Combarbalá	150	60·0	75·0
Co 9	Combarbalá	150	0·0	10·0

An important aspect that can be observed from these results is the great productive potential of some ecotypes. For example, in 1986 the ecotypes Hco5 and Hco6 produced 130 and 100 kg/fruits/tree respectively. This shows that good production could be obtained from improved and properly treated individuals. However, due to the alternate bearing habit of this species (Mooney *et al.* 1982), the production can differ completely from one season to the next. Thus, trees that have a large production in one season are likely to bear no fruits the following season.

Production Variation Between Each Period

With respect to the great variation in fruit production in different seasons, or alternate bearing, without disregarding other important factors (Monselise & Goldschmidt 1982), certain results reveal that the internal competition for reserves and other products would apparently be of importance in the regulation of fruit production in algarrobo (Wilson *et al.* 1974). According to research on the evolution of the development of algarrobo buds (Salvo 1986), both flower induction and differentiation takes place at the same time as full fruit development. This happens during December and January in central Chile. Hence, in periods of heavy fruit development, their growth will hinder good bud development and will result in decreased fruit production the following season.

Production Variation Within the Same Individual During a Season

According to the information presented above and because there is no data available for several periods, it is hard to quantify the influence of the genetic factor on production variations. However, environmental factors, such as temperature, water stress and radiation, can strongly influence productivity. Thus, in periods of water deficit during spring a large quantity of fruit is produced (Mooney *et al.* 1982).

Based on our studies in the Chacabuco zone (33° S, 71° W) radiation plays a significant role in flower and fruit production. This is particularly important in regions such as central Chile, where the sun rays fall more directly upon the branches on the north side of the tree compared with those on the south (Table 28.2).

Table 28.2: Average of Photosynthetically Active Radiation ($\mu E\ m^{-2}\ s^{-1}$) at Midday in Different Orientations in Three Algarrobo Trees during Flowering and Fruiting in the Chacabuco Zone (33° S, 71° W)

	North	South	East	West
Flowering	941·4 a	301·9 c	686·9 b	542·2 bc
Fruiting	1062·7 a	319·2 c	329·1 c	605·7 c

Same letters show there are no significant differences (P < 0·05). After Salvo 1986.

Table 28.3: Average Temperature of Algarrobo Buds in Different Orientations at Midday during the Flowering Period in the Chacabuco Zone (33° S, 71° W)

	North	South
Temperature of bud tissue (°C)	29·8	19·8 a
Environmental temperature (°C)	19·5 a	19·5 a

Same letters show there are no significant differences (P < 0·05). After Salvo 1986.

This directly influences the temperature of the tissues, which undoubtedly is higher in the north side (Table 28.3) and is probably the cause of the earlier bud burst on that side. Consequently, these buds achieve a better condition to compete for reserves and photoassimilates earlier than the less developed buds on the south side. Besides, at the beginning of spring and before bud burst, buds on the north side present better developed leaves and inflorescences than those on the south side, which often present aborted structures (Fig. 28.1). Summing up, in many Chilean algarrobo ecotypes, fruit production is greater on the north side of the tree (Table 28.4).

VARIATION OF FRUIT QUALITY

Preliminary studies have shown a very high variability in the nutritive value of algarrobo fruits. Figs. 28.2a&b present results of the nutritive value for ruminants of four different ecotypes estimated by *in vitro* gas production ($CO_2 + CH_4$) according to the method proposed by Menke *et al.* (1979). Trials made

Figure 28.1: *At the Beginning of Spring and before Bud Burst.*

Buds on the North Side (a) are much more developed than those on the South Side (b). The latter exhibit a greater number of aborted inflorescences.

 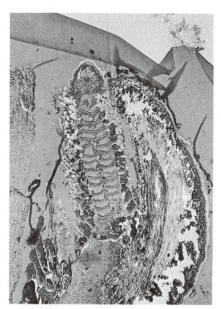

Table 28.4: *Number and Total Weight of Fruits According to the Orientation in Three Adult Algarrobo Trees in the Chacabuco Zone, during the 1986 Season*

	Tree 1		Tree 2		Tree 3	
	North	South	North	South	North	South
No. of fruits	1900	882	2400	800	608	282
Total weight of fruit (kg)	2·6	1·0	3·9	1·1	0·6	0·4

After Salvo 1986.

with ruminal juice of sheep and goats indicate the fodder quality of fruit ecotype MP8 is as good as that of alfalfa (*Medicago sativa*) whereas that of ecotypes Co16 and E3 was poor with a low gas production.

Very few studies on the fruit quality of the algarrobo have been made in Chile, and they have not considered the variation among ecotypes or seasonal variation. Besides most of the studies have been based on insufficient numbers of samples (CORFO 1985).

The conclusion that all Chilean algarrobo fruits have a high fodder quality should be carefully considered before planning a genetic improvement programme. In addition, more detailed studies are needed to characterize the real

Figure 28.2: *Nutritive Value of Fruits From Different Algarrobo Ecotypes.*

Estimated by *in vitro* gas production with ruminal juice of Goats (a) and Sheep (b).

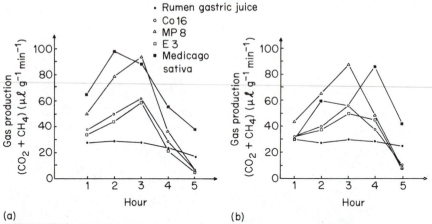

(a)   (b)

nutritive quality of fruits of diverse ecotypes growing under different ecological conditions.

Table 28.5 shows the nutritive value and principal characteristics of fruits for 70 algarrobo ecotypes collected during 1986 from ten geographical localities of northern Chile. Total N by micro-Kjehldahl, organic matter (OM) by ignition, dry matter (DM) dried by forced-air at 65 °C, neutral detergent fibre (NDF) after Van Soest (1967) and *in vitro* OM digestibility (Do) after Tilley & Terry (1963) were recorded.

Table 28.5: *Variation in Nutritive Value and Main Characteristics of Fruits of 70 Algarrobo Ecotypes*

	Average	Maximal	Minimal	Sx	Variation coefficient (VC)
Total N_2 (%)	1·8	2·8	0·9	0·36	20·4
Organic matter (%)	96·5	97·6	95·5	0·55	15·9
Dry matter (%)	88·1	93·6	79·2	2·42	2·8
Neutral detergent fibre (%)	24·1	33·6	15·4	3·78	15·7
In vitro digestibility (%)	66·6	75·8	44·4	6·0	8·9
Fruit size (cm)	13·0	21·3	7·8	2·84	–
Seeds/cm/fruit (No/cm)	1·6	2·4	1·0	0·34	–

From Table 28.5 it is concluded that, on average, algarrobo fruits have an adequate nutritive quality as fodder. The cell wall content (NDF) can be considered low; the Do (66·6%) represents a good fodder, and N content, although lower than that for high quality forage such as alfalfa, is not negligible, especially when considering the arid conditions in which algarrobo thrives.

Since almost all of these variables show a very high variation coefficient (VC), which indicates biases, these results cannot be extrapolated to other ecotypes or *Prosopis* species without incurring misleading results. Thus, in OM, N content and NDF, the VC values are 15·9, 20·4 and 15·7 respectively. These high VC values, however, could be of interest in selecting ecotypes of desirable characteristics.

The VC for *in vitro* DO was not as high as in the above variables, but there were marked differences among fruits of different ecotypes. The highest Do value was 75·8 per cent, similar to that previously reported (CORFO 1985), which indicates that, in general, algarrobo is a better forage source than other Prosopis trees such as *P. tamarugo, P. velutina* and *P. pubescens* (Becker & Grosjean 1980; CORFO 1985). In our work, however, the lowest Do value was only 44·4%.

The variation in fruit sizes and seeds/cm of fruit, which are important characteristics for the genetic improvement of this species, also presented a high VC. The seeds contain the largest proportion of fruit proteins (Becker & Grosjean 1980) and fruit size is an important agronomic characteristic for harvest management.

Results indicated that there were no important correlations among the variables, as demonstrated by the small correlation coefficient obtained between all the pairs of variables regressed (Table 28.6). Nevertheless, some of these coefficients were statistically significant such as that of OM regressed with NDF ($r = -0·47$) or with N ($r = -0·52$) ($P < 0·01$). The above would indicate that as OM increases in the fruits, the content of neutral detergent digestible solubles also increases, whereas the total N content decreases, which is unexpected since the reverse normally takes place (Minson 1982). This situation could result from the fact that the seeds contain the largest quantity of N and their number per fruit is independent from the variations in OM content ($r = -0·05$). The greatest amount of OM would then be associated with increases in the mesocarp rather than in the seeds.

Table 28.6: Correlation Coefficients among Different Quality Variables of Fruits of 70 Algarrobo Ecotypes

	DM	O	Total N	NDF	FS	Seeds/ cm/fruit
In vitro digestibility (Do)	0·16	0·17	−0·09	−0·29	0·00	0·22
Dry matter (DM)		0·01	−0·07	0·09	0·17	0·03
Organic matter (OM)			−0·52[2]	−0·47[2]	−0·15	−0·05
Total N				0·17	0·16	0·24[1]
Neutral detergent fibre (NDF)					−0·07	−0·07
Fruit size (FS)						−0·10

1. $P < 0·05$; 2. $P < 0·01$.

The correlation between OM and NDF was very low ($r = -0.29$) but significant ($P < 0.05$). This result is normally expected since with a greater fibre content digestibility usually decreases.

The fact that Do did not show any association with variables such as OM and N, as expected (Minson 1982), could be explained by the high variability of recorded values. On the other hand, this result as well as those obtained in previous work would suggest that some substances in the fruits could be masking or inhibiting the enzymatic action in the *in vitro* digestibility trials.

When the recorded variables were analysed by means of a stepwise correlation where Do was considered as the dependent variable and all the others as independent, neither the r value nor the statistical significance were improved. The equation obtained with the considered variables was as follows:

$$Do = 9.59 + 0.67 \, DM + 0.48 \, NDF + 5.7 \, seed/fruit$$
$$r^2 = 0.19; \quad P < 0.05$$

This equation would confirm the possible occurrence of some substances, apart from those of the recorded variables, that could be biasing the Do values.

DISCUSSION

The high variation in fruit production and quality shown by the Chilean algarrobo is a very useful characteristic for its genetic improvement. However, there are also some other important factors which can significantly affect this variability and which, if disregarded, can lead to erroneous extrapolation in the selection of productive ecotypes.

Apparently, environmental factors such as radiation, temperature and water availability directly regulate the productivity of this species and probably its fruit quality. With respect to the latter, there is insufficient experimental evidence, but preliminary observations have demonstrated that fruits from the same individual can exhibit very different pigmentation, shape and size from year to year (Salvo 1986). Changes in fruit shape and size can be strongly correlated with the amount of fruit produced in that season, while pigmentation changes could indicate alterations in the chemical composition that still need to be investigated.

Imagining that it is feasible to obtain the ideal type of the algarrobo with fruits of an adequate nutritive value and a yield of 50 kg/tree/year, then a community of approximately 100 algarrobo trees per ha with no more than 200 mm of rainfall would produce almost 5000 kg DM/ha/year plus 3000 kg of grass DM/ha/year (Olivares *et al.* 1983).

Another advantage presented by this species is that its fruits are produced in the summer when the natural pasture of the zone has reached the critical level of forage production.

ACKNOWLEDGEMENTS

We thank Dr. Cabrera of the Instituto de Tecnologia de los Alimentos, Universidad de Chile for the assistance in the gas analysis and the valuable English assistance provided by A. M. Espinoza.

REFERENCES

Becker, R. and O. K. Grosjean (1980) A compositional study of pods of the varieties of Mesquita (*Prosopis glandulosa, P. velutina*). *J. Agric. Food Chem.* 28: 22–25.

Corporacion de Fomento de la Produccion. Chile (CORFO) (1985) Valoración nutricional de tamarugo y algarrobo y perfiles metabólicos de ovinos y caprinos en la Pampa del Tamarugal. In: *Symposium Estado actual sobre Prosopis tamarugo,* M. Habit (Ed.), pp. 75–133. FAO. América Latina y Caribe, Producción Vegetal, *Arica, Chile.*

Felker, P. (1982) Seleção de Fenótipos de Prosopis para produção de vagens e de combustivel de madeira. In: Empressa de Pesquisa Agropecuaria de Rio Grande do Norte SA Emparn (Ed.), Algarrobo, Vol. II, pp. 7–24. EMPARN SA, Natal, Brazil.

Habit, M. A. D., D. Contreras and R. H. Gonzalez (1981) *Prosopis tamarugo: Arbusto forrajero para zonas aridas.* FAO, Santiago, Reprinted 1981 as *Prosopis tamarugo: Fodder tree for arid zones.* Plant Production and Protection Division, Paper No. 26, FAO, Rome.

Hunziker, J. H., L. A. Poccio, C. A. Naranjo, R. A. Palacios and A. B. Andrada (1975) Cytogenetics of some species and natural hybrids in Prosopis. *Can. J. Genet. Cytol.* 17: 253–262.

Menke, K. H., L. Raab, A. Salewski, H. Steingass, D. Friz and W. Schneider (1979) Gas production method. *J. Agric. Sci., Cam.* 93: 217–222.

Meyer, R. E., R. H. Haas and C. W. Wendt. (1973) Interaction of environmental variables on growth and development of honey mesquite. *Bot. Gaz.* 134: 173–178.

Minson, O. (1982) Effect of chemical composition on feed digestibility and metabolizable energy. *Nutr. Abs. Rev.* (8), 52: 592–615.

Monselise, S. R. and E. E. Goldschmidt (1982) Alternate bearing in fruit trees. *Hort. Rev.* 4: 128–173.

Mooney, H. A., B. Simpson and O. T. Solbrig (1982) Phenology, Morphology, Physiology. In: *Mesquite,* B. B. Simpson (Ed.). US/IBP Synthesis Series 4, Dowden, Hutchinson & Ross Inc., Stroudsburg, Pennsylvania.

Olivares, A., R. Cornejo and J. Gándara (1983) Influencia de la estrata arbustiva (*Acacia caven* (Mol.)) Hook. et Arn. en el crecimiento de la estrata herbacea. *Av. Prod. Anim. (Chile)* 3: 19–28.

Salvo, B. (1986) Estudio de la floración y desarrollo de los frutos en algarrobo (Prosopis chilensis (Mol.) Stuntz). Unpublished Ing. Ag. thesis, Univ. de Chile, Santiago.

Tilley, J. and R. Terry (1963) A two-stage technique for the *in vitro* digestion of forage crops. *J. Brit. Grassld. Soc.* 18: 104–111.

Van Soest, P. (1967) Development of a comprehensive system of feed analysis and its application to forages. *J. Anim. Sci.* 26: 119–128.

Wilson, R. T., D. R. Krieg and B. E. Dahl (1974) A physiological study of developing pods and leaves of honey mesquite. *J. Range Manag.* 27: 202–203.

29

Development of *Prosopis* Species Leguminous Trees as an Agricultural Crop

R. M. Saunders and R. Becker

INTRODUCTION

Prosopis spp. are perennial leguminous trees that occur in many locations around the world, ordinarily in arid and semiarid zones (Fig. 29.1). Depending upon the species and local environmental situation, the tree ranges in size from a small bush to a tree of several metres height. In the past, *Prosopis* pods were an important food for civilizations in the Americas and Asia, though nowadays tend only to be consumed by wildlife. There are some exceptions, for example in Mexico, several hundred tons are collected for animal feed, and in Argentina some pods are used to prepare a fermented beverage. *Prosopis* biology has been reviewed by Simpson (1977) and its ecology by Felker (1979).

The annual yield of pods from *Prosopis* spp. varies widely, but has been projected to be 10 000 kg/ha from cultivated trees (Felker *et al.* 1984). This encouraging yield, and the inherent ability of *Prosopis* to thrive under harsh environments has prompted a closer look at the food and feed potential of the pods. On this occasion the examination of the pods is done from a modern technological perspective. The questions being addressed are:

(a) As a food or feed, what are the properties of *Prosopis*?

(b) How does *Prosopis* compare with alternatives?

(c) How can *Prosopis* be shown, via processing if necessary, to be sufficiently competitive economically in order to convince anyone to cultivate it?

(d) Are there any special qualities about *Prosopis* which would enhance its development as a new food and/or food and/or industrial resource?

In order for *Prosopis* (or any other new crop) to assume a position as an economic crop for farmers and industry, there are a series of postharvest events which must take place. These include:

(a) Accumulation (reaching a critical mass quantity),
(b) Identity standards (chemical, physical, nutritional, appearance),
(c) Utility (price, availability), and
(d) Market demand (on-farm, local, regional, national, international, government).

Figure 29.1: Occurrence of Prosopis Species in the Americas, Africa, and the Middle East.

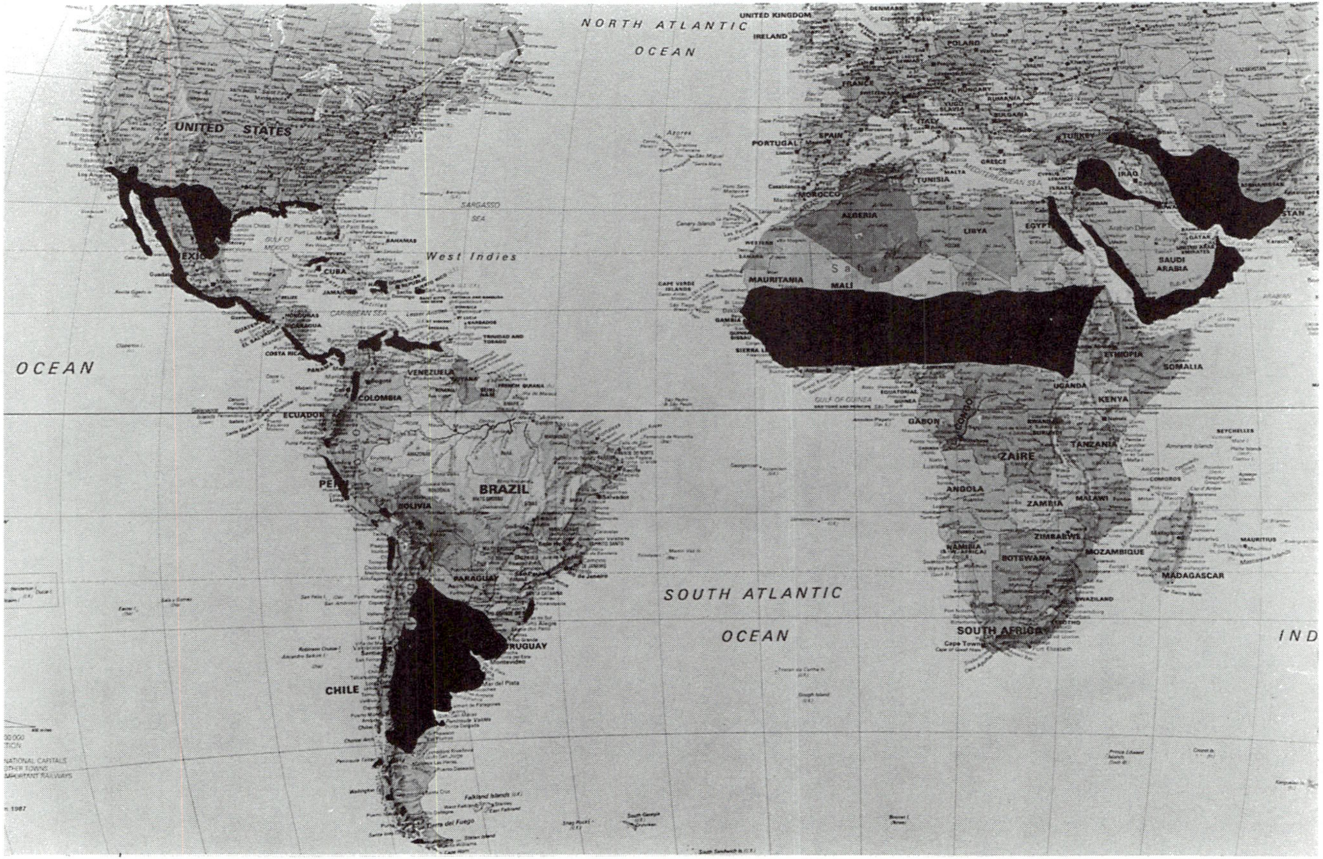

(a) *Accumulation:* In order to motivate the farmer to cultivate and harvest *Prosopis*, there needs to be a reasonable profit per hectare, and one that persuades the farmer to choose *Prosopis* over an alternative crop. This profit is after all costs associated with growing and harvesting have been considered. At the very early stages of development, *Prosopis* holds an advantage in that it is indigenously widespread, so the initial cost associated with it is primarily its harvesting cost. Because of this advantage, *Prosopis* development might be expected to move considerably faster than an alternative crop which is not already growing. However, collection of pods from wild plants has drawbacks. Wild trees are often difficult to cultivate, require pruning and are difficult to harvest. Such situations would not be expected to yield as much as a well managed mesquite orchard.

Accumulation of *Prosopis* pods necessitates a minimum critical mass. Processing of the pods requires an investment in equipment and storage facilities which must be amortized. Obviously more pods processed means less cost per unit processed.

(b) *Identity Standards:* All foods in commerce, though perhaps not always at the subsistence level, have standards of identity. These include chemical analyses (e.g. protein, sugar content), physical standards (e.g. particle size, foreign material), nutritional standards, GRAS status, and grading standards based generally upon appearance. Such standards must be set and accepted by mutual consent among the growers, processors and end-users.

(c) *Utility:* The price of *Prosopis* pod products must be competitive with other alternatives that could be grown on the same land, or not more than the cost of importing equivalent foods or feeds. There must be a supply of *Prosopis* of sufficient quantity such that the processor and/or marketing agency can depend on a consistent supply. If money is spent on developing and advertising a food or feed product, the persons responsible will only commit themselves to doing so if a guaranteed supply of raw material can be anticipated.

(d) *Marketing:* The normal progression expected for a new crop such as *Prosopis* would be as follows:

   (i) Increase on-farm usage of *Prosopis*. This implies the use of pods and its fractions as food and feed.

   (ii) Increase local usage through local product development, publicity and sales.

   (iii) Introduce on a regional and national scale through established processors and retailers.

(iv)   Government support would enhance action (3). The government would presumably benefit from development of an indigenous resource, and one that increases employment, food and feed availability, and perhaps decreasing importation of food to that region.

There are numerous problems associated with all of the points noted above, and all of the points are important, in varying degrees, for any new food or feed crop. They can come together successfully only if the bottom line shows a profit to the farmer, the processor and the retailer. A profit can only exist if *Prosopis* products are competitive with alternatives and/or show a superiority in functional performance such that a premium price can be obtained for one or more of the products. An exception to this would be if a government subsidized the development, particularly in the early stages, though one cannot rely upon government subsidies indefinitely.

This paper describes technological steps in establishing *Prosopis* as a new food and feed resource for industry. Those factors outlined above which are essential for successful introduction of a new plant resource are incorporated into the approach. By doing so, the work outlined herein is intended to serve as a model for postharvest exploitation of a new crop for food and feed industries. Parts of this work have been described elsewhere (Del Valle *et al.* 1986; Saunders *et al.* 1986).

The proximate composition of a number of *Prosopis* spp. from different countries are shown in Table 29.1. If the mean values for these four samples (protein, 11·6%; fat, 2·2%; fibre, 21·4%; ash, 4·0%; carbohydrate, 54·7%) are compared to a common grain, it is apparent that *Prosopis* pods have a considerably higher content of fibre. Even so, the total available nutrients exceed those of common grains on a yield/hectare basis if one assumes the project of Felker *et al.*'s (1984) to be correct. Materials displaying a composition like that of the *Prosopis* spp. would be expected to find immediate utility in mixed animal feed, particularly as a component of ruminant feed. Reports are available documenting feed use at 20% level of *Prosopis juliflora* for calves (Talpada *et al.* 1982), and *P. pallida* for cows, horses, donkeys, pigs and chickens (Eggler 1947). There are no detailed data available concerning feed efficiencies of *Prosopis* pods for animals, and consequently there is no exact economic value that can yet be assigned to the pod as a feed. Even so, there seems little doubt that the pods would be a profitable feed if obtainable at the yields projected by

Table 29.1: *Proximate Composition and Origin of Prosopis Pods*

Sample	Moisture	Protein	Fat	Fibre	Ash	Carbohydrate
P. velutina (USA)	4·2	11·6	2·8	22·4	3·5	55·5
P. tamarugo (Chile)	4·3	9·4	0·6	30·7	4·5	50·5
P. chilensis (Chile)	6·8	10·5	1·7	11·2	2·5	67·3
P. glandulosa (Mexico)	9·0	14·7	3·2	21·2	5·5	46·3

Felker *et al.* (1984). The actual break-even yield/hectare figure undoubtedly will vary for different species since there would be variations in feed efficiencies and costs associated with cultivation and harvesting.

The use of agricultural commodities as food rather than feed normally generates higher profit returns to the growers and therefore are more likely to motivate their development. With this fact in mind the following investigations were carried out. Technology has been developed to convert *Prosopis* pods into fractions with unique composition and functionality for potential food use.

Pod Fractionation

A scheme developed for milling and separation of *Prosopis* pods into subfractions is summarized in Fig. 29.2 (Meyer 1984). Because of the high sugar content, the pods must be dried to about 5% moisture before milling. The pods are then broken into 2–5 cm pieces using a Reitz Disintegrator, at slow speed to avoid releasing seeds. A Bauer disk mill (Model 148, size 8, disc 8114, gap-setting 40, speed 3450 rpm) proved to be the optimum system for milling the pod pieces into subfractions A plus B plus seed (Fig. 29.2). A vibrating screen sifter (Sweco, top screen, 4·8 mm round holes; bottom screen, 1·0 mm square holes) separated the milled mixture. The endocarp hulls (fraction B, Fig. 29.2) were retained on the top screen, while the seeds plus fraction A, and some pericarp fragments passed through. The seeds and pericarp fragments were retained by the lower screen whereas fraction A passed through the screen. The seeds were cleaned from pericarp fragments by separation in a cyclone (McGill Laboratory aspirator, 66–17721). Cleaned seeds at this stage have been scarified by the mill and, since this is essential for a high germination rate (>90%), are thus well suited for planting in forestry programmes.

Seed splitting was accomplished in a Udy cyclone mill (steel impeller, rough abrasive frame, 2000 rpm), or in a graintester (Strong-Scott, 17810). The milled seeds split into two fractions, the outer seedcoat/endosperm fraction C (endosperm splits) and the inner cotyledon fraction D. The economically important seed mucilage occurs in the endosperm and is later separated from the testa. About 15% of the seed splitting product is an inseparable fines mixture of fractions C (endosperm splits) and D (cotyledon) which is actually added to fraction B, whereas 85% of the seeds is converted to fractions C and D which are easily separable by air classification.

The process yield, average composition, and range in analysis of the milling fractions are listed in Table 29.2. The wide range in composition reflects the extreme diversity in pod morphology and composition encountered with *Prosopis*. For example, some pods are very rich in sugar and consequently taste very sweet whereas others are much less so.

Figure 29.2: *Flowsheet for Disc Milling and Separation of Prosopis velutina Pods.*

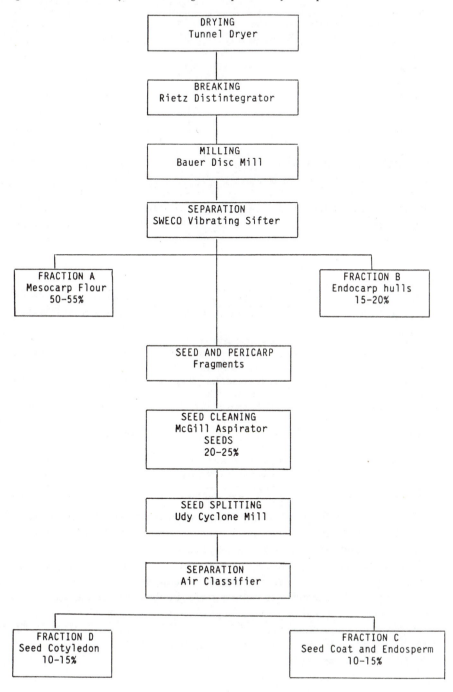

Table 29.2: *Process Yield and Composition of Prosopis velutina Pod Subfractions After Milling*

Fraction and process yield (%)	Protein (%)	Fat (%)	Fibre (%)	Ash (%)	Carbohydrate (%)
A (exomesocarp) 50–55%	9·5(6–11)	2·0(0·8–4)	16(10–12)	4·5(3·7–4·9)	48(25–58)
B (endocarp) 15–10%	6·5(4–9)	1·3(0·8–1·8)	45(39–55)	3·5(3·3–3·9)	6(3–11)
C (endosperm splits) 10–15%	5(n.a.)	0	8(n.a.)	n.a.	87·7(n.a.)
D (cotyledon)	64(59–66)	8(5–11)	4(3–6)	4·7(4–5)	n.a.

Fraction A (Exocarp Flour)

This floury fraction has a particle size of less than 1 mm, and further reduction in particle size can be obtained by milling on conventional mills. The dominant component of this fraction consists of simple sugars, which have been identified as sucrose (92%), glucose (3%) and fructose (5%) (Meyer 1984).

Table 29.3: *Use of Prosopis velutina (Fraction A) Exomesocarp Flour in Conventional Food Systems*

Food System	Observation	Conclusion
Composite flours	loaf character acceptable to 10%	could displace 10% of wheat flour
Breads	incorporation into wheat flour	
Chapati & cracker	acceptable to 50% incorporation into wheat flour	could displace 50% of wheat flour
Flakes	acceptable to 50% incorporation into wheat flour	could displace 50% of wheat flour
	35% incorporation	superior sheeting, strength, crispness
Tortilla chips	20% incorporation into corn flour	ranked superior to 100% corn chip by test panels
Fermentation	alcohol produced	liquor or energy source
Syrup extraction	flavourful sucrose syrup extractable	sweetener production

Thorough tests (Table 29.3) have been conducted on use of fraction A as a partial substitute for wheat flour in cereal-based products. Because of the absence of gluten and starch, fraction A finds use only where structural integrity of the product is not an essential characteristic. Its distinctive taste makes it an interesting component for use in sweetening and in certain baked products.

Through use of consumer taste panels, in the case of leavened bread, substitution of wheat flour at levels up to 10 per cent was judged acceptable, whereas in chapatis, crackers and flakes, up to 50 per cent was acceptable. Corn tortilla chips containing 20 per cent exocarp flour were judged to be superior to the 100 per cent corn chips.

Extraction of a sugar (sucrose) syrup from fraction A has been carried out using machinery and processing technology existing in conventional sugar plants (Meyer 1984). Using *Saccharomyces cerevisiae*, the sucrose within fraction A from *Prosopis chilensis* (Torres 1987) and *P. velutina* (Meyer 1984) has been successfully converted into ethanol in yields as high as 92 per cent. After a series of evaluations of fraction A flour, Meyer (1984) concluded that 'quantitative incorporation into composite flours is almost unlimited from the technological point of view...the taste of all samples was very pleasant'. The economic value of this fraction A flour thus would be expected to be not lower than that of wheat flour.

Fraction B (Endocarp Hulls)

Chemical analysis indicates that this material would not find use as human food. It could find use as a low-quality cattle food, or alternatively as a combustion source. Its burning value, on a dry weight basis, was found to be 16999 kJ/kg (Table 29.4; Meyer 1984), which compares favourably with an average value of 16300 kJ/kg (dry basis) for cereal crop residues, or 13860 kJ/kg for sugarcane bagasse. Possibly in a *Prosopis* processing plant, burning this material would provide energy for initial pod drying.

Table 29.4: *Burning Value of Prosopis Endocarp Hulls (dry weight basis)*

Material	Burning value kJ/kg
Endocarp hulls	16999
Sugar cane bagasse	13860
Cereal crop residue	16300

Fraction C (Seed Coat and Endosperm)

This fraction contained approximately 40% seed coat and 60% endosperm. It is unlikely that the seed coats have any unique functional or nutritional value, but the endosperm, which contains galactomannan gum, could be of considerable economic value. Galactomannan was separable from the splits by an aqueous process, producing gum of 85–95% purity, yield 60%.

In situations where lower purity (40–65%) would be acceptable, dry milling separation can be employed (Meyer 1984). Galactomannan gums are used in industries as widely diverse as oil-well drilling, pharmaceuticals, food, and pet food (Whistler & BeMiller 1959). The gum chosen for a given application depends on two criteria: functional characteristics and price. The functional quality of purified *Prosopis velutina* and *P. chilensis* galactomannan is very close to that of guar galactomannan, the most common of these gums (Fig. 29.3). In water, the *Prosopis* galactomannan causes excellent viscosity increases, and displays thermal stabilities, either alone or in a mixture with xanthan, analogous to those displayed by guar and guar-xanthan. These characteristics indicate the economic value, and utility of *Prosopis* gum should be analogous to those of guar gum. In contrast, very little galactomannan gum was found in *Prosopis tamarugo*.

Figure 29.3: *Viscosity of Prosopis Galactomannan (concentration 0·5%)*

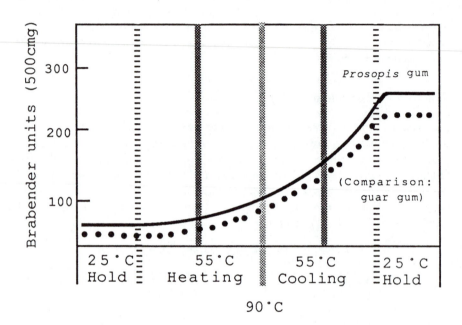

Fraction D (Seed Cotyledon)

This fraction comprises the cotyledons and germ of the *Prosopis* seed. Its main constituents are protein and fat, by virtue of the amino acid composition of its protein, it would be expected to find food use where protein content and quality need to be enhanced. In one species investigated, after defatting, *Prosopis velutina* fraction D displayed functional characteristics little different from soy protein concentrate (Meyer 1984). These characteristics, which are generally beneficial for food applications, included protein solubility, whipping properties, and emulsifying properties. Such properties indicate a high economic value and utility for *Prosopis* fraction D comparable to soy protein concentrate.

The nutritional properties of *Prosopis velutina* have been documented in studies with rats and chicks (Meyer *et al.* 1986), examples of which are shown in Tables 29.5 and 29.6. In general, the feeding value of pods and seeds are acceptable, and, for comparative purposes, tend to be similar to those existing for wheat bran. Toxic factors were found to be absent in *Prosopis velutina, P. glandulosa* and *P. chilensis*, but present (but of as yet undetermined nature) in *P.*

tamarugo (Torres 1987). It must be pointed out that the latter observation was with rats, whereas large animals do not appear to be affected. Nevertheless, it casts doubt upon the appropriateness of *P. tamarugo* as a source of food.

Table 29.5: *Nutritional Value of Prosopis velutina in Rats*

	Protein Efficiency Ratio	N Digestibility
Whole pods	0·71–1·47	44–56
Seeds	0·69–1·50	58–64
Casein	2·50	98
Wheat bran	1·70	64

Table 29.6: *Metabolizable Energy Values of Prosopis velutina Pods and Common Feedstuffs for Chickens*

Feedstuff	Metabolizable Energy kcal/gm
Prosopis velutina pods	
Uncooked, 20% of diet	1·65
Cooked, 20% of diet	0·70
Corn	3·40
Oats	2·66
Alfalfa meal	1·41
Wheat bran	1·19

It is apparent from processing, nutritional and functional observations so far that pods of three *Prosopis* spp. can be converted into four subfractions. It can be reasonably expected that fraction A has a value akin to that of wheat flour, fraction B to that of a low-grade animal feed, or firewood. Fraction C, for the major part would have a value like that of guar gum (at least ten times the value of wheat flour), and fraction D, a value similar to that of soy protein concentrate (4–5 times the value of wheat flour). The actual appearance, yield, general composition, and potential utility to the food industry of the four subfractions derived by milling and fractionation of *Prosopis velutina* are summarized in Fig. 29.4.

These figures for potential value of subfractions recoverable from *Prosopis* must be offset against the total costs involved in their growing, harvesting, storage, processing and so on. A design has been developed for a commercial plant capable of processing 1000 kg/hr of pods. An economic analysis of costs associated with such a processing plant, and the production costs that would be anticipated for fractions A, B, C and D is available (Meyer 1984). A village-scale portable unit has also been devised for subfractionation of *Prosopis* into these four fractions. In this latter case, the capital investment and associated processing costs are extremely low.

Figure 29.4: *Appearance, yield, composition and potential utility to the food industry of fractions derived by milling of Prosopis velutina pods.*

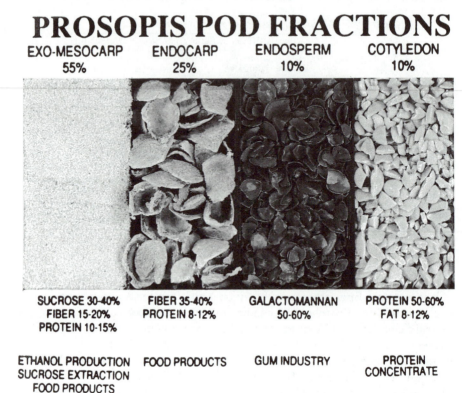

PROSOPIS POD FRACTIONS

EXO-MESOCARP	ENDOCARP	ENDOSPERM	COTYLEDON
55%	25%	10%	10%

SUCROSE 30-40%	FIBER 35-40%	GALACTOMANNAN	PROTEIN 50-60%
FIBER 15-20%	PROTEIN 8-12%	50-60%	FAT 8-12%
PROTEIN 10-15%			

ETHANOL PRODUCTION	FOOD PRODUCTS	GUM INDUSTRY	PROTEIN
SUCROSE EXTRACTION			CONCENTRATE
FOOD PRODUCTS			

Whole Pod Processing

In addition to the process described above, a second milling scheme for conversion of *Prosopis* pods into a nutritionally enriched subfraction has been developed (Del Valle *et al*. 1986). This process is considerably easier and less expensive than the first milling scheme described. In this second process, the *Prosopis* pods are first toasted, then milled to a coarse flour by grinding the pods through an Alpine pin-mill. The final ground toasted flour is separated into particles of different size through use of screens (Fig. 29.5). The finest sized particles are enriched in protein and reduced in fibre content when compared to the starting pod material. Of the processing conditions studied, toasting at 125°C for 40 minutes was found to provide the maximum yield of fines, concomitant with maximizing nutritional quality (Table 29.7). The final *Prosopis* flour (fines fraction) is a fine (less than 100 mesh) cream-coloured low-cost flour. The composition and yield of the flour from *P. glandulosa* is shown in Table 29.8.

Figure 29.5: *Flowsheet for pin-milling and fractionation of Prosopis glandulosa pods.*

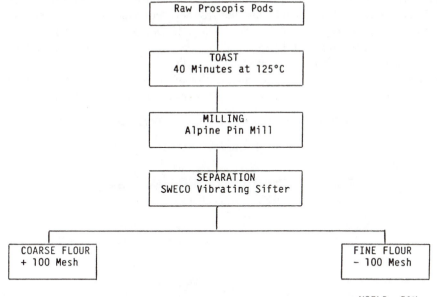

YIELD: 56%

Table 29.7: *Protein Quality of Toasted Prosopis glandulosa by Weight Gain of Red Flour Beetle (Tribolium castaneum) Larvae*

Toasting conditions	Mean weight gain/larva ± SD
Control sample,	
whole wheat flour	2·96 ± 0·6
100 °C/40 min	1·18 ± 0·12
100 °C/80 min	1·24 ± 0·01
125 °C/20 min	1·24 ± 0·05
125 °C/40 min	1·71 ± 0·13
150 °C/10 min	0·97 ± 0·13
150 °C/20 min	1·31 ± 0·09

Table 29.8: *Proximate Composition and Yield of Prosopis glandulosa Pods and Subfractions Prepared by Toasting and Pin-milling*

Component	Pods	Coarse Flour +100 mesh	Fine Flour −100 mesh
Moisture %	4·9	1·4	2·1
Protein %	11·2	9·1	13·6
Fat %	1·9	1·8	2·0
Ash %	4·6	3·6	4·6
Acid detergent fibre %	28·6	38·8	23·4
Cellulose %	20·1	29·4	16·5
Yield	100	44	56

The functionality, and therefore the intrinsic economic value of this flour is documented in Table 29.9, which summarizes situations in Mexico where many different food products benefit functionally and economically by inclusion of *Prosopis glandulosa* flour. Of eight products tested, cost estimates for six of them were less after the *Prosopis* flour was included in the formulation. Consumer evaluation of these products was positive. One product, the Peanut Butter Candy Marzipan with *Prosopis,* is now available commercially under the trade name 'DYN-DYN'. Details of its composition are listed in Table 29.10.

Table 29.9: *Cost Estimates for Products With and Without Added Prosopis Pod Flour*

Product	Price US$	
	without	with *Prosopis*
Beverage base	1·01/kg	0·94/kg
Rice 'Horchata' (dry basis)	0·29/kg	
Prosopis 'Horchata' (dry basis)		0·43/kg
Pinole	0·37/kg	0·39/kg
Peanut butter candy	0·11/25 g	0·10/25 g
Peanut butter, sweetened	3·31/kg	
Prosopis pod butter, sweetened		1·68/kg
Cream of wheat (dry basis)	1·15/kg	1·07/kg
Rolled oats (dry basis)	0·94/kg	0·90/kg
Yoghurt	1·11/litre	1·10/litre

Table 29.10: *Proximate and Nutritional Composition of Prosopis Flour – Peanut Butter Candy*

Ingredient	%
Prosopis flour	10
Peanuts	29
Sucrose	60
Salt	1
Protein	11
Fat	16
Sucrose	65
Other carbohydrates	2
Fibre	2
Calories per 25 g serving	108

These examples of incorporating *Prosopis* flour derived from inexpensive pods using an inexpensive processing technique are an example of a successful endeavour to promote a new crop as a food resource. If ingredients can be displaced from a conventional product by the new crop whereby there is a decrease in the price of the product without a negative charge in consumer appeal or nutritional value, then an economic value can be assigned to the new crop. In the examples cited in Table 29.9 it is clear that the *Prosopis* flour prepared by the process described here has a favourable utility value and cost that will spur its economic development.

As a complement to the evaluations of *Prosopis* pods and pod subfractions as food or feed, studies have been and still are underway on utilization of other parts of the *Prosopis* plant. Maximum utilization of the crop is essential in deriving favourable economics for any crop. In the case of *Prosopis*, there are at least four other avenues of utilization in addition to the pod crop. These are:

(a) Use of part of the foliage as a ruminant feed,
(b) Harvesting of gum exudate (for food use, pharmaceuticals, etc.) from the trunks and limbs of the tree,
(c) Use of the tree for lumber or firewood at some later stage in its life cycle,
(d) Arresting desertification by planting of *Prosopis* spp.

In addressing these aspects, the following conclusions have been reached or anticipated. The foliage is acceptable for ruminants, though unlikely for non-ruminants; a factor toxic to mice has been discovered in *Prosopis* foliage (Lyon *et al.* 1987). *Prosopis* lumber was widely used in the early Americas in the past. At present, it is enjoying a resurgence in demand and value because of its hardness and visual attractiveness. *Prosopis* charcoal currently enjoys a reputation as the most desirable charcoal in the United States. It retails for about US$1·50/kg.

A collection of 55 accessions of *Prosopis* are under study at a USDA plantation in Southern California for purposes of evaluating pod yield and biomass production (tree/lumber/charcoal weight), and gum exudate production.

Table 29.11: *Advantages and Disadvantages Anticipated in the Development of Prosopis spp. for Food and Feed Industries.*

Advantages	Disadvantages
Indigenous	Cultivation
Stress-tolerance	Harvesting
Biomass production (pods, firewood, forage, gum exudate)	Acceptance (?)
	Economics (?)
Subfractionation of pods (by milling)	
Functionality of milled fractions	

The advantages and disadvantages outlined for development of *Prosopis* as a new resource crop for industry are summarized in Table 29.11. Proper management of any *Prosopis* venture is critical, but use of the major part of the pods as a food resource, the minor part as a feed, in conjunction with possible harvesting of gum-exudate and lumber, supports the model for *Prosopis* development. This model becomes even more attractive in Third World situations if the aspect of environmental tolerance of the plant is included, and if one includes a factor associated with generation of employment and income.

ACKNOWLEDGEMENTS

The authors wish to thank the following people for contributions to this manuscript. F. Del Valle and E. Marco, Mexico, and M. E. Torres of Chile, for data, D. Irving, J. Hoefer and V. Breda for artwork, R. Knowles and G. Hanners for technical assistance, and A. Anderson for typing.

The work was carried out with financial assistance from the US Agency for International Development, Office of Science Advisor, S & T Small Activity Project No. 936–140–6 (3.F32).

REFERENCES

Del Valle, F. R., E. Marco, R. Becker and R. M. Saunders (1986) Preparation of a mesquite (*Prosopis* spp.) food enhanced protein reduced fiber fraction. *J. Food Sci.* 51: 1215–1217.

Eggler, F. (1947) *Arid Southeast Oahu Vegetation.* Hawaii, Ecological Monographs, No. 17.

Felker, P. (1979) Mesquite: An all-purpose leguminous arid land tree. In: *New Agricultural Crops*, G. A. Richie (Ed.), pp.89–132. Westview Press, Boulder, Colorado.

Felker, P., P. R. Clark, J. F. Osbora and G. H. Cannell (1984) *Prosopis* pod production – comparison of North American, South American, Hawaiian and African germplasm in young plantations. *Econ. Bot.* 38: 36–51.

Lyon, C. K., M. R. Gumbmann, and R. Becker (1987) Value of mesquite leaves as forage. *J. Sci. Food Agric.* (Accepted for publication).

Meyer, D. (1984) *Processing, Utilization and Economics of Mesquite Pods as a Raw Material for the Food Industry.* Ph.D. Thesis, Swiss Federal Institute of Technology, Zurich, Switzerland.

Meyer, D., R. Becker, M. R. Gumbmann, Pran Vohra, H. Neukom and R. M. Saunders (1986) Processing, composition, nutritional evaluation and utilization of Mesquite (*Prosopis* spp.) as a raw material for the food industry. *J. Agric. Food Chem.* 34: 914-919.

Saunders, R. M., R. Becker, D. Meyer, F. R. Del Valle, E. Marco and M. E. Torres (1986) Identification of commercial milling techniques to produce high sugar, high fiber, high protein, and high galactomannan gum fractions from *Prosopis* pods. *Forest Ecol. Manag.* 16: 169–180.

Simpson, B. B. (1977) *Mesquite – Its Biology in Two Desert Ecosystems.* Dowden, Hutchinson and Ross, Stroudsburg, Pennsylvania.

Talpada, P. M., M. B. Pande, J. S. Patel and P. C. Shukla (1982) Note on the utilization of pods of *Prosopis juliflora* in the ration of growing calves. *Indian J. Anim. Sci.* 52: 567–569.

Torres, M. D. (1987) INTEC-Santiago, Chile. Private communication.

Whistler, R. and J. H. BeMiller (1959) *Industrial Gums, Polysaccharides and their Derivatives.* Academic Press, New York.

30

Rattan as a Crop for Smallholders in the Humid Tropics

J. Dransfield

ABSTRACT*

Rattan is the flexible raw material for cane furniture. Its source is the stems of climbing Calmoid palms which occur in greatest abundance in Southeast Asia and Malaysia. Besides entering world trade, rattans are also intensively utilized locally for a wide range of purposes. Of the ca. 600 species only about 256 consistently enter trade. Most cane is collected from the wild, but a small but increasing proportion originates from plantations. Although many aspects of rattan cultivation have yet to be studied, it does seem feasible to cultivate the small diameter canes, *Calamus caesius* and *C. trachycoleus*. Large diameter canes, potentially of greatest value, have not yet proved to be amenable to cultivation. Small diameter rattan is an ideal crop for smallholders in the humid tropics. It grows well with relatively little attention, and is particularly well suited for planting among tree crops such as fruit trees or smallholding rubber, thus increasing considerably the cash returns from smallholdings. Unfortunately, the necessary wait of 7–10 years before the first good harvest has partly been responsible for discouraging wider planting.

---

*   The full text of this paper is in press, in Proceedings of the 1986 Palm Symposium, organized by the Society for Economic Botany. It will appear shortly in *Advances in Economic Botany*, published by the New York Botanical Garden.

31

The Pejibaye Palm: Economic Potential and Research Priorities

C. R. Clement and D. B. Arkcoll

INTRODUCTION

The pejibaye (*Bactris gasipaes*) is a caespitose palm that may attain 20+ m in height. Stem diameter varies from 15 to 30 cm and internode length from 2 to 30 cm, but becomes reduced with age after 5 years. The internodes are armed with numerous black, brittle spines, although spineless mutant occurs and have been selected for in several areas. The stem is topped by a crown of 15 to 25 pinnate fronds (Fig. 31.1), with the leaflets inserted at different angles. The inflorescences develop in the axils of the senescent fronds. After pollination the bunch (Fig. 31.2) may contain between 50 and 1000 fruit and weigh between one and 25 kg. Numerous factors may cause fruit drop: poor pollination, poor nutrition, drought, crowding, insects and diseases. The fruit (Fig. 31.3) that ripen have a starchy/oily moist mesocarp, a fibrous red, orange or yellow epicarp, and a single endocarp with a fibrous/oily white endosperm. Individual fruit may weigh between 10 and 250 g (FAO 1986).

The pejibaye was domesticated in tropical America, probably in western Amazonia (Clement, in press). The domestication process enormously increased the genetic variability of the species and will be described below with respect to fruit quality and uses. Some Amerindian groups selected the pejibaye as a starchy edible fruit, sometimes used to make flour or fermented to make alcoholic beverages (Clement, in press). In some areas the pejibaye may have been as important a staple as cassava (*Manihot esculenta*). It is much more nutritious and can produce considerable amounts of food year after year, while cassava is generally only harvested once or twice from each area on a 5–10 year cycle. Patino (1963) points out that all parts of the plant were used, although the fruit was the most important.

From an initial domestication in western Amazonia the Amerindians distributed the pejibaye throughout the tropical lowlands of South America and as far north as Honduras (Mora Urpi 1984). Figure 31.4 presents its modern distribution.

Figure 31.1: *A Three Year Old Pejibaye Palm in the INPA Germplasm Collection.*

Note the off-shoots at the base of the stem.

Figure 31.2: A 'Mesocarpa' Landrace Fruit Bunch, Weighing about Six Kilograms.

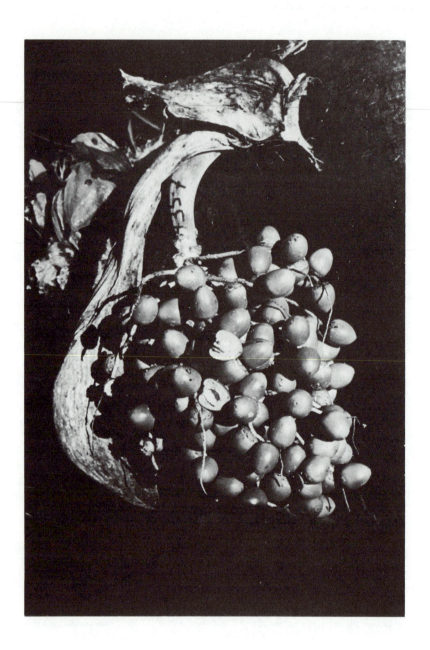

Figure 31.3: *Fruit Size and Shape Diversity from the Putumayo 'Macrocarpa' Landrace.*

Collected from the Benjamin Constant, Amazonas, Brazil population maintained at the Univ. Costa Rica Germplasm Collection.

Figure 31.4: *Approximate Modern Distribution of the Pejibaye in Latin America.*
Recent introductions to new areas have not been included.

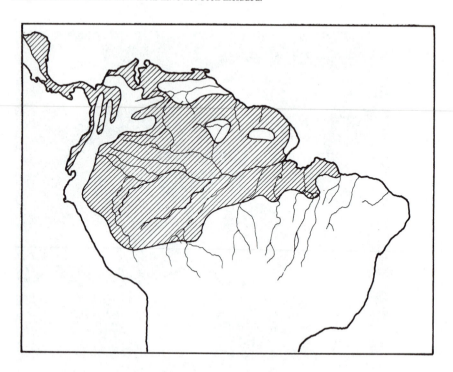

Mora Urpi & Clement (1985) described eight landraces of pejibaye found in Amazonia by germplasm collection expeditions sponsored by the US Agency for International Development. Their classification is based principally on fruit size, because this appears to have been most modified by selection during domestication. Fruit composition also varies considerably but has not yet been clearly defined. According to Mora Urpi & Clement (1985) there are three groups of landraces, based exclusively on fruit size: 'microcarpa', with small, generally more oily, fibrous fruit, averaging 20–25 g (two landraces described); 'mesocarpa', with medium, rather starchier, less oily and fibrous fruit, averaging 25–70 g (four landraces described); and 'macrocarpa', with large, very starchy fruit, low in oil and fibre, averaging more than 70 g (two landraces described). Clement (1986a) defined a ninth landrace in the 'mesocarpa' group and suggested that Mori Urpi's (1984) Occidental group of races was, in fact, one landrace within the 'mesocarpa' group. As will be made clear below, different landraces appear ideally suited for different uses.

Clement & Mora Urpi (1987) have outlined five possible uses for pejibaye that appear to have considerable economic potential: fruit for (1) direct human consumption; (2) animal ration; (3) flour; (4) oil; and (5) heart of palm

308

(palmito). These authors, and Clement (in press), have discussed these products from the breeder's point of view and have presented initial ideotypes for each product. This paper will outline the currently available results of interest and discuss some of the research priorities.

COMPOSITION

Bunch components

Recent work has shown that bunch components, especially bunch weight and fruit number (Table 31.1), vary considerably from population to population and landrace to landrace. Piedrahita & Velez (1982), working on the Colombian Pacific ('mesocarpa' group), found surprisingly low bunch component values and a very low fruit to bunch ratio. They considered these to be due to pests and diseases. Arkcoll & Aguiar (1984) studied more than 100 bunches near Manaus, AM, Brazil, with samples from different landraces and many populations; as they worked before Mora Urpi & Clement's (1985) definition of Amazonian landraces they did not distinguish among these. Their data showed low bunch weight, but high fruit number and fruit to bunch ratio, suggesting that most of their samples were from the Para 'microcarpa' landrace.

Table 31.1: *Average Bunch Composition Data from Several Sources*

Author	Bunch weight (kg)	Fruit number	Fruit/bunch ratio (%)
Piedrahita & Velez (1982)	3·3	61	87
Arkcoll & Aguiar (1984)	3·6	96	93
Clement & Mori Urpi (1985)	7·9	149	96
Clement (1986)	4·8	106	91

Clement & Mora Urpi (1985) found significant differences in bunch component values among Amazonian landraces, with the 'microcarpa' landraces presenting low weight and high fruit numbers, while the 'macrocarpa' landraces presented moderate to high bunch weights, lower fruit number and approximately equivalent fruit to bunch ratios. 'Mesocarpa' landraces were generally intermediate in bunch weights and fruit numbers although the largest bunches found were from 'mesocarpa' populations. Clement (1986a) found significant differences among three 'mesocarpa' populations in Costa Rica, both *in situ* and *ex situ*, for fruit number and bunch weight.

Piedrahita & Velez's (1982) observations in Colombia and Arkcoll & Aguiar's (1984) comments about fruit drop in Manaus show that there are already serious phytosanitary problems that must be resolved before monoculture plantations can be recommended. Pejibaye was domesticated and grown in a swidden environment (Clement, in press), which kept phytosanitary problems at acceptable levels. A change to monoculture, at least in its natural range, can be expected to exacerbate pest and disease problems.

FRUIT COMPOSITION

Fruit generally have a large amount of pulp because the seeds are small (Table 31.2). Piedrahita & Velez (1982) report typical 'mesocarpa' fruit data, with an especially high pulp to fruit ratio. Arkcoll & Aguiar (1984) report intermediate data, due to the mixture of landraces in their sample. Their slightly lower mesocarp percent and pulp percent data indicate the lower values typical of 'microcarpa' landraces.

Table 31.2: Average Fruit Composition Data from Several Sources

Author	Fruit weight (g)	Seed Weight (g)	Mesocarp %	Pulp %
Piedrahita & Velez (1982)	50	3·6	80·8	92·8
Arkcoll & Aguiar (1984)	35	2·9	76·1	90·3
Clement & Mora Urpi (1985)	58	4·1	–	91·7
Clement (1986)	42	4·6	–	88·5

Clement & Mora Urpi (1985) found significant differences among Amazonian landraces for the three components that they studied (Table 31.2). Clement (1986a) found similar differences among three Costa Rican 'mesocarp' populations. Clement (in press) suggested that pulp to fruit ratio is an important discriminant among landrace groups, with higher ratios in 'macrocarp' landraces.

MESOCARP COMPOSITION

The literature on pejibaye mesocarp composition is relatively extensive, but most early reports are based on only one or two samples, without information on collection locality or population characteristics (Clement & Arkcoll 1985).

The three more detailed studies reported in Table 31.3 found enormous variability in their different samples. While Piedrahita & Velez (1982) and Arkcoll & Aguiar (1984) worked with *in situ* samples, CIPRONA (1986) worked with *ex situ* samples from the Universidad de Costa Rica (UCR) and the Centro Agronomico Tropical para Investigacion y Ensenanza (CATIE)

Table 31.3: Average Mesocarp Composition Data from Several Sources (% dry weight)

Author	Humidity	Protein	Oils	N-free extract	Fibre	Ash
Piedrahita & Velez (1982)	49·8	9·8	11·5	73·7	2·8	2·4
Arkcoll & Aguiar (1984)	55·7	6·9	23·0	59·5	9·3	1·3
CIPRONA (1986)	56·7	6·1	8·3	79·9	3·6	2·1

germplasm banks. Arkcoll & Aguiar (1984) were looking for plants with high oil levels; they found 2 to 61·7% in the dry mesocarp, the latter in a small 'microcarpa' fruit. CIPRONA (1986) found 17·5% protein. The enormous variation shows the genetic potential available for exploitation in a breeding programme.

A number of minor components of interest have also been found in the mesocarp. These include trace amounts of B vitamins and vitamin C (INN 1959; Johannessen 1967; Leung 1961), none to enormous amounts of carotenoids (Arkcoll & Aguiar 1984) (Table 31.4), a few oxalate crystals in and just under the skin (Arkcoll & Aguiar 1984), small amounts of an alkaloid, pupunhadine (Fonsenca 1927), and a trypsin inhibitor (Murillo et al. 1983).

Table 31.4: *Amino Acid Composition Data of Pejibaye Mesocarp Protein from Several Sources (%/g N)*

Amino acid	Author: Piedrahita & Velez (1982)	Zapata (1972)	Zumbado & Murillo (1984)
Lysine[1]	4·22	4·6	4·12
Histidine[1]	2·72	2·0	1·76
Arginine[1]	7·26	9·2	5·69
Aspartic acid	4·99	4·6	n.r.
Threonine[1]	2·91	2·5	3·53
Serine	3·75	3·6	n.r.
Glutamic acid	4·68	6·3	n.r.
Proline	2·70	2·9	n.r.
Glycine[1]	3·21	4·5	5·29
Alanine	4·14	3·6	n.r.
Cystine	trace	–	n.r.
Valine[1]	2·76	2·7	3·73
Isoleucine[1]	1·96	1·7	3·14
Leucine[1]	2·63	2·6	5·49
Methionine[1]	1·46	1·3	1·57
Tyrosine[1]	1·65	1·4	2·75
Phenylalanine	1·82	1·3	2·75
Tryptophan[1]	0·92	n.r.	n.r.
Protein % (d.wt.)	9·0	5·7	5·1

1. Essential amino acid. n.r. = not reported

Most raw fruit stings the mouth, due to the oxalate crystals, but these dissolve on cooking or are removed with the skin. The trypsin inhibitor is like those commonly found in most beans and is also inactivated by cooking. The amount of these two antinutritional factors varies considerably and it may be possible to select fruit in which they are absent, thus reducing the cost of preparing for example, animal rations.

Seed composition

Arkcoll & Aguiar (1984) and CIPRONA (1986) present data on kernel composition. As in other palms this is a potential source of lauric oils. It is, however, a potentially important sub-product, providing both oil and a fibrous kernel meal. The small size of most pejibaye seed means that this will always be a minor item, but may be worth pursuing to defray costs.

Protein quality

Very little work has been done on the amino acid composition of pejibaye mesocarp protein (Piedrahita & Velez 1982; Zapata 1972; Zumbado & Murillo 1984), although some comparisons between pejibaye and other foods have been made. All essential amino acids are present, although some are at slightly lower levels than in maize. The variability found suggests that selection for limiting amino acids may be worthwhile.

Oil quality

Oil quality has been studied more than protein quality, but only CIPRONA (1986) has studied a significant number of samples. Table 31.5 outlines some of these studies and compares pejibaye with oil palm (*Elaeis guineensis*). Although Zapata (1972), Hammond *et al.* (1982) and CIPRONA (1986) report high levels of unsaturated fatty acids; these are all from 'mesocarpa' populations with relatively low oil levels. Silva & Amelotti (1983) report oil quality similar to that of oil palm from a sample that is probably from the Para 'microcarpa' landrace, where Arkcoll & Aguiar (1984) found high oil levels.

Hammond *et al.* (1982) and Silva & Amelotti (1983) characterized the fatty acids in a few samples. The results show that fatty acid contents vary a little, with oleic acid contents between 40–60%. They also showed the position of

Table 31.5: *Fatty Acid Composition Data of Pejibaye Mesocarp Oils from Several Sources (% oil)*

Fatty acid	Author: Zapata (1972)	Hammond et al. (1982)	Silva & Amelotti (1983)	CIPRONA (1986)	Oil Palm[2]
Palmitic	40·2	29·6	44·8	32·2	42·2
Stearic	0·4	trace	1·5	1·5	4·9
Total SFA[1]	40·6	29·6	46·3	33·7	47·1
Palmitoleic	10·5	5·3	6·5	8·3	–
Oleic	47·5	50·3	41·0	45·5	40·6
Linoleic	1·4	12·5	4·8	11·6	11·2
Total UFA[1]	59·4	69·9	53·3	67·4	51·8

1. SFA – saturated fatty acids; UFA –unsaturated fatty acids 2. Noiret & Wuidart (1976).

fatty acids in the triglycerides, but did not show the different types of triglycerides as is now possible. Triglyceride structure may also vary, as samples with similar fatty acid contents have different melting points. Some are liquid at room temperature or even at 8°C, while others are solid. Analysis of fatty acid composition and triglyceride structure should be included in the breeding programme, to see if fruit with more valuable oils can be identified and selected.

POTENTIAL USES AND PRODUCTIVITY

Fruit for direct human consumption

This is the traditional use for the pejibaye fruit and is the only one known in most places. The whole fruit is separated from the bunch and boiled in salted water for 30 to 60 minutes, to improve flavour and eliminate the irritating oxalate crystals and trypsin inhibitor. Piedrahita & Velez (1982) report that cooking for 15 to 20 minutes deactivates peroxidase activity. The fruit are then peeled, halved and pitted, and are ready for consumption. Pejibaye fruit are frequently consumed at breakfast, or as an appetizer before later meals. When the fruit are dry, starchy and low in oil, they are tastier when accompanied by mayonnaise or a sauce. The cooked fruit can also be used whole in stews, or ground to a flour for use in a variety of preparations and pastries. About 40 pejibaye recipes have been collected by Calvo (1981).

While the flavour of fruit are quite variable, the typical pejibaye has a distinctive bland to strong flavour of its own, depending upon carotenoid content. A liking for this is acquired fairly easily but it is not exciting enough to generate new markets without good quality control and clever marketing. Several authors have mentioned that pejibaye may taste like European chestnut (*Castanea sativa*) (Popenoe & Jiminez 1921; NAS 1975) but this is doubtful. Arkcoll & Aguiar (1984) found some similar to potato, others to maize, and some which were sweet. The latter were much more attractive than most fruit.

Mesocarp texture is also variable, ranging from that of a soggy potato to that of a good raw cashew (*Anacardium occidentale*). Texture is determined by water, starch, fibre and oil content, with watery, low starch types being soggy and starchy, low water types being floury to crunchy, even after cooking. Preferences vary from region to region; in Costa Rica the 'best' fruit are dry, high in starch and with a nutty texture; in Manaus the 'best' fruit are less dry and starchy, with a less nutty though still firm, texture. Moderate oil levels are also important to fruit quality in Manaus and influence texture.

In Costa Rica whole or halves, pitted or unpitted, peeled or unpeeled fruit are marketed in brine in 500 and 1000 g jars or cans. Quality is extremely variable but they sell well when fresh fruit are unavailable. In Columbia dehydrated fruit have been prepared for market, but have not been sampled by us.

High yields have been projected from small plots or population data. Corley (this volume) points out the risks of this type of projection. Mora Urpi (1984) reports 25 t/ha/yr of fresh bunches from non-selected, fertilized germplasm ('mesocarpa' group) growing in good edapho-climatic conditions in Costa Rica. Clement (1987) projected much lower yields from population data ('microcarpa' group) near Manaus, with 6–10 t/ha/yr for unselected, unfertilized germplasm on poor soils with three months drought. Moreira Gomes *et al.* (1987) estimated potential yields of 24 t/ha/yr from the Fonte Boa ('mesocarpa' group) population for unselected, unfertilized germplasm on poor soils with a favourable climate. Clement & Mora Urpi (1987) suggest that much higher yields should be easily attained within a cycle or two of any improvement programme, although factors causing fruit drop must be eliminated, especially on poor soils.

Given these yields and strong market acceptance where it is well known, there are good prospects for producing good quality fruit for the local and regional markets. Market saturation levels are, however, unknown. Excessive supply would lower prices from the current US$0·50–1·00 per kilogram range for best quality fruit. This might increase consumption somewhat, but could lower farm income, unless alternative uses are developed.

Fruit for animal ration

This is the major alternative use being studied at the moment. Because its dried fruit can partially or completely substitute maize for many uses, the pejibaye may be considered a tree cereal. This is especially true for animal rations, where pejibaye flour can substitute part of the maize base that is generally used. Costa Rican researchers are currently leading this effort and Murillo & Zumbado (1986) have reviewed recent work with chicken feeds. Their team (Soto 1983; Zumbado & Murillo 1984; Cooz 1984; Loynaz 1985; Facuseh 1986; Espinoza 1986) has studied pejibaye preparation (autoclaving, extrusion, sun drying, etc.) and maize substitution levels for starter and primary rations for layers and broilers.

Using second quality fruit (first quality having high market value) their results showed that pejibaye can be used as the principal energy source in primary rations, but should be used less intensively in starter rations. Heat treatment is essential, especially for starter rations, to deactivate the trypsin inhibitor. Heat treatment by extrusion is cheaper than other methods tested and is recommended. Moderate levels (30–60 per cent substitution of maize) also allows the production of a cheaper meal, both for starter and for primary rations, at Costa Rican costs for maize and second quality pejibaye.

One attractive option, that avoids drying and heat treatments, is to ensile the fruit on the farm and feed it directly to stock, possibly with a protein supplement. There is some suggestion that acid ferment may break down the antinutritional factors (Sangil 1985) and it appears to be appreciated by pigs (I. Araujo, pers. comm.). Ensiling would also be an excellent way to store cheaply pejibaye fruit for animal feeds.

Potential yields for animal ration should be similar to those for fresh fruit (previous section), especially if used in silage. For use in chicken feeds, the fruit would have to be dried, thus reducing the final product to 50 per cent of the initial yield. A breeding programme for fruit for animal ration would select for starchy, low (10%) oil fruit, from the Putumayo macrocarpa landrace for example, and should yield in excess of 25–30 t/ha/yr of fresh bunches (Clement, in press).

In many areas of the humid tropics cereals do not yield well without considerable amounts of inputs and know-how. For example, on the nutrient poor oxisols near Manaus, maize smallholders rarely obtain more than 800–1000 kg/ha/yr. This suggests that pejibaye might develop a market as a component of animal rations, if it can be produced cheaper than imported maize. Tracy (1985) was able to obtain a small profit when using sun-power to dry second quality pejibaye for animal ration in a rainy climate. A more efficient, low-cost drying method should improve profits. A 300 ha plantation in Costa Rica will soon provide information on the economics of pejibaye for fresh fruit and animal ration.

Fruit for flour

To avoid saturating the high value fresh fruit market, it is interesting to consider the potential for developing other products for human consumption. Among Calvo's (1981) 40 recipes are several for making breads, cakes and other pastries from pejibaye flour. Tracy (1986) tested pejibaye flour, mixed with wheat flour, for bread in Costa Rica. Ninety per cent wheat and 10 per cent pejibaye gave bread dough with excellent baking characteristics, slightly less protein, more energy (from the oil) and more vitamin A (beta-carotene). Eighty-five per cent wheat and 15 per cent pejibaye gave a slightly heavier dough, similar to 'natural' whole wheat bread. Both were acceptable to a small group of consumers. This product might develop well in the 'natural foods' market.

Tracy (1986) also tested several cake recipes with good results, both in the kitchen and on the market. Pejibaye flour must be mixed with wheat flour to make a good cake in Costa Rica. In the Manaus region, however, some peji-bayes can be used pure, with excellent results. This may be due to the higher oil content.

Clement & Mora Urpi (1987) point out that pejibaye flour is quite similar to yellow cassava or maize flour and could substitute these in many areas, with nutritional advantages over the cassava flour. This was one of the alternative products developed by the Amerindians, who appear to have domesticated the Putumayo and Vaupes 'macrocarpa' landraces specially for this purpose

(Clement, in press). These landraces have extremely high starch levels and low oil levels in large fruit, excellent for making flour that stores well for long periods. This flour can also be fermented to make 'chicha', a beer-gruel with good peachy flavour. Flours can also be extruded to make a variety of attractive snacks that might find a large market.

As Tracy (1985) pointed out, for regions that do not produce bread cereals, like the humid tropics, even 10 per cent substitution of wheat can have a favourable effect on the local balance of payments, by reducing imports. Yields of 10–12 t/ha/yr of dry flour are thought to be possible, if pests and diseases are controlled (Clement & Mora Urpi 1987) and would probably be economically viable. Phytosanitary quality control would be extremely important, as shown by Piedrahita & Velez's (1982) 30 per cent unacceptable fruit quality levels, as this flour would go to human consumption and must be of high quality.

Fruit for oil

Arkcoll & Aguiar (1984) were the first to point out pejibaye's oil potential. By searching for high oil fruit these authors eventually found fruit with 62% oil in the dry mesocarp and 34% oil on bunch weight. Clement & Arkcoll (1985) later pointed out that the oily fruit are more frequent in the more primitive (i.e. less selected) populations, especially in the 'microcarpa' group. This seems to be due to Amerindian selection for starch, which, being negatively correlated with oil, means that the more selected landraces have low oil levels in the mesocarp. While per hectare oil yield (extrapolated from the best plants, with *in vitro* cloning) (Clement & Arkcoll 1985) are lower than established crops, they are higher than these were at a similar stage of development, suggesting that modern breeding and biotechnology methods could produce a new oil crop quickly.

Although there are already excellent oil crops for the wet tropics (i.e. *Elaeis guineensis* and *Cocos nucifera*), the need for greater crop diversity is well demonstrated by the diseases and pests that limit or prevent their use in several countries (Meunier 1976). Pejibaye has an advantage over these two in that it will also provide a significant quality of a high quantity meal after oil extraction, suitable for humans and animals in regions where this would otherwise be imported.

Unsaturated fatty acids are more common in pejibaye mesocarp oil than in oil palm (Table 31.5), although the caiaué (*Elaeis oleifera*) has similar oils and is now being introgressed with oil palm to improve oil quality in that species (Hartley 1977). The fact remains, however, that unsaturated fats have a good market value at present, and are interesting from both a nutritional and industrial point of view.

As pointed out by Arkcoll & Aguiar (1984) and emphasized by Clement & Arkcoll (1985), most pejibayes have oil separation problems when pressure extracted. The oil, starch and water form an emulsion that must be solvent extracted. There are plants that have good separation characteristics, but using only these would severely reduce the genetic base of any improvement programme. The oiliest pejibayes are relatively dry, however, so that separation may not be a severe problem with improved materials.

Clement & Arkcoll (1985) suggest that yields of 2–3 t/ha/yr of oil are immediately feasible with *in vitro* cloning and could easily be raised to 5 or more tons in an improvement programme. The germplasm is available (Clement & Arkcoll 1985) and trial crosses have been made, but long-term financing is required for this breeding programme. The problem of fruit drop must also be resolved before this product can become commercially viable.

Palmito

Palmito, or palm hearts, are the only product for which pejibaye is currently grown on a commercial scale. There are more than 2000 ha planted in Costa Rica (Clement & Mora Urpi 1987), although the market appears to be very volatile because they must compete with palmito from the acai palm (*Euterpe oleracea*). The latter are extracted nearly free of cost from enormous natural populations in the estuary of the Amazon river. Quality control varies considerably from company to company and their intensive exploitation is devastating the natural populations. Both of these factors may open a larger market for pejibaye, as plantation quality control has been good and plantations are managed rather than devastated.

Processing technology for the pejibaye palmito has been developed in Costa Rica and in Brazil. The most detailed studies are by Ferreira *et al.* (1982a & b) at the Instituto de Tecnologia de Alimentos (ITAL), in Campinas, Sao Paulo, Brazil. V. Ferreira (pers. comm.) has observed that the Costa Rican palmito may be stored without the salt solution changing its appearance, while the Brazilian palmito becomes turbid within a short time. Research is underway to identify the causes of this turbidity, so as to be able to select against it in an improvement programme. Clement *et al.* (in press) have recently reviewed the research results and necessities with respect to palmito of pejibaye and suggest an improved numerical idiotype.

Most pejibayes are spiny, both on the trunk and on the leaf petiole and rachis, which complicates extraction of the palmito. Several spineless populations have been found in western Amazonia, especially around Yurimaguas, Peru (Clement *et al.* in press). Many other western Amazonian populations show variable frequencies of spineless or reduced spininess. There is also a spineless population in Costa Rica. This germplasm will form the genetic base for the improvement programmes currently planned in Brazil and Costa Rica.

Moriera Gomes & Arkcoll (in press) report yields of 1·2 t/ha of market quality palmito on oxisols in Manaus at the first harvest at 2 years, falling to between 600 and 900 kg/ha at subsequent harvests. This decline is probably due to management of the plantation, rather than biological factors, as it has not been reported from Costa Rica. Zamora (1985) reported 3 t/ha/yr of field harvested palmito in Costa Rica, of which 20 to 30 per cent is of cannable market quality. Zamora (1985) also reported on a density trial whose highest yields were above 3·5 t/ha/yr. With selected germplasm and good agronomic practices it may be possible to attain nearly 2 t/ha/yr of market quality palmito.

Both the apical and basal residue from palmito extraction have some potential use and require more intensive research to find new uses. The basal residue, just below the apical meristem, is very tender and has a crispy texture. This could be made into a cream soup or be thin-sliced (transversely) as a substitute for bamboo shoots or deep fried to make chips. The apical residue is slightly fibrous leaf and petiole material, which can serve as a vegetable. Use of the rest of the tree as a forage is also possible (Moreira Gomes & Arkcoll, in press). With the use of both of these residues palmito costs could be lowered, and may eventually compete with acai.

RESEARCH PRIORITIES

In the previous section, utilization and processing have been identified as major limiting factors for the expansion of pejibaye as a crop. Although this is a major area that needs research, there are two other areas that should be receiving priority attention: phytosanitary problems and plant physiology for improvement. These priority areas are currently receiving some attention, but funds for research are meagre in Latin America at present, so that these efforts are far from ideal. Clement (in press) has recently reviewed breeding objectives and presented ideotypes for each major potential use. Space does not permit a detailed analysis of research necessities, so only a short outline will be presented here and the discussion will be divided in two: fruit and palmito.

For fruit production, research into alternative uses for the fruit, flour and residues from oil extraction is necessary. Processing of the whole fruit and the flour will be necessary to open export markets for pejibaye, as fresh fruit export from the humid tropics is unlikely to develop beyond the curiosity level. The 'natural foods' market should become an initial target and will require food technology aimed specifically at these consumers. Economic elimination of the trypsin inhibitor must be developed. AT CIPRONA, drum driers are being studied as a way of producing flour. Early results indicate elimination of the inhibitor but require careful economic analysis, as costs must remain competitive with maize and wheat. Ensiling whole or mashed fruit must be studied as a means of storing and treating pejibaye for animal rations.

A detailed analysis of chemical variation within and among landraces will guide the breeder in his selection of a genetic base for each improvement programme. Especially important are starch, protein, oil and carotene quantities and qualities. The nature of the oil/starch complex must be defined and

318

economic means of extraction on an industrial scale developed. Detailed analysis of the starch might suggest new uses for the flour.

Phytosanitary problems, both insect and diseases, must be surveyed so that potential economic pests can be identified before they become a problem. The problem of fruit drop is especially serious in the Amazon, where poor soils surely play a part in lowering plant defences to pest attack. This suggests that plant micro-nutrient needs must be linked with phytosanitary research. On better soils, fruit drop does not appear to be a problem, but other insect and disease pests may be. *Phytophthora* species are currently becoming a problem in Costa Rica, and leaf mites are a problem in some areas of Brazil and Costa Rica. Biological control of pests, agro-ecological control of diseases and genetic resistance to both must be researched as major priorities, especially where pejibaye may become a monoculture crop.

Growth and physiology of the plant, bunch and fruit, both pre- and post-harvest, require attention, especially to support the improvement programmes. Corley (1983) has emphasized the importance of growth and physiological parameters in the improvement of yield of tropical crops and this methodology is currently being adapted for use with pejibaye. There are, however, many aspects of bunch and fruit growth and physiology that require equally urgent attention.

Through the use of the Bunch and Harvest Indexes it will be possible to select for dwarf plants, as large annual height increment during the early years makes the pejibaye difficult to harvest. Reduced height increment is essential for monoculture plantations, but may not be so important for agroforestry systems (Clement 1986b). The efficiency with which the plant absorbs, uses and recycles nutrients must be studied, as genetic variation of these qualities exists and will certainly be important selection criteria in improvement.

For palmito production the need for food technology research is less urgent, as the process is quite simple and well studied. The turbidity in canned palmito from the Amazon must be studied to determine the causes and whether or not these are amenable to selection. Alternative methods for preparing market quality palmito might widen the market. Uses for the apical and basal residues must be found and marketed.

The production of palmito is a simpler system than the production of fruit, as the plant is always in a juvenile stage, i.e. only the vegetative aspects are important, and because the economic product is also vegetative. Growth and physiological parameters must be studied and selected for in the opposite direction to that for fruit production. More vigorous plants, with larger leaves and, especially petioles, are required, while for fruit production dwarf characters will be favoured.

Because the palmito is basically a leaf crop, leaf and stem pests and diseases must be intensively studied. Leaf mites have already been observed in Brazil and Costa Rica and were used as population descriptors in Clement (1986a). Apparent genetic variation for resistance to these pests has been observed. Some

Amazonian populations appear to be especially susceptible to leaf mite in Costa Rica, so much so that the major spineless population from the Amazon (Yurimaguas, Peru) can probably not be used in the Costa Rican improvement programme.

For most of the research necessities listed here, there are able researchers ready to work. The major problem is that financial resources are scarce and are not generally available for the development of new crops. In order for the pejibaye, and other crops mentioned at this symposium, to develop to the point of entering into the national or world markets, international financial assistance will be necessary.

REFERENCES

Arkcoll, D. B. and P. L. Aguiar (1984) Peach palm (*Bactris gasipaes* H. B. K.), a new source of vegetable oil from the wet tropics. *J. Sci. Food Agric.* 35: 520–526.

Calvo, I. M. (1981) *Usos culinarios del chontaduro.* Inst. Ciencias del Valle Caucano, Cali, Colombia.

CIPRONA (1986) *Aprovechamiento industrial del pejibaye (Bactris gasipaes).* Rep. Invest. Centro de Investigaciónes de Productos Naturales/Univ. Costa Rica, San José.

Clement, C. R. (1986a) *Descriptores minimos para el pejibaye (Bactris gasipaes H. B. K.) y sus implicaciónes filogenéticas.* Thesis, Univ. Costa Rica, San José, Costa Rica.

Clement, C. R. (1986b) The pejibaye palm (*Bactris gasipaes* H. B. K.) as an agroforestry component. *Agroforestry Systems* 4: 205–219.

Clement, C. R. (1987) A pupunha, uma arvore domesticada. *Ciencia Hoje* 5, 29: 42–49.

Clement, C. R. (in press) Domestication of the pejibaye palm (*Bactris gasipaes*): past and present. *Advances in Economic Botany.*

Clement, C. R. and D. B. Arkcoll (1985) El *Bactris gasipaes* H. B. K. (Palmae) como cultivo oleaginoso: potencial y prioridades de investigación. In: *Informe del Seminario-Taller sobre Oleaginosas Promisorias,* L. E. Forero P. (Ed.), pp. 160–179. Programa Interciencias de Recursos Biologicos, Bogota, Colombia.

Clement, C. R. and J. Mora Urpi (1985) Phenotypic variation of peach palm observed in the Amazon basin. In: *Final report: Peach palm (Bactris gasipaes H. B. K.) germplasm bank,* pp. 92–106. US Agency of International Development, San José, Costa Rica.

Clement, C. R. and J. Mora Urpi (1987) The pejibaye (*Bactris gasipaes* H. B. K. Arecaceae); multi-use potential for the lowland humid tropics. *Econ. Bot.* 41: 302–311.

Clement, C. R., W. B. Chavez F. and J. B. Moreira Gomes (in press) Copnsiderações sobre a pupunha (*Bactris gasipaes* H. B. K.) como produtora de palmito. In: *Anais do Primeiro Encontro de Pesquisadores em Palmito.* Centro Nacional de Pesquisas Florestais/EMBRAPA, Curitiba, Brazil.

Cooz S., A. (1984) *Efecto de la sustitución de maiz por harina de pejibaye en dietas para pollas de reemplazo durante la etapa de iniciación.* Thesis, Univ. Costa Rica, San José.

Corley, R. H. V. (1983) Potential productivity of tropical perennial crops. *Exper. Agric.* 19: 217-237.

Espinoza J., A. (1986) *Sustitución del maiz por harina de pejibaye tratada térmicamente en dietas para gallinas ponedoras.* Thesis, Univ. Costa Rica, San José, Costa Rica.

The Pejibaye Palm

FAO (1986) *Food and Fruit-bearing Forest Species. 3. Examples from Latin America.* Forestry Paper 44/3, FAO, Rome.

Facuseh J., E. (1986) *Efecto del tiempo de almacenamiento, tratamiento termico y suplementación energetica de la harina de pejibaye (Bactris gasipaes) en dietas para pollos parrilleros.* Thesis, Univ. Costa Rica, San Jose, Costa Rica.

Ferreira, V. L. P., M. Graner, M. L. A. Bovi, I. S. Draetta, J. E. Paschoalino and I. Shirose (1982a) Comparação entre os palmitos de *Guilielma speciosa* e *Euterpe edulis*. I. Avaliações fisicas, organolepticas e bioquimicas. *Coletanea do Instituto de Tecnologia de Alimentos* 12: 255–272.

Ferreira, V. L. P., M. Graner, M. L. A. Bovi, I. B. Figueiredo, E. Angelucci and Y. Yokomizo (1982b) Comparacão entre os palmitos de *Guilielma speciosa* e *Euterpe edulis*. II. Avaliações fisicas e quimicas. *Coletanea do Instituto de Tecnologia de Alimentos* 12: 273–282.

Fonseca, E. T. (1927) *Oleos vegetais brasileiros.* Instituto de Tecnologia de Alimentos, Campinas, Brazil.

Hammond, E. G., W. P. Pan and J. Mora Urpi (1982) Fatty acid composition and glyceride structure of the mesocarp and kernel oils of pejibaye palm (*Bactris gasipaes* H. B. K.). *Rev. Biol. Trop.*, 30: 91–93.

Hartley, C. W. S. (1977) *The Oil Palm.* Longmans, London.

INN (1959) *Tabla de composición de alimentos Colombianos.* Inst. Nacional de Nutrición, Bogota.

Johannessen, C. L. (1967) Pejibaye palm: physical and chemical analysis of the fruit. *Econ. Bot.* 21: 371–378.

Leung, W.-T. (1961) *Tabla de composición de alimentos para uso en América Latina.* INCAP, Ciudad de Guatemala.

Loynaz, B., A. (1985) *Utilizacion de la harina de pejibaye extrusada bajo diferentes temperaturas en dietas de iniciación de pollos de engorde.* Thesis, Univ. Costa Rica, San José, Costa Rica.

Meunier, J. (1976) Prospectives for the Palmae. A necessity for the improvement of oil yielding palms. *Oleagineaux* 31: 156–7.

Mora Urpi, J. (1984) El pejibaye (*Bactris gasipaes* H. B. K.): origen, biologia floral y manejo agronomico. In: *Palmeras poco utilizadas de América tropical,* pp. 118–160. FAO/Centro Agronomico Tropical de Investigación y Ensenanza, Turrialba, Costa Rica.

Mora Urpi, J. and C. R. Clement (1985) Races and populations of peach palm found in the Amazon basin. In: *Final report: Peach palm (Bactris gasipaes H. B. K.) germplasm bank,* pp. 107–141. US Agency of International Development. San José, Costa Rica.

Moreira Gomes, J. B. and D. B. Arkcoll (in press) Estudos iniciais sobre a produção de pupunha (*Bactris gasipaes*) em plantações. In: *Anais do Primeiro Encontro de Pesquisadores em Palmito.* Centro Nacional de Pesquisas Florestais/EMBRAPA. Curitiba, Brazil.

Moreira Gomes, J. B., C. R. Clement, S. A. N. Ferreira and C. E. L. Fonseca (1987) *Variação fenotipica de pupunha (Bactris gasipaes H. B. K.) selecionada da população de Fonte Boa, Amazonas.* Rep. Invest. Instituto Nacional de Pesquisas da Amazonia, Manaus, Brazil.

Murillo R., M. and M. Zumbado A. (1986) *Composición quimica y valor nutritivo de la harina de pejibaye en la alimentacion de las aves.* Rep. Invest. Univ. Costa Rica/ CONICIT, San José.

Murillo R., M., A. Kroneberg, J. F. Mata, J. G. Calzada and V. Castro (1983) Estudio preliminar sobre factores inhibidores de enzimas proteolíticas en la harina de pejibaye (*Bactris gasipaes*). *Rev. Biol. Trop.* 31: 227–231.

NAS (1975) *Underexploited Tropical Plants with Promising Economic Value.* National Academy of Sciences, Washington, DC.

Noiret, J. M. and W. Wuidart (1976) Possibilitiés d'amélioration de la composition en acides gras de l'huile de palmae. Résultats et perspectives. *Oleagineaux* 31: 465–474.

Patino, V. M. (1963) Plantas cultivadas y animales domesticos en América Equinoccial. Imprenta Departmental, Cali, Colombia.

Piedrahita G., C. A. and C. A. Velez P. (1982) *Metodos de obtención y conservación de las harinas obtenidas a partir de los frutos de la palma de chontaduro (Bactris gasipaes H. B. K.).* Rep. Invest. Univ. del Valle, Cali, Colombia.

Popenoe, W. and O. Jimenez (1921) The pejibaye, a neglected food plant of tropical America. *J. Heredity* 12: 154–66.

Sangil, J. (1985) *Evaluación del efecto de diferentes procedimentos quimicos y biologicos sobre la actividad inhibidora de enzimas proteoliticas presentes en la harina de pejibaye (Bactris gasipaes) utilizada en alimentacion animal.* Thesis, Univ. Costa Rica, San José, Costa Rica.

Silva, W. G. and G. Amelotti (1983) Composizione della sostanza grassa del fruitto di *Guilielma speciosa* (Pupunha). *Riv. Ital. Sostanza Grasse* 60: 767–70.

Soto T., S. (1983) *Utilización de la harina de pejibaye en dietas para pollos de engorde.* Thesis, Univ. Costa Rica, San José, Costa Rica.

Tracy, M. D. (1985) *The pejibaye fruit: problems and prospects for its development in Costa Rica.* Thesis, Univ. Texas at Austin, Austin, Texas.

Tracy, M. D. (1986) Harina de pejibaye, una opción prometadora. *Bulletin RETADAR* 23, 3.

Zamora F., C. (1985) Densidades de siembra de pejibaye para palmito con tallo simple. In: *Diversificación Agricola,* pp. 75–78. Asociación Bananera Nacional, San José, Costa Rica.

Zapata, A. (1972) Pejibaye palm from the Pacific coast of Colombia (a detailed chemical analysis). *Econ. Bot.* 26: 156–159.

Zumbado, A., M. and M. Murillo (1984) Composition and nutritive value of pejibay *(Bactris gasipaes),*in animal feeds. *Rev. Biol. Trop.* 32: 51–56.

32

Native Neotropical Palms: a Resource of Global Interest

M. J. Balick

INTRODUCTION

Palms are important sources of food, fibre, fuel, shelter and medicine, and greatly enhance the physical environment in areas where they are found in abundance. Several important species of palms come to mind in any consideration of this family, namely, the date palm (*Phoenix dactylifera*), coconut palm (*Cocos nucifera*), and African Oil Palm (*Elaeis guineensis*). These are the so-called major species of great value to the cash economy of vast regions of the tropics, with significant contributions to the gross national product of many countries. For example, global production of African palm oil exceeds four million tons annually, and over one million tons of coconut oil are exported each year. The African oil palm is a relatively recent domesticate, taken from the wild and planted in the Old World around the turn of the century. Yields of this species have risen dramatically as compared to its original yields in the wild.

There are several hundred other species of palms in the tropical and subtropical regions of the world that provide a great variety of commercial and subsistence commodities of benefit to people. These regional resources are of such value that they should be considered of global interest, and be properly inventoried and evaluated for use before the rainforest habitats where they are found are destroyed.

One of the great advantages of the so-called minor species of useful palms is that they grow in habitats considered marginal for conventional forms of agricultural exploitation. For example, one of the great palm resources of Amazonian Peru, *Mauritia flexuosa,* grows primarily on swampy or flooded land. Local people harvest these palms on a regular basis, obtaining food, fibre, oil, thatch, insect protein and other useful commodities from this species. There are almost no alternative uses for the land on which this palm grows, if the options are limited to conventional crops. Another major advantage of the minor species of palms is their multi-use, as mentioned above, providing a variety of products from a single plant.

It is extremely difficult to ascertain the true value of the native palm species as currently utilized. There are few if any accurate estimates of the impact of palms on the subsistence economies of tropical-dwelling peoples. It would be very useful to try and calculate the replacement value of the hundreds of plants, including palms, used by an Amazonian Indian or colonist as part of their life-style. Only then will we begin to have an idea about the true value of these native plants. One of the better examples for estimating the value of palms in a neotropical setting is found in a government report from Brazil (IBGE 1982) and data compiled from this are presented in Table 32.1. In this example from a single country the cash value of the harvest and sale of 14 local palms is estimated at over 100 million dollars. As such estimates are based on field interviews with local collectors and industries, it is often the case that the true economic impact is much greater. Our own work with one of the species in Table 32.1, *Orbignya phalerata*, has shown it to have an economic impact in the commercial sector alone over twice that listed in Table 32.1, and of even greater importance to the subsistence-based peoples whose livelihood depend heavily on its products.

Table 32.1: Economic Value and Quantity of Native Palm Products Produced in Brazil – 1980

Product	Species (Common Name)	Quantity (Tons)	Value (1980 $)
Wax	*Copernicia cerifera* (Carnauba)	18 857	16 426 054
	Syagrus coronata (Licuri)	10	5652
Fibre	*Mauritia flexuosa* (Buriti)	614	446 862
	Copernicia cerifera (Carnauba)	1399	53 521
	Syagrus sp. (?) (Butia)	1186	60 886
	Attalea funifera (Piaçava)	55 939	12 717 109
	Astrocaryum tucuma (Tucum)	102	48 534
Oils	*Orbignya phalerata* (Babaçu)	250 951	60 432 775
	Syagrus coronata (Macauba)	195	19 441
	Astrocaryum sp. (Tucum)	8381	1 519 625
	Astrocaryum murumuru (Murumuru)	10	440
Other Foods	*Euterpe oleracea* fruits (Açai)	59 591	7 939 527
	Euterpe spp. and others (Palmito)	114 408	6 777 029

At the current time there is a resurgence of interest in the use of native palm resources. A number of international foundations are supporting research on the evaluation of the economic impact of these species, with the idea that the resources should be identified and saved from extinction either through improved utilization of existing stands (management) or domestication for use elsewhere in the tropics and subtropics. Stimulated by increased research activity, scientists in a number of countries have formed networks, either formal or informal, to advance knowledge about and protection of the palm resources. The first major symposium devoted to useful neotropical palms was held at CATIE, Costa Rica in 1983, and the resulting proceedings are an action plan for the future utilization and conservation of tropical American palms (FAO 1983). A more recent symposium held at the New York Botanical Garden in 1986, supported in part by the World Wildlife Foundation-US, explored the value and conservation needs of under-utilized palm species around the world (Balick, in press). In addition, workshops on individual neotropical palms, such as *Jessenia bataua*, *Bactris gasipaes* and *Orbignya phalerata* in Brazil and Colombia have helped stimulate activity in improving utilization and protection of these species.

There is a great volume of literature on the uses of the lesser-known palms, and the reader should refer to bibliographies on the various species (e.g. Guerro & Clement 1982; Balick & Beck, in press) for more specific information. The purpose of the present contribution is to provide a very brief summary of the range of uses of the neotropical palms, and some thoughts on the need for their increased rational exploitation in a way that also ensures the long-term conservation of this resource.

PRODUCTS FROM NEOTROPICAL PALMS

This section is devoted to the commodities provided by neotropical palms, and their impact on commercial and subsistence economies.

Oil

Neotropical palms are especially high in the diversity of oils produced. Two types of oils are common, that found in the mesocarp and another found in the endosperm. Neotropical genera that produce oil include *Acrocomia*, *Astrocaryum*, *Bactris*, *Butia*, *Elaeis*, *Jessenia*, *Manicaria*, *Mauritia*, *Oenocarpus*, *Orbignya*, *Scheelea* and *Syagrus*. The chemical characteristics of some of these oils, as measured by their fatty acid composition, are outlined in Table 32.2. It is evident from this table that neotropical palm oils can range from lauric acid types (*Orbignya*, *Manicaria* similar to coconut oil to oleic acid types *Jessenia* similar to olive oil). Interestingly enough, the greatest concentration of oleic

Table 32.2: *Native Neotropical Oil Palms – Fatty Acid Composition*

Fatty acid (% in oil)	Astrocaryum murumuru (Kernel)	Astrocaryum tucuma (Kernel)	Manicaria saccifera (Mesocarp)	Jessenia bataua (Mesocarp)	Elaeis oleifera (Mesocarp)	Orbignya phalerata (Kernel)	Bactris gasipaes (Mesocarp)	Bactris gasipaes (Kernel)
Myristic	36·8	21·6	18·9		0·1–0·2	15·4		28·4
Palmitic	4·6	6·4	8·2	9·2	18·8–24·2	8·5	29·6	10·4
Palmitoleic					1·2–1·5		5·3	
Stearic	2·2	1·7	2·4	5·9	0·6–2·2	2·7		3·1
Oleic	10·8	13·2	9·7	81·4	63·0–67·0	16·1	50·3	18·2
Lauric	42·5	48·9	47·5	3·5		44·1		33·3
Linoleic					5·8–15·9	1·4	12·5	
Linolenic	0·4	2·5	1·4		0·5–0·6		1·8	5·1
Capric	1·6	4·4	6·6			6·6		0·6
Caprilic	1·1	1·3	5·3			4·8		0·5

Sources: Balick (1982); Mora-Urpi (1983).

acid-producing palms is in the Amazon Basin, an area where thousands of tons of olive oil are imported each year to meet domestic needs.

Among the higher yielding taxa of oleaginous palms are *Acrocomia totai*, *Bactris gasipaes*, and *Orbignya phalerata*. Each of these can produce thousands of tons of oil per hectare, either in plantation rows or through the management of wild or semi-wild populations. Both *A. totai* and *O. phalerata* are aggressive species, often taking over an area after the forest is cleared and becoming difficult to eradicate once established.

The present limitations to expanding the use of these palms include a lack of knowledge of the basic biology of the species, as well as the lower yields when compared with some of the current alternative crops. However, as seen in the example of the African oil palm, when a promising tree is domesticated, yields increase dramatically and the species can become competitive. One specific example of how the lack of basic biology hinders proper use is in the area of symbiont relationships. Some of the species, such as *Jessenia bataua* and *Orbignya phalerata* have mycorrhizal relationships in their wild habitats. When seeds are planted in sterilized nursery soils, growth is retarded and maturity delayed (Balick & T. V. St. John, unpubl. data). Other species have complex pollinator relationships (Henderson 1986) and, if the pollinators are not present at the site where the palms are introduced, fruit set is inhibited and yields are reduced.

Fruit

One of the better examples of fruit production from native neotropical palms is in the genus *Euterpe*, where almost 60 000 tons of *E. oleracea* fruit are harvested annually in Amazonian Brazil for production of a nutritious beverage and sherbet. Ripe panicles of fruit are cut from the trees, which often grow in seasonally inundated habitats or in swampy areas, and the mesocarp removed and mixed with water to produce the dark purple beverage. While the nutritional composition of the beverage is not overwhelming, it is an important adjunct to the regional diet as well as the economy.

Fruits of many neotropical palms are eaten locally, as the panicles ripen and the fruits fall to the ground. The mesocarp of *Acrocomia totai*, *Mauritia flexuosa*, *Aiphanes caryotifolia*, *Jessenia bataua* and *Oenocarpus bacaba* are favourite foods in the areas where the palms are found. Many of these mesocarps are brightly coloured – orange, yellow and red – a characteristic presumed to aid in seed dispersal. Our analyses of these fruits show the mesocarp pulp to be rich in vitamin A. For example, the vitamin A (from carotene) content of *Mauritia flexuosa* is 7190 IU/100 g (wet weight basis) and that of *Aiphanes caryotifolia* 16 000 IU/100 g (wet weight basis) (Balick, unpubl. data; Balick & Gershoff, in press). While not always appreciated by outsiders, the nutritional benefits derived from these fruits are important supplements to the local diet. In some areas the fresh fruits are sold commercially (Padoch 1987) while elsewhere they may be preserved or made into a sweet paste.

Fibre

The fibre produced from palms is defined as a structural fibre, as it is derived primarily from the supporting tissue of the plant. Schultes (1977) outlined a number of important fibre palms of the Amazon valley, including *Astrocaryum tucuma, A. vulgare, Euterpe oleracea, Mauritia flexuosa* and *Oenocarpus bacaba*. Fibre produced from *Astrocaryum* species is quite durable, and during the last century was used in rigging for naval vessels. The Museums of Economic Botany at Kew contain rope samples produced from this genus that were collected by Richard Spruce during his lengthy explorations of the Amazon Valley in the 1800s. Today the fibre is used to produce hammocks, nets, fishing lines, bow strings, fishing spears and handicrafts for trade or sale. As shown in Table 32.1, in 1980 Brazil reported commercial production of 102 tons of fibre from *Astrocaryum tucuma*.

Leopoldinia piassaba is an important fibre palm used for brush making and brooms. It is harvested from the leaf petiole that produces long strands of fine rounded fibres 1·5–1·8 m long. The fibres are harvested by hand from wild stands of palms, cleaned and bundled into cone-shaped bunches ca. 1·2 m tall. With shortages in broom fibre from other natural sources, *Leopoldinia piassaba* could play a minor role in meeting this need.

An additional example of a promising fibre palm is the genus *Desmoncus*, the only group of vining palms in the Neotropics. These 'New World rattans', as they are sometimes called, grow to 40 metres or more into the forest canopy. The cane diameters range from ca. 1–2·5 cm, and are covered with spiny sheath and petiole bases. In the Amazon Valley, baskets are commonly made from these palms. One of the more unusual baskets is the *tipi-tipi*, or cassava press, a ca. 2 m long cylindrical basket press used to process fermented, grated cassava into meal. During a recent trip to Belize, I noted that the abundant populations of *Desmoncus* species, locally called bayal, are used extensively for handicrafts. Clearly the potential of this New World rattan needs further investigation.

Fuel

Approximately 2·5 billion people in the developing world use wood or wood products to meet 50 per cent or more of their energy needs (NAS 1980). Although palm 'wood' is not a common fuel, as it is composed primarily of starchy ground tissue, palm fruits and leaf petioles are used for this purpose. For example, in many areas of South America, the solid endocarps of *Attalea, Scheelea* and *Orbignya* species are either burned when dry or converted to charcoal. Charcoal produced from *Orbignya phalerata* burns with more energy and less polluting fumes than mineral coal. In northeastern Brazil where these palms are common, the charcoal is used to fuel many industrial factories and also for household cooking. The charcoal produced from the endocarps is the natural complement to oilseed production which only uses the oleaginous kernels for oil and press cake. Thus, three important industrial commodities are produced from the fruit of this single palm.

In Bolivia the leaf rachis and petiole of *Orbignya phalerata* are used to fuel bread ovens. Local people are of the opinion that this fuel source burns evenly and for a maximum amount of time, and thus is the fuel choice for this purpose. Because the trees are not destroyed to harvest the leaves for this purpose, rachis/petiole firewood appears to be a most reasonable solution for meeting selected fuel needs in this area.

Medicines and Chemical Products

While over 25 per cent of the prescriptions written for medicines in the US are based on a higher plant product (Farnsworth & Morris 1976), and perhaps half of all prescription medicines have some type of plant in their developmental history, palms are not usually thought of for their medicinal or chemical value. Plotkin & Balick (1984) did a survey of the folk uses of new world palms and found that at least 48 species had references to medicinal value. Various palm parts, such as fruits, leaves, stems and roots are purported to have medicinal value, and are employed in the preparation of teas, poultices, curative baths and therapeutic drops. Some of these uses are presented in Table 32.3. Table 32.4 presents information on the biological activities of selected palm extracts. This information has been gathered from a literature survey of palms from around the world. In general, few palms have been tested for biological activity; the more common ones that have been investigated have proven of interest. It is recommended that additional research be carried out in this area.

Table 32.3: Some Ethnomedicinal Uses of South American Palms

Species	Use	Part
Acrocomia sclerocarpa	Increase fertility	Fermented sap
Astrocaryum ayri	Taenifuge	Dried endosperm
Bactris minor	For snakebite	Fruits
Copernicia cerifera	Diuretic, depurative	Dried roots
Desmoncus rudentum	Eczema	Roots
Euterpe oleracea	Anti-diarrhoea	Oil from fruits
Bactris insignis	Anti-rheumatic	Oil from seeds
Jessenia bataua	Anti-tubercular	Oil from mesocarp
Manicaria saccifera	Cough remedy	Liquid endosperm
Maximiliana maripa	For colds	Fruits
Scheelea princeps	Hair conditioner	Endosperm oil
Syagrus picrophylla	Diuretic	Fermented mesocarp

One new line of investigation that is proceeding is the collection of New World palms by the New York Botanical Garden for testing for anti-cancer and anti-AIDS activity, as part of a more general collection program under contract with the National Cancer Institute in Washington DC. Because this program is at a very early stage, no conclusions can be reported at this time.

Table 32.4: *Biological Activities of Selected Palm Extracts*

Species	Part	Activity	Reference
Borassus flabellifer	dried shoots	cytotoxic	Greig *et al.*(1980)
Borassus flabellifer	dried shoots	general toxicity	Greig *et al.*(1980)
Calamus floribundus	aerial parts	hypotensive	Dahr *et al.*(1973)
Calamus rotang	aerial parts	antitumor	Dahr *et al.*(1968)
Calamus rotang	aerial parts	antispasmodic	Dahr *et al.*(1968)
Cocos nucifera	seed oil	oestrogenic	Booth *et al.*(1960)
Cocos nucifera	dried shell	hypoglycemic (weak)	Morrison & West (1982)
Daemonorops draco	resin	antituberculosis	Rao *et al.* (1982) (weak)
Elaeis guineensis	seedling	oestrogenic	Butenandt & Jacobi (1933)
Hyphaene thebaica	fruit	ganglionic blocking	Sharaf *et al.*(1972)
Hyphaene thebaica	fruit	hypotensive	Sharaf *et al.*(1972)
Hyphaene thebaica	fruit	cardiotonic	Sharaf *et al.*(1972)
Hyphaene thebaica	fruit	smooth muscle stimulant	Sharaf *et al.* (1972)
Hyphaene thebaica	fruit	uterine relaxation	Sharaf *et al.* (1972)
Phoenix dactylifera	kernel	oestrogenic	Butenandt & Jacobi (1933)
Phoenix dactylifera	dried pollen	oestrogenic	El Ridi & Wafa (1947)
Phoenix dactylifera	pollen	gonadotropin synthesis inhibition	Soliman & Soliman (1957)
Phoenix dactylifera	pollen	gonadotropin synthesis stimulation	Soliman & Soliman (1957)
Phoenix dactylifera	pollen	gonadotropin release stimulation	Soliman & Soliman (1957)
Phoenix sylvestris	leaf	hypothermic	Dahr *et al.* (1973)
Phoenix sylvestris	leaf	hypotensive	Dahr *et al.* (1973)
Phoenix sylvestris	fresh leaf	antiascariasis	Chaturvedi & Tiwari (1981)
Sabal serrulata	dried fruit	anti-inflammatory	Wagner & Flachsbarth (1981)
Serenoa serrulata (= *S. repens*	fruit	oestrogenic	Elghamry & Hansel (1969)
Washingtonia filifera	dried pericarp	plant germination inhibition	Khan (1982)

IMPROVING RESEARCH NETWORKS

Because the rates of tropical deforestation are increasing yearly, greater and greater areas of palm habitats are falling under the axe or chainsaw each year. There is now the realization that global resources once taken for granted, such as fuel and industrial feedstocks, are truely finite and may well be exhausted within the next century. Thus, there is the need for increasing the level of commitment to investigating the palm resource before it is destroyed through overexploitation and ignorance. As mentioned previously, research could proceed along two lines, domestication of species for agriculture and improving management techniques for the rational exploitation of native palm populations.

It has been gratifying to see programs developing in Latin America along both of these lines. For example, the Museu Paraense Emilio Goeldi in Belém, Brazil has a program to study the ecology of native populations of *Euterpe oleracea* in the area and develop management strategies for improving the harvest of palm heart and fruit. The Instituto Nacional de Pesquisas Amazonicas (INPA) in Manaus, Brazil has a program to domesticate the *Bactris gasipaes* palm that involves a multinational effort to collect as much germplasm over the range of its distribution as possible, for evaluation in a plantation or smallholder setting. This particular endeavour is discussed elsewhere in this volume. Collaboration between a number of Brazilian agencies such as Centro Nacional de Recursos Genéticos/Empresa Brasileira de Pesquisa Agropecuária (CENARGEN/ EMBRAPA), Unidade de Execução de Pesquisa de Ambito Estadual-Teresina (EUPAE), Empresa Maranhense de Pesquisa Agropecuária (EMAPA), and the New York Botanical Garden since 1980 has resulted in much new information about *Orbignya phalerata* as well as the creation of trial plantings in several sites to improve this important resource. A recent program undertaken by CENARGEN/EMBRAPA is working towards the improved utilization of *Acrocomia totai*, which is estimated to have some of the highest yields of any of the neotropical oil palms.

All of these previous examples are evidence of the resurgent interest in the palm resource. However, it is essential if these efforts are to succeed that stable sources of funding be found to ensure the long- term research crucial to such an effort. This can be through international agencies or the collaboration of the private sector, as has happened in Brazil with some of the species being studied. Conferences that continue to build and strengthen the research networks are an essential part of this endeavour, and the past few years have seen several major convocations towards this end. There is now a source of written documentation to distribute research results, the newsletter produced by CENARGEN/ EMBRAPA sponsored by the FAO entitled *Useful Palms of Tropical America*. All of this leads me to be somewhat optimistic that, if current rates of increasing interest hold, many of these resources will receive at least a cursory evaluation before the balance of the genetic material is destroyed through deforestation. An analysis of the diversity of New World palms by Lleras *et al.* (1983) outlined 14 areas of high diversity of palm species. Unfortunately, these areas are also some of the areas of highest rates of destruction of tropical forest and colonization programs. It is important to realize that, even if germplasm of selected useful crop species is protected, the only real way of preserving it is through *in-situ* conservation, the creation of biological reserves. Clearly, we are in a race against time.

CONCLUSION

This paper has provided only the briefest overview of some of the present and potential uses of neotropical palms. The need for both basic and applied research on the palm source continues, as does the need for more aggressive conservation efforts. Through a serious investment in research on this family, many new species of value to both the industrial and subsistence sectors could be developed for use around the globe.

REFERENCES

Balick, M. J. (1982) Palmas neotropicales: nuevas fuentes de aceites comestibles. *Interciencia* 7, 1: 25–29.

Balick, M. J. (in press) The Palm – Tree of Life. Proceedings of a symposium on the biology, utilization and conservation of palms. *Advances in Economic Botany* 6.

Balick, M. J. and H. T. Beck (Eds) (in press) *An Annotated Bibliography of the Useful Palms of the World.* Colombia University Press, New York.

Balick, M. J. and S. N. Gershoff (Eds), (in press) A nutritional study of *Aiphanes caryotifolia* (Palmae) fruit, an exceptional source of vitamin A and high quality protein from Tropical America. *Advances in Economic Botany* 7.

Booth, A. N., E. M. Bickoff and G. O. Kohler (1960) Estrogen-like activity in vegetable oils and mill by-products. *Science* 131: 1807.

Butenandt, A. and H. Jacobi (1933) The determination and crystallization of plant tokokinins (thelykinins) and their identification as alpha-follicular hormones. *Z. Physiol. Chem.*

Dahr, M. L., M. M. Dahr, B. N. Dhawan, B. N. Mehrotra and C. Ray (1986) Screening of Indian plants for biological activity. Part 1. *Indian J. Exp. Biol.* 6: 232–247.

Dahr, M. L., M. M. Dahr, B. N. Dhawan, B. N. Mehrotra, R. C. Srimal and J. S. Tandon (1973) Screening of Indian plants for biological activity. Part IV. *Indian J. Exper. Biol.* 11: 43.

El Ridi, M. S. and M. A. Wafa (1947) An oestrogenic substance in palm pollen grains of the date palm. *J. Egypt. Med. Assoc.* 30: 124.

Elghamry, M. I. and R. Hansel (1969) Activity and isolated phytoestrogen of shrub palmetto fruits (*Serenoa repens*), a small estrogenic plants. *Experimentia* 25: 828.

FAO (1983) *Palmeras poco utilizadas de América Tropical.* Food and Agricultural Organization of the United Nations. Imprenta LIL, San José, Costa Rica.

Farnsworth, N. R. and R. W. Morris (1976) Higher plants – the sleeping giant of drug development. *Amer. J. Pharm.* 148: 46–52.

Greig, J. B., S. J. E. Kay and R. J. Bennetts (1980) A toxin from the palmyra palm, *Borassus flabellifer*: partial purification and effects in rats. *Food Cosmet. Toxicology* 18: 483-488.

Guerro, A. M. A de and C. R. Clement (1982) *Pejibaye (Bactris gasipaes) Bibliografía parcialmente anotada*, CATIE, Turrialba, Costa Rica.

Harishanker, H. C. Chaturvedi and P. V. Tiwari (1981) Incidence of ascarisis in children (1–2 years) and its treatment by certain drugs. *J. Sci. Res. Plants Med.* 2, 4:86–90.

Henderson, A. (1986) A review of pollination studies in the Palmae. *Bot. Rev.* 52(3) 221–259.

IBGE (1982) *Produção extrativa vegetal.* Fundação Instituto Brazileiro de Geografia e Estatística – IBGE, Rio de Janeiro.

Khan, M. I. (1982) Allelopathic potential of dry fruits of *Washingtonia filifera*: inhibition of seed germination. *Physiol. Plant.* 54: 323–328.

Lleras, E., D. C. Giacometti and L. Coradin (1983) Areas criticas de distribucion de palmas en las Americas para colecta, evaluacion y conservacion, In: *Palmeras poco utilizadas de América Tropical*. Food and Agricultural Organization of the United Nations. Imprenta LIL, San José, Costa Rica.

Mora-Urpi, J. (1983) El Pejibaye (*Bactris gasipaes* H. B. K.): origen, biologia floral y manejo agronómico. In: CATIE/FAO, Palmeras poco utilizadas de América Tropical. Imprenta LIL, S. A., San José, Costa Rica.

Morrison, E. Y. S. A. and M. West (1982) A preliminary study of the effects of some West Indian medicinal plants on blood sugar levels in the dog. *West Ind. Med. J.* 31: 194-197.

NAS (1980) *Firewood crops. Shrub and tree species for energy production*. National Academy of Sciences, Washington, DC.

Padoch, C. (1987) Risky business. *Nat. Hist.* 96(10): 56–65.

Plotkin, M. J. and M. J. Balick (1984) Medicinal uses of South American palms. *J. Ethnopharm.* 10: 157–179.

Rao, G. S. R., M. A. Gerhart, R. T. Lee IV, L. A. Mitscher and S. Drake (1982) Antimicrobial agents from higher plants, Dragon's Blood resin. *J. Nat. Prod.* 45: 646–648.

Schultes, R. E. (1977) Promising structural fiber plants of the Colombian Amazon. *Principes* 21: 72–82.

Sharaf, A., A. Sorour, N. Gomaa and M. Youssef (1972) Some pharmacological studies on *Hyphaene thebaica* fruit. *Qual. Plant Mater. Veg.* 22: 83.

Soliman, F. A. and L. Soliman (1957) The gonadotropic activity of date palm pollen grains. *Experientia* 13: 411–412.

Wagner, H. and H. Flachsbarth (1981) A new antiphlogistic principle *Sabal serrulata*. I. *Planta Medica* 41: 244–251.

33

Food Production by Selective Algal Biomass in the Desert

K. Shinohara, Y. Zhao and G. H. Sato

INTRODUCTION

The mass culture of microalgae has fascinated people for many years because of their extremely high productivity and high nutritional value, compared to terrestrial plants (Oswald 1980). The well-known examples are *Chlorella* and *Spirulina*. At present, some companies in several countries are producing about 10–300 tons dry algae in year-round operations in artificial open ponds (Ciferri & Tiboni 1985). However, the utilization of these algae for food is mostly limited to relatively expensive health foods, although a few products are used for feeding fish.

There are some factors for the limited use of algae. One of the major factors being the contamination of algal ponds with other organisms, resulting in the reduction of algal yield (Burlew 1953). This problem can be solved by selection of algae, especially the selection of thermophilic and halophilic (salt tolerant) algae which can grow in the harsh conditions that would be lethal to contaminants. Such harsh conditions can be found in the arid or desert area where there is sufficient sunlight necessary for the growth of plants and the space available for large scale agriculture or aquaculture. The only available water resource in such areas is sea water or brackish water, as many deserts are close to the sea. Shallow algal ponds in the desert would reach temperature extremes in which no potential contaminants can grow. The logical organisms then would be thermophilic and halophilic.

We have isolated some thermophilic and halophilic blue-green algae which have a high growth rate in the hot sea water in the desert areas (Shinohara *et al.* 1983). We are also developing the production of food by using the combination of sea water, sunlight, algae, brine shrimps, shrimps and fishes. This research is known as the 'Manzanar Project' and is directed by the W. Alton Jones Cell Science Center, Inc., in Lake Placid, New York, USA.

SELECTION OF THERMOPHILIC AND HALOPHILIC ALGAE

Selection of thermophilic and halophilic algae was carried out on algae which were collected from hot springs at temperatures of 60 °C. The algae were cultured in artificial sea water supplemented with an aqueous extract of sewage sludge. The extract was prepared by mixing five grams of sewage sludge with 10 ml of water. The mixture was boiled for ten minutes, filtered and the filtrate added to one litre of sea water as the final growth medium. Cultures were maintained at 45–50 °C under fluorescent illumination. The algae which grew were cloned in 1·5% agar in a medium composed of artificial sea water supplemented with an extract of sewage sludge; the clonal strains were grown in the sea water medium at 45–50 °C without CO_2 gas. By repeating these procedures, we isolated two kinds of thermophilic and halophilic blue-green algae, *Synechococcus elongatus* var. *vestitus* and *Spirulina subsalsa* var. *crassior*.

Synechococcus elongatus var. *vestitus* are blunt-end unicellular rods 1·8–2·4 μm wide and 4·0–7·6 μm long. Their growth temperature range is 25–60 °C; the optimal pH for growth is between 7·5 and 8·5. Halophilic *Synechococcus elongatus* var. *vestitus* showed no growth in fresh water medium, as shown in Fig. 33.1. *Spirulina subsalsa* var. *crassior* has spiral-shaped chains ca. 2 μm wide and ca. 100 μm long. Unlike other *Spirulina, S. subsalsa* var. *crassior* has no septa. The genus *Spirulina* is the most popular blue-green algae for algal production, especially *S. platensis* and *S. maxima*, which are currently exclusively used on an industrial scale because they form floating mats on the surface of ponds; therefore it is quite easy to harvest by filtration with simple nets. Generally, *Spirulina* grows very well in the alkaline medium and requires bicarbonate ions for growth. We also confirmed that *S. subsalsa* var. *crassior* grows in sea water supplemented with an extract of chicken manure in the presence of sodium bicarbonate without CO_2 gas. The temperature, pH and salinity range for its growth was 25–43 °C, 7·5–10·5 and 0–5%, respectively.

ALGAL PRODUCTION IN THE DESERT

Medium sized cultures of the clonal strains were grown in sea water medium in 14 litre carboys with bubbled air to increase gas exchange and agitation, and kept outdoors in the sun in the Imperial Desert area of California, USA. When the cultures reached maximum density ca. 4–8 g wet weight/litre, the entire 14 litres was added to a shallow pond (15·5 × 3 m) which had been filled to a depth of 20 cm with natural salt water supplemented with the extract of sewage sludge.

The algal yields varied with the different water temperatures. Figure 33.2 shows the relation between the yield of *S. elongatus* var. *vestitus* and the culture temperature in the pond. An aliquot (1 litre) of the culture was centrifuged and the dry sedimented algae weighed to give the yield. This weight was plotted against the water temperature at the time of sampling. The yield of algae increased with increasing water temperatures. High water temperatures were achieved by covering the pond with 6 mm clear polyethylene, but comparable temperatures could be achieved in hotter climates in uncovered ponds.

Figure 33.1: *Growth of Halophilic Strain of Synechococcus elongatus var. vestitus in Sea Water Medium and Fresh Water Medium.*

The sea water medium was composed of sea water supplemented with an aqueous extract of sewage sludge. The fresh medium used was that of Cassel & Hutchinson 1954. The water temperature was 45 °C, and the cultures were constantly illuminated with fluorescent lights.

▲ growth in sea water medium; △ growth in fresh water medium.

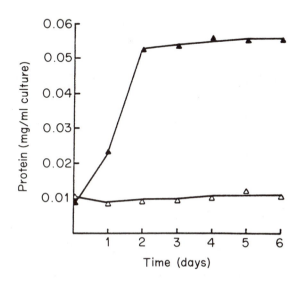

Composition of major components of *Synechococcus elongatus* var. *vestitus* and *Spirulina subsalsa* var. *crassior* are listed in Table 33.1, indicating that these algae contain about 60% protein. Amino acid contents of these proteins were comparable to the standard elaborated by the Food and Agriculture Organization and that of *Spirulina platensis*.

The growth of these algae was rapid at optimum temperature. Fig. 33.3 shows the amount of *Synechococcus elongatus* var. *vestitus* per unit volume on successive days. The cell density of algae doubled 24 hours after inoculation. The yield of *S. elongatus* var. *vestitus* was 30–60 g dry weight/m²/day which is equivalent to 50–100 tons dry weight/acre/year, while the yield of *Spirulina subsalsa* var. *crassior* was 20 g dry weight/m²/day (about 30 tons/acre/year). These yields were higher than those of conventional crops giving 5–30 tons dry weight/acre/year.

Figure 33.2: *Relation of Growth of Synechococcus elongatus var. vestitus and Water Temperature.*

Half of the culture in the pond was replaced daily with fresh sea water medium. Yields were calculated from the differences of algae content of the ponds after replacement of the contents 24 hours later.

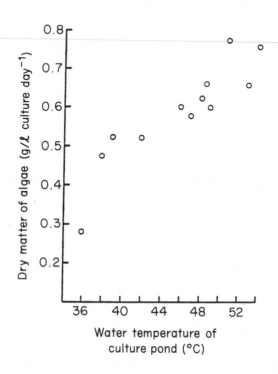

Table 33.1: *Composition of Major Components of Synechococcus elongatus var. vestitus and Spirulina subsalsa var. crassior*

%	Synechococcus elongatus[1]	Spirulina subsalsa[2]	Spirulina subsalsa[3]	Spirulina platensis[3]
Protein	59·3	61·7	64·6	62·4
Lipid	7·3	12·6	12·1	14·3
Carbohydrate	19·2	9·4	8·4	8·8
Ribonucleic acid	1·2	3·6	3·6	4·1
Ash	4·0	4·9	4·5	5·5

Media: 1. Sea water supplemented with 0·25% sewage sludge extract. **2.** Synthetic medium containing 26·0 g $NaHCO_3$, 0·1 g $MgSO_4$, 0·5 g K_2HPO_4, 0·04 g $CaCl_2$, 2·5 g $NaNO_3$, 0·01 g $FeSO_4$, 1·0 g NaCl, 0·08 g EDTA, 1·0 ml A_5 solution and 1 litre water. **3.** Sea water supplemented with 10·0 g $NaHCO_3$ and extract of 2·5 g chicken manure.

Figure 33.3: *Yield of Synechococcus elongatus var. vestitus per 24 hours after Inoculation.*

From the culture pond in May and June 1982; water temperature 41–45 °C.

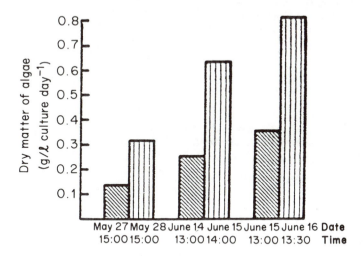

Microscopic examination showed these outdoor cultures to be almost pure. The dual selective conditions of high temperature and high salinity seem effectively to exclude other organisms. We have also fed brine shrimp, young carp, *Tilapia* spp., mice and chicken on these algae and found good maintenance of growth rates.

As stated above, we demonstrated that the selection of thermophilic and halophilic blue-green algae and the combination of these algae with desert, sea water, sunshine and sewage sludge or animal manure are quite useful for the large-scale production of pure inexpensive and highly proteinaceous food in the desert area.

AQUACULTURE IN THE DESERT

In the case of *Synechococcus elongatus* var. *vestitus*, because it is unicellular, it is difficult to harvest. We have developed the necessary aquacultural techniques in the desert area (Manzanar Project). This involves growing in the laboratory algae that thrive in hot sea water where competitive organisms are killed by heat and salt. The algae are fed to brine shrimp or *Tilapia* which are then fed to larger shrimps or fish. The method of food production is simple enough for people with limited training easily to maintain the unit. Recently, scientists from the Manzanar Project set up an actual production unit at Iquique in the Atacama Desert, Chile, one of the driest deserts in the world, with the cooperation of researchers

Figure 33.4: *Growth of Human Myeloma Cell Line, RPMI 8226 Cells in the Presence of Algal Extract.*

▲———▲ basal medium;

●———● basal medium supplemented with algal extract;

o———o basal medium supplemented with fetal calf serum.

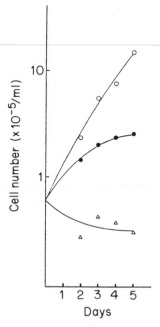

from the Arturo Prat University, Chile and another unit in Putian, Fujian Province, China. These facilities are actually producing brine shrimp, local shrimps and fish in two-acre ponds. The conversion ratio for algae fed to *Tilapia* was 1·4 which is comparable to that when commercial dry feed was used (McLarney 1984). This project will also provide economic opportunities in the desert area as well as nutritious and plentiful food.

USEFUL SUBSTANCES FROM ALGAE

Algal biomass can also be used as a source of chemicals and other materials for the food industry, medicine or research. Pigments such as phycocyanins, β-carotene, and xanthophylls have already been produced and marketed for food additives or feed supplement. The very high content of α-linolenic acid in algae, especially *Spirulina*, has also attracted attention recently since it is an essential fatty acid for animals and humans. We have also found the existence of biologically active compounds in blue-green algae. We found that extracts of *Synechococcus elongatus* var. *vestitus* and *Spirulina subsalsa* var. *crassior*

Figure 33.5: *Production of Mucus Polysaccharides by Spirulina subsalsa var. crassior, using 100 ml of culture medium.*

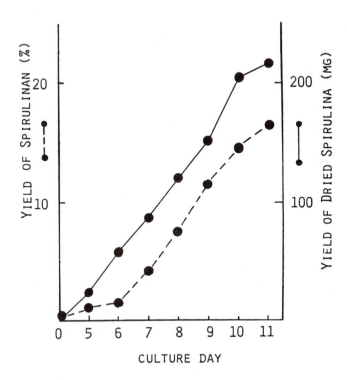

promote the growth of animal cells, especially lymphocyte cells which are involved in immunization as shown in Fig. 33.4 (Shinohara *et al.* 1986). One of the active compounds may be phycocyanins. It is also possible that the algal extracts or certain polysaccharides may have anticarcinogenic and anticataractic effects. *Spirulina subsalsa* var. *crassior* were also found to produce relatively high amounts of mucous polysaccharides into the medium (Fig. 33.5). The polysaccharides were extractable with hot water and formed a gel in the presence of K^+, Na^+, Ca^{++} and Mg^{++}. These properties are similar to those of carrageenans and alginic acid, which are widely used as food additives or chemicals.

We conclude that the algal biomass is a valuable source of biologically active materials as well as producing useful materials for the food industry.

REFERENCES

Burlew, J. S. (Ed) (1953) *Algal Culture from Laboratory to Pilot Plant*. Carnegie Institute of Washington, Washington DC.

Cassel, W. A. and W. G. Hutchinson (1954) Nuclear studies on the smaller Myxophyceae. *Exper. Cell Research* 6: 134–150.

Ciferri, O. and O. Tiboni (1985) The biochemistry and industrial potential of *Spirulina*. *Ann. Rev. Microbiol.* 39: 503–526.

McLarney, W. J. (1984) *The Freshwater Aquaculture Book: Feeding rates, techniques and strategies.* Hartly & Marks, New York.

Oswald, W. J. (1980) *Algae Biomass: Algal production-problems, achievements and potential.* Elsevier/North Holland Biomedical, Amsterdam.

Shinohara, K., Y-H. Zhao and G. H. Sato (1983) *Advances in Gene Technology: Molecular genetics of plants and animals.* Academic Press, New York.

Shinohara, K., Y. Okura, T. Koyano, H. Murakami, E-H. Kim and H. Omura (1986) Growth-promoting effects of an extract of a thermophilic blue-green alga, *Synechococcus elongatus* var. on human cells. *Agric. Biol. Chem.* 50: 2225–2230.

SCIENCE AND MISCELLANEOUS

34

Plant Introduction and International Responsibilities

J. T. Williams

INTRODUCTION

In any consideration of 'new' crops a clear distinction is needed between those that are totally new to mankind and those that are new on the world scene but are well known in restricted areas and those that are new in specific areas and represent aspects of cropping diversification. In parallel the use of the particular species or varieties need consideration, whether for food, feed or industrial purposes. Such distinctions are necessary because the research agendas differ markedly and affect funding priorities and availability as well as how important such agendas are in an international context.

Against these considerations there are a number of scientific implications, from which emerge a number of important principles which may well form focal points in the discussion during this symposium.

PLANT RESOURCES AND PLANT INTRODUCTION

The programme of this symposium is wide ranging in relation to tapping the diversity of the plant world. We must bear in mind, at an international meeting, a number of pressing topics. Firstly, who will benefit from any new research attempts on whatever plants we discuss and in this respect we must balance the advantages to affluent countries and industries with their over-production in agriculture and the advantages to the rural poor in vast areas of the world who are likely to continue to have to rely on the health and income generation of local agriculture, and also to the increasing urban poor in the developing world who have for a long time posed problems for agricultural research and poverty alleviation programmes. It is an over-simplification to think that crops are produced to feed such peoples or generate income to buy food. Secondly, the whims of individual scientists who think they have a potential crop, for whatever purpose, usually ignore the aspects of collecting diversity and moving it elsewhere and do not consider that the diversity in effect belongs generically to mankind. Hence a number of considerations require sensitivity, particularly at the present when there are loud spokesmen arguing in effect for property rights on plants within national boundaries. If scientists had been more sensitive, and if their use of such materials were to be balanced by effective aid-packages (as for

345

instance the genetic resources collection work of the Consultative Group on International Agricultural Research balanced by the free provision of new breeding materials) we would be in a better situation to defuse numerous arguments being voiced. Thirdly, programmes need to be developed which tap the diversity of plants from all parts of the world and the knowledge base on which we build is indeed elementary when an inordinately high percentage of plants from the remoter tropical areas have not yet even been described and might become extinct without any potential use having been tested or documented. Hence there must be a very strong link between any new crops programmes or projects and conservation aspects. We should therefore be critical in the choice of research agendas, and avoid the patronizing attitude all too common in telling others what they should do and what they should grow especially if in the case of some minor food crops the option has been open to them for generations and they have rejected it. What I am trying to say is that simply producing new crops or better crops of lesser-known species may not reduce the number of hungry, may throw into imbalance the money supplies and there are other pitfalls which can indeed be avoided if thought about from the outset and some policy research is included with the scientific work.

I have used the phrase plant introduction in the title of this paper because the vast majority of agricultural and industrial crop plants have been moved from one part of the world to another from time immemorial. Assembling collections of plants from foreign lands and keeping them in botanic gardens is ancient, and religion and mythology have often motivated their construction (Plucknett *et al.* 1987). The need for plants for medicine and plantation crops were added requirements. The role such institutions played in developing industries based on plants is legion for such crops as rubber, vanilla, cinchona, oil palm, sugarcane, coffee, banana, as well as for the more staple food crops. In fact in a post-colonial world some authors have tried to impugn illegal behaviour on some of these past activities although, as in the case of rubber transferred from the Amazon through the Royal Botanic Gardens, Kew, to South East Asia, the operations were quite above board (Hepper 1982) and the results benefited peoples in the poorer countries.

Over the past century the activities of plant collectors and botanical institutions gave way to plant introduction programmes largely as agricultural operations of governments. Some of these have become the basis for much of the scientific background of the programmes now dealing with the collection, conservation and use of a wide range of diversity particularly for agriculture. Here I refer to genetic resources programmes and two historical examples suffice. Firstly, the work of the Russian plant explorer Nikolai Vavilov, the centenary of whose birth is celebrated later this year (1987), and the UK Commonwealth Potato Collection built up over decades and still being used for plant breeding in association with other major international collections.

Plant introduction programmes, since they are related largely to agriculture, have tended, in parallel with the development of sophisticated and effective plant breeding, to develop into genetic resources programmes. This has been a perfectly logical evolution in international research centres, developed countries and those developing ones with plant breeding and plant improvement capabilities (Williams 1984) but it is a pity that the poorer countries often do not have aggressive plant introduction programmes dealing with the introduction of promising varieties (rather than genetic resources for plant improvement). It seems that international concern over crop genetic resources conservation and use has caused bandwagons (even in the wider conservation movement) which indiscriminately mix up the need for plant resources of immediate potential use with collection of wide variation which need a tremendous scientific input to produce useful forms. Even the use of plant resources requires careful screening of potential. After a decade of jojoba planting and use a report called for the better identification of quality specimens (NAS 1985).

Let me now examine the plant resources for new crops for which carefully planned plant introduction programmes are needed – and in some cases, depending on the proposed use of the resources and the time scale envisaged, more comprehensive genetic resources programmes. I have based my comments on the range of interests represented by speakers at this symposium, and have excluded existing crops which are discussed in the concept of the production of 'better' crops not 'new' crops. There are the following categories:

(a) *Non-domesticated species used by man.* These are, in effect, wild species, some of which, e.g. tree species and many species for pasture and range use may have been extensively planted. In many cases these plantings have been in countries remote from where the species occur naturally. These wild species also include countless species of economic importance as medicinal plants, the plants or plant parts being gathered from the wild. In most cases the genetic reservoirs of these species remain unknown, largely untapped by man and most species will require minimal plant improvement efforts to enhance their usefulness. At present many of these wild species are under threat, particularly tree species and medicinals which are over-collected.

(b) *Wild species with potential use.* Much has been written and said, often on the basis of slender evidence, of thousands of wild species which might become useful in the future. Certainly it is unlikely that this vast group will yield any major new food crop but many could be useful to industries as the source of secondary products, but for limited periods until such products are synthesized chemically. Our hope is that nature conservation will maintain sufficient species-rich areas so that such wild species can be examined when needed. The International Board for Plant Genetic Resources has made the point that reserve areas need scientists who can act as curators, monitor population dynamics and supply

347

materials; however, the conservationists are only just getting around to producing inventories of plants in reserves and we cannot expect accurate data for years yet. Moreover, there should not be extravagant claims made on the value of species in reserves until such claims have been substantiated by research. If such claims continue we will see perpetuated claims by journalists that breeders want a perennial maize which they do not or that all the reserves have plants which can produce products which will form billion-dollar industries.

(c) *Wild relatives of crops.* Although these are not domesticated and they do have potential use they form a separate category because of the time-frame for their use. My own organization, IBPGR, is actively pursuing research on crop genepools and, depending on the level of breeding activity on the crop, extends its scientific work narrowly or more widely in relation to the closeness of relationships between species (IBPGR 1987). For instance in many crops we are involved with only a few closely related species; but in others, e.g. wheat, barley and pearl millet, we take into account quite distantly related species. The point here is that we do know from scientific research the overall species relationships in the genepools and the broad patterns of genetic variation and research targets are set and funded to examine other crop genepools where information is more scanty.

(d) *Keystone species.* Frankel & Soule (1981) recognized wild species, the preservation of which *in situ* are necessary for ecosystem stability. Gilbert (1980) recognized 'mobile link' species and 'keystone mutualists'; mobile links are animals which play a major role in the life processes of plants, e.g. for seed dispersal or pollination, and keystone mutualists are plants which support complexes of mobile links. Such species need to be recognized for the maintenance of the species complexes in the ecosystem, many of which include plant resources used by man or are potentially useful.

Depending on the priority for exploitation, therefore, research on new crops must be built on preliminary decisions as follows.

(a) Does the species need to be brought into cultivation? If so is production of the plant adequate without a major programme of improvement? Will the time-scale of cultivation be such that insurance has to be made against potential genetic vulnerability due to diseases and pests by developing genetic resources programmes?

(b) If the time-scale appears very long term some preliminary research to domesticate the plant seems called for. This is not as easy as it sounds (Krochmal & Krochmal 1977). Also a wider spectrum of diversity will be needed in case there may be unforeseen dangers. This presupposes that adequate research has already been carried out on the breeding system (which may change drastically if the plant is being cultivated far from its natural habitats) and that other forms are held in readiness to replace existing planted forms if necessary and quickly.

(c) What decisions on conservation need to be made to back-up research and ensure a supply of plants?

Whereas in the past we have witnessed screening programmes and even database compilations the basic aspects I have just outlined have been ignored. Their consideration will provide a better rationale for funding operations, will provide an estimated time horizon for the research, will almost certainly need cooperative links, e.g. on conservation, and will avoid the ill-fated short-term project approach which is often doomed at the outset.

INTERNATIONAL RESPONSIBILITIES

A logical extension of the above points would be the establishment of an information centre which maintains data on all existing and proposed programmes. No existing database fulfils this need and it must be an international responsibility to avoid the criticism that it can be used unfairly. It needs to include data willingly given by the scientific community on unpublished work and negative results. Overall information needs to be freely available to all with sensitivity to the scientists' unpublished data.

This proposal may seem idealistic but in fact there is a prototype. IBPGR maintains such a computerized database on the research culled from over 1300 entries from scientists in about 70 countries on over 300 genera of plants on *in vitro* culture related to genetic conservation of important crop plants. This database is of the utmost importance in identifying research gaps, in initiating collaborative and more effective research and in monitoring the state of the art (see Withers 1984 for a discussion of the database and its use in planning).

The cost aspects for such an international facility are trivial and such costs are offset by the goodwill generated in providing free user services by the ability to query by crop species or relevant techniques.

This type of facility could provide the logical basis for cooperative and enhanced international attention to the problems of new crops, avoid the whims of individual scientists competing for limited funds and help those responsible for allocation of funds to assess the value of proposed research. Because of the interest of the industrial sector it would require their cooperation so that clearer partnerships emerge between public and private sector work and factual, non-political information in such a facility would remove areas of suspicion all too frequently present due to lack of information and understanding. It needs to be a new concept not an expansion of any existing effort which might be faltering in terms of financial support and it needs to be governed by an international organization or committee.

The success of research will, of course, depend on the continuing availability of plant resources from all parts of the world. Many of you will be aware of supposed controversies raging over recent years related to genetic resources where the world's scientific community deprecates the agitation of non-scientists, particularly ill-informed journalists, and the potential implications of political decisions in an issue muddled with the role of multinational corporations and the weaknesses of plant introduction programmes. To a large degree the scientific community worldwide is at fault for not explaining the realities. Any enhanced work on new crops must realize these tensions, be sensitive to the needs and wishes of developing countries and forge agreeable contracts so that indeed exploitation of plants benefits others. The aim must be firmly the exploitation of plants, not the exploitation of peoples or countries. This does not mean that competitive research is not needed; it is often a good thing and the information facility I have proposed would go a long way to ensuring the lack of suspicion as to motives, remove isolated scientists from their ivory towers and positively link public and private sector work wherever. Such work can and probably will have to have an international aspect in order to combine availability of materials, links with local scientists and with high-tech research elsewhere.

I am not proposing the information facility as an alternative to existing programmes of which many are of relevance to this symposium. A few examples would certainly include the International Legume Database and Information Service located here in the institution hosting this symposium, the Commonwealth Science Council's project on life-support plant species, the SEPASAL project at the Royal Botanic Gardens, Kew, UK, the database established by a past programme of the United States Department of Agriculture on new crops research (Princen 1977), data on medicinals collated through the activities of IUBS or the World Health Organization, the proposed database of the IUCN on economic plants or the activities of IBPGR on genetic resources of forages worldwide.

Many of these activities are relatively new, few are well funded and cooperative networking for many of them is a new concept. The proposed facility would complement their work and fill important gaps but more importantly could be a mechanism to enhance research in the neglected area of economic botany.

Where do we go from here?

The deliberations of this symposium will be crucial in drawing attention, yet again, to the need for enhanced work on plant exploitation. I have pointed out the need for careful and rational planning supported by, and in cooperation with, conservation efforts due to the rapid loss of diversity in many parts of the world. This stresses the need to incorporate into exploitation programmes the need for the survival and continued availability of material with adequate adaptation in maybe limited populations in nature. However, when plants become actively improved through plant breeding this needs the conservation of genepools as genetic resources where the objective is not only survival but the optimization of genetic representation to provide the building blocks for the breeder (Frankel 1983). In such cases exploitation programmes must be backed by genetic resources programmes.

It must be realized that effective work on conservation *in situ*, which will be of great importance to those species of potential use and those which act as ecosystem stabilizers, requires trained botanists of which there is a dearth in many developing countries. Species in other categories, e.g. those directly used by man, can be dealt with differently in the sense that more is known about them and there are complementary *ex situ* conservation methods which, at present, act as a safeguard to the weak *in situ* conservation aspects for these genepools. Such considerations are essential to planned plant introductions and the continued availability of plant resources.

REFERENCES

Frankel, O. H. (1983) Genetic principles of *in situ* preservation of plant species. In: *Conservation of Tropical Plant Resources,* S. K. Jain and J. K. L. Mehr (Eds), 55–56. Botanical Survey of India, New Delhi.

Frankel, O. H. and M. E. Soule (1981) *Conservation and Evolution.* Cambridge University Press. Cambridge.

Gilbert, L. E. (1980) Food web organization and the conservation of neotropical diversity. In: *Conservation Biology: an evolutionary-ecological perspective*, M. E. Soule and B. A. Wilkox, (Eds), pp. 11–34. Sinauer. Sunderland, Mass.

Hepper, F. N. (1982) *Royal Botanic Gardens, Kew: Gardens for Science and Pleasure.* HMSO, London.

IBPGR (1987) *Annual Report for 1986.* International Board for Plant Genetic Resources, Rome.

Krochmal, A. and C. Krochmal (1977) Potential for development of wild plants as row crops for use by man. In: *Crop Resources*, D. S. Seigler (Ed), pp. 75–77. Academic Press, New York.

NAS (1985) *Jojoba. A new crop for arid lands. Raw material for industry.* National Academy Press, Washington DC.

Plucknett, D. L., N. J. H. Smith, J. T. Williams and A. N. Anishetty (1987) *Gene Banks and the World's Food.* Princeton University Press, Princeton.

Princen, L. H. (1977) Potential wealth in new crops: Research and development. In: *Crop Resources*, D. S. Seigler (Ed.), 1–15. Academic Press, New York.

Withers, L. A. (1984) Germplasm conservation *in vitro*: present state of research and its application. In: *Crop Genetic Resources: Conservation and Evaluation*, J. H. W. Holden and J. T. Williams, (Eds), 138–157. Allen & Unwin, London.

Williams, J. T. (1984) A decade of crop genetic resources research. In: *Crop Genetic Resources: Conservation and Evaluation*, J. H. W. Holden and J. T. Williams (Eds), pp. 10–17. Allen & Unwin, London.

35

Kairomones – Chemical Signals Related to Plant Resistance Against Insect Attack

H. Rembold

INTRODUCTION

It can safely be stated that two thirds of all animal species are insects. This enormous abundance implies the existence of twice as many insect species as plants, and twenty times as many as mammals. The great number of about one million insect species known at present, their tremendous reproductive power which fairly often enables a female to lay more than 1000 eggs within 24 hours, with the consequence of another generation starting just a few weeks later, plus their long evolutionary history, with the oldest insect records going back almost 400 million years to fossils from the Middle Devonian period, enables them to occupy practically every environmental niche. By virtue of this adaptive power some insects have become transmitters of harmful diseases, while others are important agricultural pests. Most of them can still be controlled by the relatively cheap but toxic organochlorine, organophosphorus or carbamate compounds. However, a rapidly increasing resistance against insecticides, along with their acute off-target neurotoxicity, makes their future application an impending threat to ecology and human health.

Due primarily to their high reproductive capacity, more than 350 pest species have become partially or completely resistant against one or more of about 50 commercial insecticides. The increasing number of resistant pest insects and the problems of environmental hazards have led to a search for more selective insecticides and to studies on host-plant resistance, insect attractants, use of natural enemies, autocidal techniques, and the integration of complementary techniques such as agricultural practices. The final aim of such an integrated pest management strategy is to apply the chemical insecticide not at the beginning of pest control as a precaution against a possible build-up of a pest population, but in a more defensive way and *ultima ratio*, when the more soft methods have been exhausted.

Breeding of new crop varieties is usually done with an impetus to high yields and therefore under the pesticides umbrella. It becomes an important question therefore, whether with this breeding strategy natural resistance factors might unwittingly have become suppressed and consequently the resistance genes controlling their synthesis becoming less effective. By new breeding strategies they should then again be depressed, if the need for application of chemicals has to be reduced. The development of such new strategies is a tremendous challenge for basic research in general, especially when insect biochemistry is so poorly investigated in its applied direction, as is the case at present.

Having this situation in mind, manipulation of the endocrine control of growth, development, reproduction and behaviour through plant-specific metabolites appears as one promising complement to the knockdown chemicals at present exclusively available for the protection of mass cultures. The importance of such an approach becomes immediately understandable, if the early dynamics of a pest population is also taken into consideration. Indigenous resistance factors of a plant can only be effective before the explosion of an insect population becomes evident. Therefore, manipulation of the endocrine system of any organism cannot show an immediate effect, as is the case after application of a knockdown insecticide. In the context of natural resistance factors, sometimes physical factors like leaf hardiness may be enough to protect a plant to some extent from insect attack. Another possibility is the search for plant-borne gene products which directly interfere with the regulation of the insect hormone system. Although not much is known in this field, quite a few promising candidates may already exist in the gene pool of our crop plants resulting from a breeding program for pest resistance. Such an example from our cooperation with the International Crops Research Institute for the Semi-Arid Tropics (ICRISAT) will be discussed later on.

Endocrine regulation also works in a hierarchical manner in the insects. This means that, in close analogy to the hypothalamo-hypophysial system of mammals, neuropeptides from the central nervous system control the synthesis of steroid (moulting) and sesquiterpenoid (juvenile) hormones in peripheral glands. Attention falls on the question of how such a network of hormonal control can be influenced by external chemical factors. One possibility is an interference with the peripheral hormones – the juvenile and the moulting hormones primarily. There are indeed some plants which are producing homologues of the insect specific juvenile and moulting hormones, i.e. juvenoids of phytoecdysteroids. Even the insect moulting hormone ecdysone itself is present in several plant species. Application of synthetic juvenoids is also following the same line of competing with the natural hormone for insect control. Our own concern, however, is with another class of promising candidates. We are searching for such secondary metabolites of plant origin which either interfere with insect growth or with their behaviour. Such semiochemicals are interesting for the plant breeder in his search for less susceptible crop varieties as well as for use as trap plants. And they are also of interest for the chemist in his search for new bioactive structural elements which can be used in tailoring new types of pest.

THE AZADIRACHTINS AS INSECT GROWTH INHIBITORS

In the course of their evolution, a number of plants have learned to protect themselves against pest insects by chemical means. So-called secondary metabolites act as antifeedants or deterrents, or by disrupting insect growth and metamorphosis. Practitioners in agriculture have long been aware of such phenomena. Parts of the Indian neem tree (*Azadirachta indica*), for example, have traditionally been used as cheap remedies against various ailments, as well as to keep insects away from stored agricultural products: neem leaves are immune even against an insect as voracious as the migratory locust!

Chemical Structure of Azadirachtins

Butterworth & Morgan (1968) were the first to isolate a substance with high biological activity from neem seeds which they named azadirachtin. For its purification they had used the feeding inhibitory effect in the desert locust, *Schistocerca gregaria*. Tests under their standard conditions with the pure compound gave complete inhibition of feeding response at the remarkably low concentration of 5 mg/l. Insect growth inhibitors without any feeding inhibition were then detected and their purification from neem seed monitored by use of the *Epilachna varivestis* bioassay (Rembold *et al.* 1980). The most effective factors were identified as a whole group of triterpenoids which inhibit pupation of the *Epilachna* larvae even in the ppm range. Such natural products are interesting for plant protection purposes because, due to their not deterring the insect from feeding, they can go unnoticed with the food and then unfold their whole growth inhibiting capacity in low concentrations. A whole series of pure and nontoxic compounds were found present in the polar, methanol soluble extract from neem seeds. In the test larvae most of these substances induce the formation of reddish-brown spots in the dorso-lateral zones of the thorax. The treated larvae survived as larvae or larval-pupal intermediates for about three more weeks after appearance of the brown spots, most of them without further metamorphosis. These need seed fractions also disrupt growth in other insects, again without inducing any feeding inhibition. This type of compound is, as already stated, of special interest for the chemist from the point of his search for new chemical structures aiming at alternative plant protection strategies. Due to their interference with insect growth and development, the target of such compounds would be the insect hormone system directly.

Azadirachtin was shown to be the main component within this class of insect growth inhibitors (Rembold & Schmutterer 1981). Some of the other growth inhibitors isolated from neem seeds are isomeric in their chemical structure with azadirachtin (Rembold *et al.* 1984). They are only present in need seeds in trace amounts, as shown by Table 35.1.

Table 35.1: *Pure Azadirachtins as Isolated from 27 kg Neem Seeds (Forster 1987)*

A	3500 mg
B	700
C	4·7
D	3·8
E	9·4
F	4·5
G	3·1

These naturally occurring azadirachtins are all of the tetranortriterpenoid structure as shown for the main compound, azadirachtin A (Fig. 35.1). Its structure has recently been reassigned by three laboratories (Bilton *et al.* 1987;

Figure 35.1: *Azadirachtin*

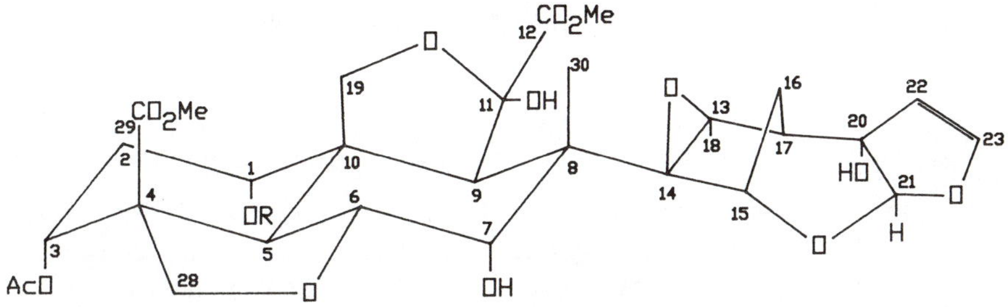

R = tigloyl

Figure 35.2: *Minimal Structure for Insect Growth Inhibitor.*

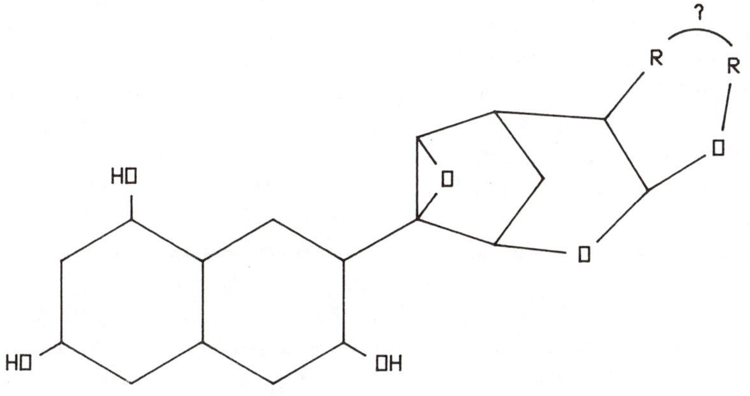

Kraus *et al.* 1987; Turner *et al.* 1987) and now unequivocally gives the basis for the structural elucidation of other azadirachtin isomers by nmr spectroscopy. They all share a very similar chemical structure and biological activity. With the exception of one (azadirachtin C, for which only a partial structure could be given), a structure based on nmr data can be proposed. In addition, by cautious chemical modifications of the mother compounds (azadirachtins A and B) nine more isomers were prepared. Based on the structures of these 16 highly active insect growth inhibitors, a minimal structure (Fig. 35.2) can now be proposed which is common to all of them (Forster 1987).

Activity in the *Epilachna varivestis* bioassay

All the azadirachtins induce at about the same concentrations three different effects, depending on the amount of substance applied to the fourth instar *Epilachna* larva under standard conditions:

(a)   Toxic effects if applied in high concentrations (>1000 ppm). Under these conditions, the larvae survive for only a few hours.

(b)   In concentrations beginning from 10 to 100 ppm, the azadirachtins are very active phagodeterrents, combined with growth disrupting activity.

(c)   In concentrations between 1 and 10 ppm, all the azadirachtins are growth inhibitors without any phagodeterring effect. This is exactly the concentration range where they can be described as ideal growth inhibitors.

Mode of azadirachtin action

As discussed in the preceding paragraph, the azadirachtins are very potent feeding inhibitors, depending on the concentration applied to the larval food. Ignoring the Janus-faced character of feeding, some of the physiological effects of growth inhibitors can therefore be subject to misinterpretation. Salannin for example, another compound from neem seeds, is a highly potent feeding deterrent for house flies (Warthen *et al.* 1978). However, it does not interfere with *Epilachna* growth and development even at high concentrations, whereas azadirachtin does both, dependent on the amount of substance taken up by the treated insect. On an average, azadirachtin seems to inhibit feeding at much lower doses in the hemimetabolous than in the holometabolous insects. However, there are exceptions from this rule on both sides.

What is the growth disrupting mode of azadirachtin action? A detailed study of the effects on the hormone system of the last larval instar of *Locusta migratoria* showed a typical dose dependence of the responding animals (Sieber & Rembold 1983). At a dose of 0·6 µg/g azadirachtin, only 10 per cent of the animals showed a reaction whereas 2 µg/g already elicited a maximal response and no larva was able to undergo or terminate ecdysis. An even more dramatic effect was induced in *Rhodnius prolixus* after the substance was taken up with a blood meal (Garcia & Rembold 1984). The effective dose that prevented ecdysis in 50 per cent of the insects was 4×10^{-4} µg/ml blood and doses higher than 1 µg/ml inhibited ecdysis by 100 per cent. Such reactions clearly indicate an interaction of azadirachtin with the hormone system of the treated larvae.

There is a pronounced effect on control of the ecdysteroid titre as first demonstrated for the 5th-instar *Locusta migratoria* (Sieber & Rembold 1983). The authors explain this effect as an interference of the compound with the larval neuroendocrine system. This argument is supported by histologic studies which clearly show an increase of the paraldehyde-fuchsin stainable material in the neurosecretory cells of the *Pars intercerebralis* of azadirachtin-treated last instar locusts. Concomitantly with the ecdysteroid, the juvenile hormone synthesis is also affected by azadirachtin (Rembold 1984). The dose necessary for inducing such effects seems to be much lower than indicated by the ED_{50} value, as found with tracer studies (Rembold *et al.* 1984). Tritium labelled azadirachtin was to a great extent excreted again within seven hours, most of it unchanged. Such a result speaks in favour of a high-affinity azadirachtin binding protein which still has to be found and which is not identical with an ecdysone binding site.

Although the molecular mechanism of azadirachtin action is still unknown, one can generalize from present knowledge that the compound irreversibly, or at least for an extended period of time, blocks and sometimes changes developmental programmes. There are species specific differences, however, in the insects' reaction to azadirachtin treatment, some of the treated insects even not reacting at all. The fact that a secondary plant product like the group of isomeric azadirachtins interferes with the endocrine regulation of insect growth and development and that until now neither acute nor chronic mammalian toxicity could be found, makes them a promising candidate in new plant protection strategies. Whether azadirachtins besides those in the Meliaceae are also present in other plant species is still unknown.

The neem tree, an economic tropical plant

The neem tree, *Azadirachta indica*, has already gained an important function in reforestation programs as a fast growing tree for tropical and subtropical areas (Schmutterer & Ascher 1984). Breeding for increased seed production will further broaden its use in plant protection programmes as a cheap natural insecticide. However, the neem tree is also a promising candidate for many other uses, including the production of fuelwood, timber, technical oils, tannins, or medicinal products. As another interesting application, the pressed cake from neem seeds is already being used as an effective fertilizer and nematocide. Primarily limited by its distribution to subtropical and tropical areas, the neem will rapidly gain an increasing economic importance.

KAIROMONES IN GRAIN LEGUMES

Plants and insects have evolved in the Devonian period and from that time onwards mutual resistance or a more or less pronounced tolerance have coevolved. The natural relationship between the structural cum chemical diversity of plants and the richness of the insect kingdom during a period of about 400 million years has produced the rich genetic pool of our wild plants. However, from the beginning of civilization onwards man has propagated those plants which he considers better in yield or in a few other qualities. In combination with mass production of a limited number of staple crops, he interfered with the established ecological balance and created, through modern agricultural practices, his new and often artificial world. In this environment, the stimulation of primary metabolic pathways – which is identical with breeding for high crop yields – has priority over the semiochemicals which are secondary metabolites. These semiochemicals (from the Greek *semeon*, a mark or signal) however, are an important component of plant resistance against phytophagous insects. For a better understanding of plant-insect relationships we must come to a better biochemical understanding of the selection pressures arising from crop monocultures, which finally establishes mass populations of those few main pests which are rapidly becoming resistant to our traditional insecticide pest management.

The semiochemicals are divided into two major groups, depending on whether the interactions between organisms are intraspecific (pheromones) or interspecific (allelochemicals). To the latter group of metabolites belong such allelochemicals represented by the terms arrestant, attractant, repellent, stimulant and deterrent. An alternative form of terminology that has been reviewed by Norlund *et al.* (1981) may also be used. An allomone is a substance which is produced by an organism of species A. If it is received by species B, it causes a reaction in B that is favourable to A and not to B. The meaning of a kairomone is similar to an allomone, but the reaction in B is favourable to B; some examples will be considered below. The meaning of a synomone is similar to an allomone, but the reaction in B is favourable to both A and B. Finally, an apneumone is a substance emitted by nonliving material, favourable to some species and detrimental to others.

Selection and breeding of varieties for resistance against a wide range of pests and diseases is one important way for competing with problems arising from crop plant monocultures. At ICRISAT, Hyderabad, India, various cultivars of chickpea (*Cicer arietinum*) and pigeonpea (*Cajanus cajan*) are grown and exposed to insect attack in special fields which have never been treated with insecticides before. The two major insect pests of these grain legumes are *Heliothis armigera* and *Melanagromyza obtusa*. Breeding programmes for selection of resistant varieties have been successful and a wide range of more or less susceptible pedigrees can be offered to chemists interested in the chemical basis of plant vs. insect relationships.

Allomones as nonvolatile exudates

Leaves and pods of both chickpea and pigeonpea are covered with hairs secreting an exudate. That of *Cicer arietinum* appears in the form of droplets whereas *Cajanus cajan* secretes only a thin film which can be washed off with organic solvents. The chickpea secretes a very acidic material with a pH ca. 1 and dry matter percentage between 12 and 75%. This may be responsible for the relatively small range of insect pests found on this crop, of which the greatest contributor to pod damage is *Heliothis armigera*. The main component of all the chickpea exudates, in highly resistant as well as in the susceptible ones, is malic acid, which accounts for 70–100% of the dry matter. There is a clear correlation between malic acid content and borer damage. The threshold of low borer damage was found in this comparative study around 250 mg malate per ml chickpea exudate (Rembold 1981).

In contrast to chickpea, the surface in pigeonpea is velvety, but without visible droplets. Here again, by use of analytical chemical methods using gas chromatography (GC) and high performance liquid chromatography (HPLC), first results demonstrate a correlation between exudate quality and the resistance of pigeonpea cultivar against *Heliothis* attack. The main attracting components in the exudate were the contact-perceivable sugars whereas other compounds seem to counteract. Malate obviously plays a subordinate role in pigeonpea exudates, since its concentration is rather low (Rembold & Winter 1982).

Volatile kairomones from chickpea and pigeonpea

Heliothis armigera is a night-active moth. One important fact when searching for a host plant on which it could deposit one of its 500–1000 eggs concerns the plant itself. The insect is very selective in determining its host, although this changes during the season. Concerning our two grain legumes, *Heliothis* is first found in the chickpea fields and later in the pigeonpea. The question arises, whether a certain kairomone is common to these two host plants and whether there are qualitative or quantitative differences between the cultivars of different susceptibility to *Heliothis* attack. Furthermore, its larvae start migrating when searching for food. Are they also following a kairomone trail and is this eventually identical with the one which attracts the adult female for oviposition?

There are very different signals controlling insect behaviour in the field. Some of them are the shape and colour of the host plants, its stage of maturation including flowering; others are site factors like humidity, temperature or soil quality. For a kairomone bioassay under laboratory conditions all these factors, other than the olfactory signal, must be avoided.

In the oviposition assay a rectangular chamber with glass walls on its sides and wire-nets at its two opposite ends is used. The tunnel is kept in a dim light and seedlings or a plastic disc, either containing the material to be tested or empty, is fixed to the wire-net. Any physical contact is avoided and also any other contact-perceivable signals. Two egg-laying *Heliothis* moths are kept in this chamber overnight. Eggs are laid on the net and can be counted either at regular time intervals or after exposure overnight. If the seedling from the pigeonpea variety which is fairly resistant against *H. armigera* attack is placed behind a net on one side of the tunnel, and a more susceptible variety on the other side, the number of eggs is much higher on the net adjacent to the more susceptible variety.

The kairomones can be extracted from the pigeonpea leaves and can be fractionated by vacuum distillation. Analysis of the active fraction by GC-MS shows that most of the compounds are aliphatic aldehydes and monoterpenes followed by sesquiterpenes. Such a fraction has also been used for characterizing different varieties through their typical GC-aromagrams. Clear differences can become evident by this method (Rembold & Tober 1985). In the field *Heliothis* females preferentially lay their eggs near to the pigeonpea flowers. If flower extracts are tested in the oviposition bioassay, a strong stimulatory effect is found. However, an extract from leaves is also attractive and the stimulus is even stronger from the steam distillate of pigeonpea leaves (Rembold & Tober 1987). It is evident from these results, that under field conditions a whole series of signals, depending also on their quantity, becomes involved before the insect starts laying its eggs. And it is also evident from the results mentioned, that it is possible to isolate biologically active kairomone components from the host plant by use of a laboratory assay and thus correlate the susceptibility of a pigeonpea variety to chemical signals.

The same arguments hold true for larval attractants; for a first screening, Saxena & Rembold (1984) used powdered chickpea seed. Larval response to volatiles was followed in a 'cross-track' and in a 'glass-tunnel' test. The seeds contain kairomones which strongly attract *H. armigera* larvae regardless as to whether they are newly emerged and unfed in the first instar or in the last instar and reared on pigeonpea leaves. As in the oviposition attractant from pigeonpea leaves, the attractants from chickpea seed are found in the hexane-soluble fraction, which could be further fractionated by HPLC into a highly active subfraction. The larval attractance bioassay was modified and as such improved by using a trident olfactometer with remote gas supply, two tubes serving the control gas streams, and the third coupled with a Tenax glass tube carrying the volatiles to be tested. Velocity of gas flow and humidity can be kept at constant level with this equipment and all plant-born signals, except for the volatiles, are excluded. Here too, either first or last instar larvae are used for the test.

Synthetic kairomones

With the improved method of olfactometer assay it has been possible to identify out of 200 individual components the four most active ones from the chick pea seed aromagram shown in Fig. 35.3 (Wallner 1987). Here too, as in the oviposition attractant, the biological activity is primarily contained in the monoterpene fraction. This group of chickpea seed volatiles was fractionated by capillary gas chromatography into its aromagram which can concomitantly be used as an analytical fingerprint for the characterization of chick pea varieties.

Figure 35.3: *Aromagram of the Terpenoid Fraction from Chickpea (Cicer arietinum) Seed Powder Volatiles, Separated by Capillary Gas Chromatography.*

From a total of 200 peaks, four (characterized by the fully dark signals) are components of the chickpea kairomone (Wallner 1987).

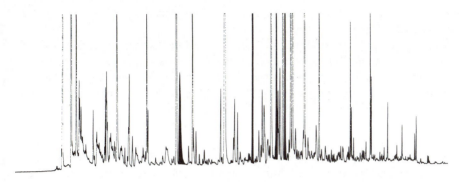

The combination of GC with a mass spectrometer equipped with data bank and subsequent comparison with authentic compounds made it possible to identify 80 of the most prominent peaks in the chick pea aromagram (unpubl. results). Out of these 80 pure substances, 16 were individually tested in the olfactometer assay and four of them were found positively to induce larval migration in the trident olfactometer. Also with a mixture of all four active compounds, in the same proportion as in the natural aromagram, the test larvae demonstrate a similar behaviour to that with volatiles from the chick pea seed powder. The synthetic kairomone is also active in the flight tunnel oviposition assay and induces a high preference behaviour in the egg laying *Heliothis* moth. A few nanograms, desorbed by several litres of air streaming through the tunnel, are sufficient for inducing a positive response by the moth or its larvae.

DISCUSSION

With the coevolution of plants and insects, the semiochemicals have gained an important role in the host finding process. This holds true for the interaction of the foraging honey bee as well as for the phytophagous parasite. First data indicates an even more sophisticated interaction between the host plant and *Heliothis* as the insect pest. Mating, as in most other species of moths, is mediated by the production and release by the females, of a sex pheromone to which conspecific males are attracted. Sex pheromone production in *H. zea* is controlled by a pheromone biosynthesis activating neuropeptide (Raina & Klun 1984). It appears that *Heliothis* species have evolved a mechanism which stops pheromone production and release, and thus prevents the possibility of mating in absence of a host plant for the resulting progeny. Such a control is achieved through a signal in the form of a volatile factor from the host plant that causes the release of the pheromone biosynthesis activating neuropeptide, which in turn initiates pheromone production. Consequently, after mating the female follows the kairomone signal and thereby finds its host plant for oviposition. With these findings in mind, it might even be possible that the oviposition kairomone described above for *H. armigera*, and the plant signal which switches on the sexual behaviour of the female moth, are identical.

There are many possibilities for making use of the kairomone work described above. One most important aspect is the quantitative analytical information given to the plant breeder, which can help him in tailoring new crop varieties. A better understanding of the influence of different habitats on intensity and quality of the kairomone and their influence on resistance and susceptibility is to be expected. The synthetic kairomone may also be used as an attractant in poisonous baits or in traps. Last but not least, the question can now be asked, whether the same volatile signal is present in all the different host plants which are attacked by this polyphagous insect and whether the insect then demonstrates preference behaviour for its selected host plant.

The development of crop plant varieties with high resistance against insect pests, fungi, diseases, etc., has several advantages compared to the classical pest control strategy. The application of pesticides may be reduced or even eliminated, resulting in decreased costs, which is most important in developing countries where subsistence farmers cannot adopt complicated technology. Another very important point is that knowing more about the chemical basis of plant resistance is a prerequisite for any gene technological approach to improvement of our existing crops. Although far from practical use, it is important to know of such compounds as the isomeric azadirachtins, which could be good candidates in breeding programmes for resistance.

ACKNOWLEDGEMENTS

Unpublished results are acknowledged as part of the doctoral theses work of A. Schroth, H. Tober and P. Wallner. Expert assistance in planning the field experiments by Dr. S. S. Lateef ICRISAT, Hyderabad, India, is also gratefully acknowledged. My own results presented on kairomones are part of a project financially supported by Deutsche Gesellschaft fur Technische Zusammenarbeit (GTZ), Eschborn, FR Germany.

REFERENCES

Bilton, J. H., H. B. Broughton, P. S. Jones, S. V. Ley, Z. Lidert, E. D. Morgan, H. S. Rzepa, R. N. Sheppard, A. M. Z. Slawin and D. J. Williams (1987) An X-ray crystallographic, mass spectroscopic, and nmr study of the limonoid insect antifeedant azadirachtin and related derivatives. *Tetrahedron* 43: 2805–2815.

Butterworth, J. H. and E. D. Morgan (1968) Isolation of a substance that suppresses feeding in locusts. *Chem. Commun.* 23–24.

Forster, H. (1987) Struktur und biologische Wirkung der Azadirachtine, einer Gruppe insektenspezifischer Wachstumshemmer aus Neem (*Azadirachta indica*). Dr. Thesis. Univ. Munich.

Garcia, E. S. and H. Rembold (1984) Effects of azadirachtin on ecdysis of *Rhodinius prolixus*. *J. Insect Physiol.* 30: 939–941.

Kraus, W., M. Bokel, A. Bruhn, R. Cramer, I. Klaiber, A. Klenk, G. Nagl, H. Pöhnl, H. Sadlo and B. Vogler (1987) Structure determination by nmr of azadirachtin and related compounds from *Azadirachta indica* A. Juss. (Meliaceae). *Tetrahedron* 43: 2817–2830.

Norlund, D. A., R. L. Jones and W. J. Lewis (Eds) (1981) *Semiochemicals: their role in pest control.* Wiley, New York.

Raina, A. K. and J. A. Klun (1984) Brain factor control of sex pheromone production in the female corn earworm moth. *Science* 225: 531–532.

Rembold, H. (1981) Malic acid in chickpea exudate – a marker for *Heliothis* resistance. *Chickpea Newsletter* 4: 18–19.

Rembold, H. (1984) Secondary plant products in insect control, with special reference to the azadirachtins. In: *Advances in Invertebrate Reproduction* 3, W. Engels (Ed.), pp. 481–491. Elsevier Science Publishers, Amsterdam.

Rembold, H. and H. Schmutterer (1981) Disruption of insect growth by neem seed components. In: *Regulation of Insect Development and Behaviour*, F. Sehnal, A. Zabza, J. J. Menn and B. Cymborowski (Eds), pp. 1087–1090. Techn. Univ. Press, Wrozland, Poland.

Rembold, H. and H. Tober (1985) Kairomones as pigeonpea resistance factors against *Heliothis armigera*. *Insect Sci. Applic.* 6: 249–252.

Rembold, H. and H. Tober (1987) Kairomones in legumes and their effect on behaviour of *Heliothis armigera*. In: *Insects – Plants*, V. Labeyrie, G. Fabres and D. Lachaise (Eds). W. Junck, Dordrecht, Holland.

Rembold, H. and E. Winter (1982) The chemist's role in host plant resistance studies. In: *Proc. Int. Workshop Heliothis Management*, W. Reed (Ed), pp. 241–250. ICRISAT, Patancheru, India.

Rembold, H., H. Forster, Ch. Czoppelt and K.-P. Sieber (1984) The azadirachtins, a group of insect growth regulators from the neem tree. In: *Natural Pesticides from the Neem Tree and other Tropical Plants,* H. Schmutterer and K. R. S. Ascher (Eds), pp. 153–162. GTZ, Eschborn, F. R. Germany.

Rembold, H., G. K. Sharma, Ch. Czoppelt and H. Schmutterer (1980) Evidence of growth disruption in insects without feeding inhibition by neem seed fractions. *J. Plant Dis. Prot.* 87: 290–297.

Saxena, K. N. and H. Rembold (1984) Attraction of *Heliothis armigera* (Hubner) larvae to chickpea seed powder constituents. *Z. Ang. Ent.* 97: 145–153.

Schmutterer, H. and K. R. S. Ascher (Eds) (1984) *Natural Pesticides from the Neem Tree and other Tropical Plants.* Proc. Second Int. Neem Conference, Rauischholzhausen, FR Germany, 25–28 May, 1983. GTZ, Eschborn, F. R. Germany.

Kairomones

Sieber, K. P. and H. Rembold (1983) The effects of azadirachtin on the endocrine control of moulting in *Locusta migratoria*. *J. Insect Physiol*. 29: 523–527.

Turner, C. J., M. S. Tempesta, R. B. Taylor, M. G. Zagorski, J. S. Termini, D. R. Schroeder and K. Nakanishi (1987) An nmr spectroscopic study of azadirachtin and its trimethyl ether. *Tetrahedron* 43: 2789–2803.

Wallner, P. (1987) Flüchtige Staffe aus *Cicer arietinum* – Samen. Identifizierung und Einfluss aus das Verhalten von *Heliothis armigera* – Larven. Munich, F. R. Germany: Univ., Dr. Thesis.

Warthen, J. D. Jr., R. E. Redfern, E. C. Uebel and G. D. Mills, Jr. (1978) An antifeedant for fall armyworm larvae from neem seeds. *USDA, Sci. & Educ. Adm., ARR-NE-1*.

36

Wild Plants a Source of Novel Anti-insect Compounds: Alkaloidal Glycosidase Inhibitors

M. S. J. Simmonds, L. E. Fellows and W. M. Blaney

INTRODUCTION

Today we are faced with increasing concern about the use of persistent pesticides, especially the narrow range of synthetic insecticides used in modern agricultural practices. The problems are great and are exacerbated by the number of insect species that have developed resistance to these insecticides.

The persistence on the planet of wild plants, despite predation by insects and other organisms, is largely attributable to their possession of defensive chemicals which have evolved over millennia as 'biodegradable pesticides'. No less interesting than the structures of these compounds are their sites and modes of action, which exploit many different aspects of the insect's life cycle and physiology. Biochemists and entomologists are now pooling their resources to determine what can be learned from these natural systems which might assist in the development of new crop protection strategies. At the same time, it is becoming clear that the demonstration of an economic use for a novel compound in a wild plant can give that species itself potential crop status.

At Kew we are privileged to have in the Gardens and in the Herbarium a wealth of plant specimens and botanical knowledge which allow us to undertake chemotaxonomic studies of plant material and to investigate the biological function of the chemicals found. Thus we are able to take a plant with known biological activity, identify the active component and then look at related plants for the presence of the same or similar types of chemicals.

In practice, many difficulties hamper the search for novel plant compounds which might find a place in an integrated pest control programme. The cost of screening extracts against a range of insects is high and the chemical composition of these extracts can vary depending on the condition of the plant (due to factors such as stress or seasonal and diurnal variation), how it was harvested and how the material was kept prior to extraction. Despite these problems, industrialists are still interested in novel compounds discovered by plant biochemists in the course of chemotaxonomic studies.

ALKALOIDAL GLYCOSIDASE INHIBITORS

As a result of a study at Kew of the generic complex *Lonchocarpus/Millettia/ Derris* (Leguminosae), best known for its content of insecticidal rotenoids, two alkaloids have been isolated which proved to be inhibitors of glycosidase enzymes: one, DMJ (for details of structures see legend to Fig. 36.1) was a structural analogue of 1-deoxy-α-D-mannose; the other, DMDP, was an analogue of β-D-fructofuranose. In both cases the ring oxygen was replaced by nitrogen (Fellows *et al.* 1979; Evans *et al.* 1985a&b), DMJ was technically a polyhydroxy derivative of the 6-sided piperidine ring, DMDP of the 5-sided pyrrolidine ring. Further studies of alkaloids in genera of the tribe Sophoreae (Leguminosae) led to the isolation of a second polyhydroxy derivative of pyrrolidine, AB1, related to DMDP by the loss of one hydroxymethyl group, from *Angylocalyx* spp. (Jones *et al.* 1985) and to a glucoside of fagomine, XZ1, from *Xanthocercis* sp. (Evans *et al.* 1985c). Free fagomine, a piperidine derivative, was first found in buckwheat (Kyoma & Sakamura 1974). It is related to DMJ by the loss of a hydroxy group (at C2 if considered as a sugar, at C5 as a piperidine). A further type of glycosidase-inhibiting alkaloid, castanospermine, was found in *Castanospermum australe*, a monospecific Sophoreae genus from Australia. This compound contained a polyhydroxy-substituted indolizidine (octahydroindolizine) nucleus, essentially the fusion of piperidine and pyrrolidine rings to give a bicyclic structure (Hohenschutz *et al.* 1981). Castanospermine subsequently proved to be a particularly potent glucosidase inhibitor (Fellows 1986).

The first alkaloidal glycosidase inhibitor to be detected in nature was nojirimycin, isolated from a *Streptomyces* sp. (Ishida *et al.* 1967) and shown to be an analogue of α-D-glucose with nitrogen replacing oxygen in the ring. The 1-deoxy derivative, deoxynojirimycin, (DNJ), was first formed from it by chemical reduction, but later shown to occur naturally in bacteria (Schmidt *et al.* 1979; Murao & Miyata 1980) and also in the mulberry *Morus nigra* (Moraceae) (Yagi *et al.* 1976). By analogy, the 1-deoxymannose analogue DMJ is conveniently referred to as 1-deoxymannojirimycin.

From the later 1970s on, researchers in many parts of the world began to discover alkaloidal glycosidase inhibitors in other plants and microorganisms and it is likely that these compounds are widespread in nature. Today, approximately 13 such compounds have been reported, each belonging to one of the three categories, polyhydroxy derivatives of piperidine, pyrrolidine or indolizidine (Fellows & Fleet 1988; Fig. 36.1). Of these, the best-known is probably swainsonine, an indolizidine mannosidase inhibitor isolated from *Swainsona* and *Astragalus* (Leguminosae) and responsible for the poisoning of cattle grazing on these plants (Colegate *et al.* 1979; Molyneux *et al.* 1982). It also occurs in fungus, *Metarhizium* sp. (Hino *et al.* 1985). Although the majority of these compounds were initially isolated from the Leguminosae, they are not restricted to this plant family; AB-1 has been found in the fern *Arachnoides*

Figure 36.1: *Structures of the Alkaloidal Glycosidase Inhibitors.*

DMPD = 2R,5R-dihydroxymethyl-3R,4R-dihydroxypyrrolidine; AB-1 = 1,4-dideoxy-1,4-imino-D-arabinitol; CY-3 = (2R,3S)-2-hydroxymethyl-3-hydroxypyrrolidine; DNJ (Deoxynojirimycin) = 1,5-dideoxy-1-5-imino-D-glucitol; DMJ (Deoxymannojirimycin) = 1,5-dideoxy-1,5-imino-D-mannitol; Fagomine = 1,2,5-trideoxy-1,5-imino-D-arabino-hexitol; XZ-1 = 4-0-(β-D-glucopyranosyl)-fagomine; BR-1 = 2S-carboxy-3R,4R,5S-trihydroxypiperidine; Castano-spermine = (1S,6S,7R,8R,8aR)-1,6,7,8-tetrahydroxyoctahydro-indolizine; Swainsonine = (1S,2R,8R,8aR)-1,2,8-trihydroxyoctahydroindolizine

DMDP   AB-1   CYB-3

DNJ   DMJ   BR-1   Fagomine   XZ-1

Swainsonine   Castanospermine

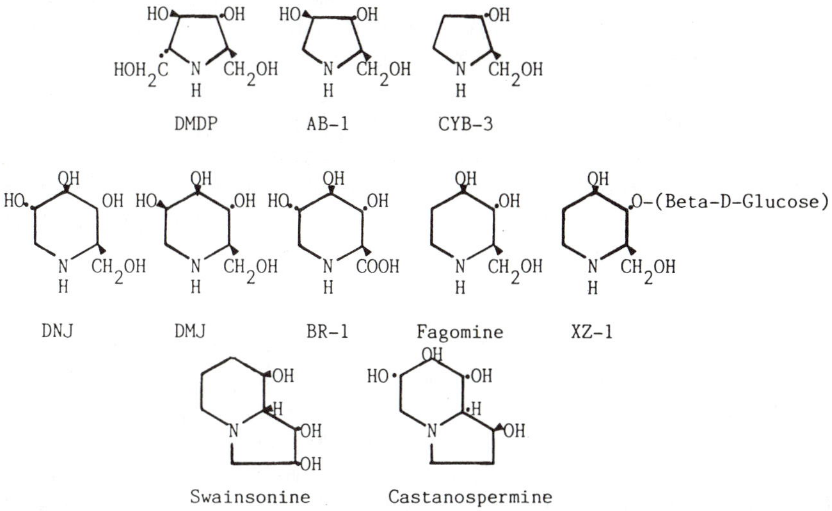

standishii (Aspidiaceae) (Fleet *et al.* 1985), DMDP in the Euphorbiaceae (Fellows *et al.* 1988).

Interest in these compounds centres on their ability to inhibit specific glycosidases (Fellows *et al.* 1986). So far it is not possible to predict in advance whether a particular enzyme will be susceptible to inhibition by a given compound since enzymes having similar activity but from different organisms may differ greatly in their response. For example, DMDP has almost no effect on mammalian digestive α-glucosidase but strongly inhibits the enzyme from the larvae of the bruchid beetle *Callosobruchus maculatus* (Evans *et al.* 1985a). DMJ and swainsonine are both α-mannosidase inhibitors, but DMJ inhibits only mannosidase 1 and swainsonine only mannosidase 11. Neither has been shown to inhibit glucosidases (Fuhrmann *et al.* 1985; Szumilo *et al.* 1986). However, it is these empirically determined differences which are proving invaluable in that inhibitors can be used to distinguish between different types of glycosidase activity in a way which was not previously possible. Also differences in susceptibility of organisms to a given compound may be exploited in pesticide formulation.

Medical Application

One area of possible medical application is the modulation of blood glucose levels in diabetes. Several alkaloids, particularly DNJ and castanospermine, inhibit intestinal glucosidases which perform the final stages in the digestion of carbohydrates and liberate free glucose: patents have been filed on both compounds as antidiabetic agents (Scofield *et al.* 1986). Another area of interest involves the selective action of these alkaloids on enzymes of glycoprotein processing, which can be used to alter the structure of the oligosaccharide side chains of the glycoproteins and consequently modify the biological activity of the complete molecule. Glycoproteins are known to perform important but little understood roles in many biological processes, such as cell-cell recognition, cancer transformation and metastasis, and the immune response. Swainsonine and castanospermine have been shown to inhibit pulmonary colonization by B16-F10 murine melanoma cells (Humphries *et al.* 1986a&b). Castanospermine prevents the expression of the transformed phenotype in feline sarcoma-transformed rat and cat embryo cells (Hadwiger *et al.* 1986), and swainsonine that of the transformed phenotype in NIH 3T3 cells transfected with human tumor DNA at non-toxic concentrations (De Santis *et al.* 1987). Swainsonine enhances the antibody response to foreign cells in immunodeficient mice (Kino *et al.* 1985), reduces trypanosome/host cell association (Villalta & Kierszenbaum 1985) and increases the susceptibility of some tumour cells to the anti-proliferative effects of interferon (Dennis 1986).

Recently, castanospermine, DNJ and DMDP have been shown to reduce the infectivity of the AIDS virus, HIV, in cultured lymphocytes at millimolar, but non-toxic, concentrations (Tyms *et al.* 1987; Gruters *et al.* 1987; Walker *et al.* 1987). The mechanism of action is still under scrutiny, but evidence so far suggests that changes to the glycoprotein on the surface of the virus render it incapable of binding to the receptor sites on further cells and thereby infecting them. Should those compounds prove useful in the treatment of AIDS patients, demand for further natural sources of these compounds will be heavy.

Anti-insect activity

Some of the alkaloidal glycosidase inhibitors have detrimental effects on insects. Campbell *et al.* (1987) found that castanospermine differentially inhibited enzymes causing the hydrolysis of the five disaccharides, lactose, cellobiose, maltose, sucrose and rehalose, known to occur in the midguts of the 19 species (representing 12 different families) of insects tested. Castanospermine inhibited all five disaccharidases in the five species of Homoptera tested, including those of the mealybug *Pseudococcus longispinus*, known to feed on plants containing castanospermine. This result appears anomalous because sucrose is present in the phloem and has to be hydrolysed to glucose for absorption, yet the insect's sucrase activity was inhibited by castanospermine. Campbell *et al.* (1987)

suggest that sucrase might be localized within the gut in such a way that it does not contact castanospermine. Other than in the Homoptera, sucrase was not inhibited; maltase was only inhibited in two species of Lepidoptera, *Manduca sexta* and *Heliothis zea*; cellobiase was inhibited in the other fourteen species, as was lactase, with the exception of the bruchid *Acanthoscelides aureolus*. This bruchid develops in the seeds of the locoweed *Astragalus oxyphysus* which contains swainsonine, and might have physiological adaptations enabling it to counteract some of these glycosidase inhibitors. Nevertheless, this bruchid was the only one of the five coleopterans tested in which trehalase was inhibited. Trehalase was also inhibited in six of the eight species of Lepidoptera and in the one species of Diptera tested. The importance of this differential inhibition of trehalase is that trehalase is the main storage sugar and source of energy in insects, and any compound that inhibits this enzyme could have potential use as an insecticide.

Nash *et al.* (1986) found that at 2×10^{-3} M castanospermine was a potent inhibitor of α- and β-glucosidase (52% and 85%, respectively) but not α- and β-galactosidase or α-mannosidase in larvae of the bruchid *Callosobruchus maculatus*. In *Tribolium confusum* it inhibited β-glucosidase (98%), α-galactosidase (59%) and to a lesser extent α-glucosidase (13%), β-galactosidase (8%) and α-mannosidase (5%). Starch is the main source of carbohydrate for these insects and to hydrolyse starch to glucose for absorption, the insect requires α-amylase together with α-glucosidase. Any inhibition of these enzymes could be deleterious to the insect. In experiments carried out *in vitro*, castanospermine did not inhibit porcine α-amylase (Nash, pers. comm.). However, it did inhibit insect α-glucosidase and was 39% more active against the α-glucosidase in *C. maculatus* than in *T. confusum*. This could explain why, in development studies, the former species is more susceptible to castanospermine (Nash *et al.* 1986).

In contrast to castanospermine, DMDP inhibits α-glucosidase more than β-glucosidase in both *Callosobruchus maculatus* and *Tribolium confusum*. However, the α-glucosidase in larvae of the bruchid *Ctenocolum tuberculatum*, which develops in seeds of some *Lonchocarpus* spp. containing DMDP, is 100% less sensitive to DMDP at 3×10^{6} M than that of *C. maculatus* (Fellows *et al.* 1988). The pyrrolidine AB-1, which is structurally related to DMDP but lacks one hydroxymethyl group (Fig. 36.1), inhibits α-glucosidase but not β-glucosidase in the larvae of both *Callosobruchus maculatus* and *Ctenocolum tuberculatum* (Nash, unpublished results).

As the above *in vitro* experiments would predict, some of the compounds increase insect mortality when ingested. DMDP caused mortality in two species of *Spodoptera* but not in two species of locusts, *Locusta migratoria* and *Schistocerca gregaria*. *Spodoptera littoralis* was affected by lower concentrations (0.005% w/w in an Agar-cellulose diet) of DMDP than *S. exempta* (0.02% w/w). This is due to the former species consuming more treated food than the latter and being poisoned by the amount of DMDP eaten (Blaney *et al.* 1984). The same occurs with castanospermine (Table 36.1). Of the compounds tested in

Table 36.1: *% Mortality Resulting From Exposure of Three Species of Lepidoptera and Two Species of Coleoptera to Different Concentrations of the Alkaloidal Glycosidase Inhibitors.*

Compound:	DNJ		Fagomine		DMDP		AB-1	CYB-3		CAST		SWAIN		Ref
Conc[b]:	20	200	20	200	20	200	20	20	200	20	200	20	200	
Lepidoptera:														
Spodoptera														
littoralis	10	20	0	0	5	50	20	5	35	10	45	5	15	S/B
S. exempta	0	10	0	5	–	30	30	0	20	40	5	–	–	S/B
Heliothis														
virescens	0	20	0	5	5	40	10	–	5	55	–	–	–	S/B
Conc.:	100	1000	100		100	300	200	1000	10	100	100	1000		
Coleoptera:														
Tribolium														
confusum	30	10	10		–	–	2	11	30	40	37	90		N/S/B
Callosobruchus														
maculatus	20	55	35		30	95	19	32	25	55	55	40		N/S/B

a. Reference to unpublished work by: S/B = Simmonds & Blaney; N/S/B = Nash, Simmonds & Blaney.

b. Conc (Concentration) = weight of compound/weight of diet (ppm).

Bioassays – Lepidoptera: 3rd stadium larvae were exposed for 5 days to a diet containing the test compounds (20 replicates/conc). Mortality was recorded after the treatment period.

Coleoptera: Adults were exposed to diet containing the test compound (see Nash *et al*. 1986 for details). Mortality represent the difference in the number of adults produced in the tests relative to those produced in the control (5 replicates/conc.).

developmental bioassays, castanospermine and DMDP caused the greatest mortality (Table 36.1). The fact that the compounds were less active against the locusts could be due to the gut in these insects acting as a barrier, preventing the compounds reaching their target enzymes as has been described in relation to tannins by Bernays & Chamberlain (1980).

Many of the alkaloidal glycosidase inhibitors deter feeding (Blaney *et al*. 1984; Dreyer *et al*. 1985; see Table 36.2). The pea aphid *Acyrthosiphon pisum* was deterred from feeding by castanospermine (50% inhibition at 1×10^{-4} M) but the aphids *Schizagraphis graminum* and *Myzus persicae* were not affected (inactive at 5×10^{-2} M) (Campbell *et al*. 1987). These authors found that only one of the three aphids, *M. persicae*, was deterred from feeding by DNJ (50% inhibition at 2.5×10^{-3} M). The pea aphid, which feeds on the locoweed *Astragalus lentiginosus*, was not deterred from feeding by swainsonine when it was incorporated in an Akey synthetic diet at 0.05% (Dreyer *et al*. 1985).

In recent studies we have evaluated the antifeedant activity of DNJ, DMJ, fagomine, DMDP, CYB-3, AB-1, castanospermine and swainsonine against a range of chewing insects, which included three species of Lepidoptera and two species of locusts (Table 36.2). In these bioassays individual insects were exposed to pairs of glass fibre discs (a control and a treatment disc). The discs were made palatable with sucrose (10–15% of disc dry weight). The treatment discs had the further addition of a test compound at concentrations which represent 0.001 to 1% of disc dry weight. The Antifeedant Index ((C – T)/(C + T) %) was calculated using the amounts (i.e. weight) of both the control © and treatment (T) discs eaten during a 4–8 hour period. As expected, the compounds differ in their antifeedant activity: the most active were castanospermine and DMDP; castanospermine was an antifeedant against all of

Table 36.2: *Inhibition of Feeding by the Alkaloidal Glycosidase Inhibitors*

Insect Compound[c]	Spodoptera littoralis[a]	Spodoptera exempta[a]	Heliothis virescens[a]	Locusta migratoria[b]	Schistocerca gregaria[b]
DNJ	*	**	–	NA	*
DMJ	*	*	–	–	–
Fagomine	NA	NA	–	–	–
XZ-1	NA	–	–	–	–
BR-1	NA	NA	–	–	–
DMDP	**	***	NA	***	***
AB-1	***	–	–	***	–
CYB-3	NA	NA	–	**	*
CAST	***	***	**	***	*
SWAIN	***	–	–	–	–

Inhibition of feeding based on Antifeedant Index (see text). *** = 75–100%; ** = 50–74%; * = 25–49%; NA = 0–24%; – = Not tested. a = 10–20 replicates/experiment; b = 10 replicates/experiment; c = concentration = 0.01–0.02 disc dry weight.

the five species tested; DMDP was an antifeedant against all of the species except *Heliothis armigera*; swainsonine and AB-1 were significantly active against *S. littoralis*; DNJ and DMJ were most effective against the oligophagous *S. exempta*; fagomine, BR-1 and XZ-1 were not active.

Hansen (1974) has suggested that a glucosidase may be involved in host plant selection as these enzymes are present in the insect's taste sensilla and could release free sugars from plants during testing of potential food. Any compound, therefore, that inhibits glucosidases could affect feeding. Thus, the mechanism of antifeedant action elicited by the piperidines (DNJ and DMJ), pyrrolidines (DMDP and AB-1) and indolizidines (castanospermine and swainsonine) might be related to their ability to inhibit enzymes in the insect's taste sensilla.

However, Blaney *et al.* (1984) suggested that the antifeedant effect of DMDP was not related to its ability to inhibit glycosidases in a taste sensillum but to its structure as an alkaloid. Recent electrophysiological studies have shown that DMDP and other sugar analogues stimulate two neurones in the lateral maxillary styloconic taste sensilla of *Spodoptera littoralis* larvae; one neurone normally responds to sugars and the other neurone responds to alkaloids, such as quinine, sparteine and dopamine (Simmonds & Blaney, unpubl.).

This electrophysiological bioassay is used to investigate how insects perceive these chemicals and has helped to explain some of the anomalies encountered in the antifeedant bioassays. The electrophysiological bioassay is a very powerful tool in helping to establish the role of any chemical (natural or synthetic) in insect behaviour, especially feeding or oviposition behaviour. Furthermore, the bioassay requires only a fraction of the test compound that would be required in a more conventional bioassay.

The compounds are dissolved in an electrolyte and the solution used to stimulate sensilla on the insect's mouthparts. Each lepidopteran larva has four sensilla which are critically involved in food selection, the two medial and two lateral maxillary styloconic sensilla, and each sensillum contains four chemosensory neurones. By recording the action potentials elicited from these neurones when stimulated by a compound, we can identify which neurone responds and the frequency of the response. This information can be correlated with the resulting behaviour, enabling us to interpret gustatory codes associated with compounds that modify feeding behaviour.

Figure 36.2 shows the result of correlating electrophysiological data with behavioural data. The compounds evoked two distinctly different categories of behaviour; at high concentrations swainsonine, AB-1, castanospermine and DMDP were antifeedant, whereas fagomine and XZ-1 did not affect feeding. Despite this behavioural difference, the overall firing rate did not differ between the two groups. However, the information conveyed to the insect's brain in the two cases differs because each group preferentially stimulated different neurones. The distinction between neurones that respond to antifeedants and those that respond to phagostimulants is not absolute and some compounds may stimulate both types of neurones. Thus the resulting behaviour will depend on the ratio of firing in these neurones.

The antifeedant effect of DMDP is thought to be partially due to the fact that it stimulates the 'alkaloid-sensitive neurone' while at the same time interacting with the 'sugar-sensitive neurone' (Blaney *et al.* 1984). This 'sugar-sensitive neurone' has at least two types of receptor sites; one responding to glucose and one to fructose. When the lateral styloconic sensillum of *Spodoptera littoralis* is stimulated sequentially with glucose, fructose and DMDP each compound initiates a similar rate of firing (Fig. 36.3). When the sensillum is stimulated with DMDP and then restimulated with a sugar, some type of interaction is evident, resulting in a decrease in the firing rate compared to the initial response to that sugar. Over the next hour the ability to respond to glucose returns and the original firing rate is restored, but with fructose recovery is much slower. This may occur because DMDP is a fructose analogue and may therefore block the fructose site. The antifeedant activity ascribed to DMDP could result from two different mechanisms: by stimulation of a deterrent neurone and/or by interference with the functioning of the fructose sites on the 'sugar-sensitive neurone'.

The specificity of these compounds against insects needs further study but the fact that the insects differ in their susceptibilities could be of use in the design of insect control programmes (Fellows 1986).

In addition to the uses described in this paper, these compounds are also being used in the study of genetic diseases, such as mannosidosis (Dorling *et al.* 1983), and in affinity chromatography (Osiecki-Newman *et al.* 1986). As the demand for these compounds is likely to increase it is important to find new sources. A survey of plants, especially wild plants, for alkaloidal glycosidase

Figure 36.2: *Correlation of Behavioural and Electrophysical Responses of Spodoptera littoralis Sixth Stadium Larvae to Test Compounds.*

<u>Behavioural response</u> is the Antifeedant Index ($(C - T)/(C + T)$ %), n = 20.
<u>Electrophysiological response</u> is the % increase in total firing, in the first second of stimulation, of the lateral and medial maxillary styloconic sensilla compared with the response to the control (0.05 M) NaCl, n = 5–10.

Concentrations (mM)	50	20	10	5	1	0.5	0.2	0.1	0.05	0.01	0.005
swainsonine, AB-1			a		b			c		d	
castanospermine			a	b	c		e	f		d,g	
DMDP	a	b	c		d			e			
CYb-3				a	b			c	d	e	
fagomine and XZ-1			a	b	c	d	e	f		g	h

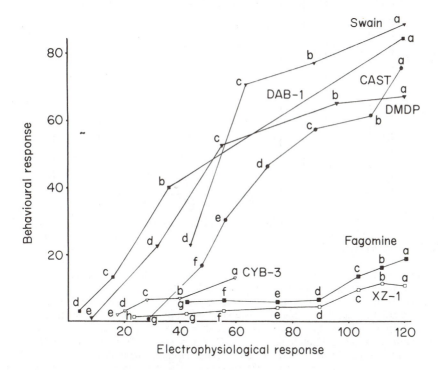

inhibitors should be encouraged, and might generate unexpected sources of income in many areas of the world where such plants grow, but are not currently harvested. The well publicized destruction of natural habitats, such as tropical rain forests, adds a very real sense of urgency to the need to investigate wild plants in this way, while they are still available to us.

Figure 36.3: *Effect of Stimulating the Lateral Maxillary Styloconic Sensilla of the Sixth Stadium Larvae of Spodoptera littoralis Sequentially with NaCl (0.05M) Glucose (0.1M), Fructose (0.1M) and DMDP (0.001M).*

Each point represents the total mean firing rate obtained in the first second of stimulation (n = 10 larvae).

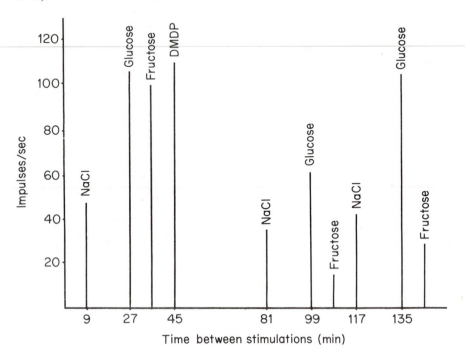

REFERENCES

Bernays, E. A. and D. J. Chamberlain (1980) A study of tolerance of ingested tannin in *Schistocerca gregaria. J. Insect Physiol.* 26: 1–18.

Blaney, W. M., M. S. J. Simmonds, S. V. Evans and L. E. Fellows (1984) The role of the secondary plant compound 2,5-dihydroxymethyl 3,4-dihydroxypyrrolidine as a feeding inhibitor for insects. *Entomol. Exper. Appl.* 39: 209–216.

Campbell, B. C., R. J. Molyneux and K. C. Jones (1987) Differential inhibition by castanospermine of various insect disaccharidases. *J. Chem. Ecol.* 13: 1759–1770.

Colegate, S. M., P. R. Dorling and C. R. Huxable (1979) A spectroscopic investigation of swainsonine: an α-mannosidase inhibitor isolated from *Swainsona canescens. Austr. J. Chem.* 32: 2257–2264.

Dennis, J. W. (1986) Effects of swainsonine and polyinosinic: polcytidylic acid on murine tumor cell growth and metastasis. *Cancer Research* 46: 5131–5136.

De Santis, R., U. V. Santer and M. C. Glick (1987) NIH 3T3 cells transfected with human tumor DNA lose the transformed phenotype when treated with swainsonine. *Biochem. Biophys. Res. Comm.* 142: 348–353.

Dorling, P., C. Huxable, I. Cenci di Bello and B. Winchester (1983) Swainsonine-induced mannosidosis: a reversible chemically inducible phenology of an inherited disorder. *Biochem. Soc. Trans.* VIII: 717–718.

Dreyer, D. L., K. J. Jones and R. J. Molyneux (1985) Feeding deterrency of some pyrrolizidine, indolizidine, and quinolizidine alkaloids towards pea aphid (*Acyrthosiphon pisum*) and evidence for phloem transport of indolizidine alkaloid swainsonine. *J. Chem. Ecol.* 11: 1045–1051.

Evans, S. V., A. M. R. Gatehouse and L. E. Fellows (1985a) Detrimental effects of 2,5-dihydroxymethyl-3,4-dihydroxypyrrolidine in some tropical legume seeds on larvae of the bruchid *Callosobruchus maculatus*. *Entomol. Exper. Appl.* 37: 257–261.

Evans, S. V., L. E. Fellows, T. K. M. Shing and G. W. J. Fleet (1985b) Glycosidase inhibition by plant alkaloids which are structural analogues of monosaccharides. *Phytochem.* 24: 1952–1955.

Evans, S. V., A. R. Hayman, L. E. Fellows, T. K. M. Shing, A. E. Derome and G. W. J. Fleet (1985c) Lack of glucosidase inhibition by, and isolation from *Xanthocercis zambesiaca* (Leguminosae) of 4-0-(β-D-glucopyranosyl)-fagomine [1,2,5-trideoxy-4-0-(β-D-glucopyranosyl)-1,5-imino-D-arabino-hexitol], a novel glucoside of a polyhydroxylated piperidine alkaloid. *Tetrahedron Letters* 26: 1465–1468.

Fellows, L. E. (1986) The biological activity of polyhydroxyalkaloids from plants. *Pesticide Science* 17: 602–606.

Fellows, L. E. and G. W. J. Fleet (1988) Alkaloid glycosidase inhibitors from plants. In: *Natural Products Isolation*, G. H. Wagman and R. Cooper (Eds). Elsevier, Amsterdam (in press).

Fellows, L. E., S. V. Evans, R. J. Nash and E. A. Bell (1986) Polyhydroxy-alkaloids as glycosidase inhibitors and their possible ecological role. In: *Natural Resistance of Plants to Pests, Roles of Allelochemicals.* M. B. Geen and P. A. Hedin (Eds), pp. 72–78.

Fellows, L. E., E. A. Bell, D. G. Lynn, F. Pilkiewics, I. Miura and K. Nakanishi (1979) Isolation and structure of an unusual cyclic amino alditol from a legume. *J. C. S. Chem. Comm.* 977–978.

Fellows, L. E., C. H. Doherty, J. M. Horn, G. C. Kite, R. J. Nash, J. T. Romeo, M. S. J. Simmonds and A. M. Scofield (1988) Distribution and biological activity of alkaloidal glycosidase inhibitors from plants. In: *Proceedings of 1987 Swainsonine Conference, Utah*, L. F. James (Ed.) (in press).

Fleet, G. W. J., S. J. Nicholas, P. W. Smith, S. V. Evans, L. E. Fellows and R. J. Nash (1985) Potent competitive of α-galactosidase and α-glucosidase activity by 1,4-dideoxy-1,4-iminiopentitols: synthesis of 1,4-dideoxy-1,4-iminoarabinitol. *Tetrahedron Letters* 26: 3131.

Fuhrman, U., E. Bause and H. Ploegh (1985) Inhibitors of oligosaccharide processing. *Biochim. Biophys. Acta* 825: 95–110.

Gruters, R. A., J. J. Needfjes, M. Tersmette, R. E. Y. de Goede, A. Tulp, H. G. Huisman, F. Miedema and H. L. Ploegh (1987) Interference with HIV-induced syncytium formation and viral infectivity by inhibitors of trimming glucosidase. *Nature* 330: 74–77.

Hadwiger, A., H. Niemann, A. Kabisch, H. Bauer and T. Tamura (1986) Appropriate glycosylation of the fms gene product is a prerequisite for its transforming potency. *EMBO* 5: 689–694.

Hansen, K. (1974) Alpha-glucosidases as sugar receptor proteins in flies. In: *25 Mosbascher Colloquium Ges. Biol. Chem.*, L. Jaenicke (Ed.), pp. 207–232. Heidelberg: Springer Verlag.

375

Hino, M., O. Nakayama, T. Tsurumi, K. Adachi, T. Shibata, H. Terano, M. Kohsaka, H. Aoki and H. J. Imanaka (1985) Studies of an immuno-regulator swainsonine. 1. Enhancement of immune response by swainsonine *in vitro*. *J. Antibiotics* 38: 926–935.

Hohenschutz, L. D., E. A. Bell, P. J. Jewess, D. P. Leworthy, R. J. Pryce, E. Arnold and J. Clardy (1981) Castanospermine, a 1,6,7,8-tetrahydroxyoctahydroindolizidine alkaloid, from seeds of *Castanospermum australe*. *Phytochem*. 20: 811–814.

Humphries, M. J., K. Matsumoto, S. L. White and K. Olden (1986a) Inhibition of experimental metastasis by castanospermine in mice: blockage of two distinct stages of tumor colonisation by oligosaccharide processing inhibitors. *Cancer Research* 46: 5215–5222.

Humphries, M. J., K. Matsumoto, S. L. White and K. Olden (1986b) Oligosaccharide modification by swainsonine treatment inhibits pulmonary colonisation by b16-f10 murine melanoma cells. *Proc. Nat. Acad. Sci.* 33: 1752–1756.

Ishida, N. K., T. Kumagai, T. Niida, T. Tsurvoka and H. Yumoto (1967) Nojirimycin, a new antibiotic. II Isolation characterisation and biological activity. *J. Antibiotics*. Ser A20: 66–71.

Jones, D. W. C., R. J. Nash, E. A. Bell and J. M. Williams (1985) Identification of the 2-hydroxymethyl-3,4-dihydroxypyrrolidine (or 1,4-dideoxy-1.4-iminopentitol) from *Anglocalyx boutiqueanus* and from *Arachnoides standishii* as the (2R,3R,4S)-isomer by the synthesis of its enantiomer. *Tetrahedron Letters* 26: 3125–3126.

Kino, T., N. Inamura, K. Nakahara, S. Kijoto, T. Goto, H. Terano, M. Kohsaka, H. Aoki and H. Imanaka (1985) Studies of an immunoregulator swainsonine. 2. Effect of swainsonine on mouse immunodeficient system and experimental murine tumor. *J. Antibiotics* 38: 936–940.

Kyoma, M. and S. Sakamura (1974) The structure of a new piperidine derivative from buckwheat seeds (*Fagopyrum esculentum* Moench). *Agric. Biol. Chem.* 38: 1111–1112.

Molyneux, R. J. and L. F. James (1982) Loco intoxication: Indolizidine alkaloids of spotted locoweed (*Astragalus lentiginosus*). *Science* 216: 190–191.

Murao, S. and S. Miyata (1980) Isolation and characterization of a new trehalase inhibitor, S-GI. *Agric. Biol. Chem.* 44: 219–221.

Nash, R. J., K. A. Fenton, A. M. R. Gatehouse and E. A. Bell (1986) Effects of the plant alkaloid castanospermine as an antimetabolite and a feeding deterrent to phytophagous insects. *Entomol. Exper. Appl.* 42: 71–77.

Osiecki-Newman, K. M., D. Fabbro, T. Dinur, S. Boas, S. Gatt, G. Legler, R. J. Desnick and G. A. Grabowski (1986) Human acid β-glucosidase: affinity purification of the normal placental and Gaucher disease splenic enzymes on N-alkyl-deoxy-nojirimycin-sepharose. *Enzyme* 35: 147.

Schmidt, D. D., W. Formmer, L. Muller and E. Truscheit (1979) Glucosidase-inhibitors aus Bazillen. *Naturwiss.* 66: 584–585.

Scofield, A. M., L. E. Fellows, R. J. Nash and G. W. J. Fleet (1986) Inhibition of mammalian digestive disaccharidases by polyhydroxy alkaloids. *Life Science* 39: 645–650.

Szumilo, T., G. P. Kaushal, H. Hori and A. D. Elbein (1986) Purification and properties of a glycoprotein processing α-mannosidase from mung bean seedlings. *Plant Physiol* 81: 383–389.

Tyms, A. S. *et al.* (1987) Castanospermine and other plant alkaloid inhibitors of glucosidase activity block the growth of HIV. *Lancet* October 31st: 1025–1026.

Villalta, F. and F. Kierszenbaum (1985) The effect of swainsonine on the association of *Trypanosoma cruzi* with host cells. *Molecular Biochem. Parasit.* 16: 1–10.

Walker, B. D., M. Kowalski, W. G. Goh, K. Kozarsky, M. Krieger, C. Rosen, L. Rohrschneider, W. A. Haseltine and J. Sodroski (1987) Inhibition of human immunodeficiency virus syncytium formation and virus replication by castanospermine. *Proc. Nat. Acad. Sci.* 84: 8120–8124.

Yagi, M., T. Kouno, Y. Aoyagi and H. Murai (1976) The structure of moranoline, a piperidine alkaloid from *Morus* species. *Nippon Nogei Kagaku Kaishi* 50: 571–572 (*Chem Abstr.* 86: 167851, 1976).

37

Bioactive Phytochemicals – the Search for New Sources

P. G. Waterman

INTRODUCTION

Throughout recorded history man has relied heavily on the plant kingdom for biologically active compounds to treat illness, to assist in the capture of animal protein and to protect crops against spoilage and the depredations of herbivores and disease vectors. In this paper I concern myself primarily with plants as sources of compounds with bioactivity of potential value to the treatment of diseases in man but the arguments presented apply equally to the search for compounds of value in agriculture.

THE SITUATION IN THE WORLD TODAY

The Developed World

In this country the importance of natural products as sources of bioactive compounds has been challenged by the advent of synthetic substances. However, natural products have proved remarkably resistant to displacement, even in the developed world. To illustrate this point I have compared the officially recognized plant and fungus species and their products included in the British Pharmacopoeias of 1932 and 1980 (Table 37.1). This reveals that over a period of about 50 years there has been an actual increase in numbers from 94 monographs in 1932 to 98 in 1980. This is due primarily to the advent of antibiotics; in 1932 ergot and its alkaloids was the only fungal material included, whilst in 1980 fungal and bacterial sources accounted for 22 of the monographs. Thus monographs for higher plants and their products have declined from 93 to 76. A breakdown of these figures for higher plants reveals that 57 monographs are common to the two sources, 36 that were present in the 1932 volume have been dropped and 19 new monographs have been adopted.

In one respect, however, the above figures are somewhat misleading. While many of the plant drugs lost since 1932 were biologically active most of the new higher plant products can be regarded as biologically inert diluents or flavours; the alkaloids of *Rauvolfia* spp. (Apocyanaceae) and *Catharanthus roseus* (Apocyanaceae) are obvious exceptions. The major reason behind this decline in the use of bioactive plants has been the enormous expansion in the role of

Bioactive Phytochemicals

Table 37.1: A Comparison of the Crude Drugs and Derivatives from Fungi and Higher Plants Official in the British Pharmacopoeias of 1932 and 1980

Type of Substance	1932 only	Present in 1932 and 1980	1980 only
Fixed oils	5	4	3
Esters/acids	2	3	0
Carbohydrates	1	3	4
Phenolics	6	6	2
Volatile oils	8	17	4
Resins	6	3	1
Saponins/Cardioactive steroids	1	5	0
Terpenes	2	1	1
Alkaloids	5	16	3
Vitamin preparations	0	0	1
Antibiotics	0	0	21

Figure 37.1: Vincritine.

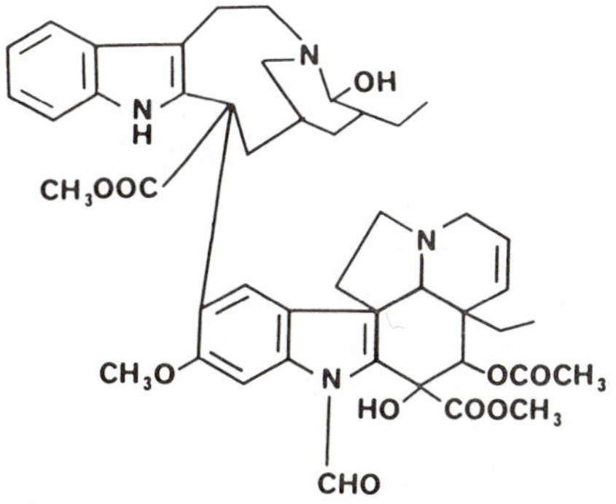

synthetic drugs (Sneader 1985) which were very few in number in 1932. Wholly synthetic substances have definite practical advantages over whole plant drugs in terms of quality assurance (absence of biological variation, the consequent accurate dosage, simpler identification and quality control procedures, ease of manipulation for optimal methods of administration). Such advantages are shared by pure natural products isolated from living organisms and the semisynthetic derivatives obtained from them and numerous examples in this category exist, such as the cytotoxic alkaloid vincritine (Fig. 37.1) from *Catharanthus roseus* and the cardiotonic glycoside digoxin (Fig. 37.2) from *Digitalis lanata* (Scrophulariaceae).

Yet while diminished in number in comparison with previous times the cost of prescription drugs derived solely from plant sources was still estimated as $8 billion in the USA in 1980 (Balandrin *et al.* 1985) and the percentage of prescriptions based primarily on natural products including antibiotics has

Figure 37.2: *Digoxin.*

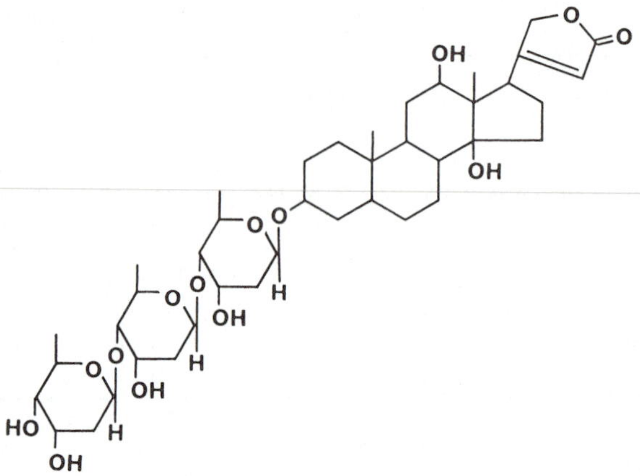

remained fairly constant at about 25 per cent for a number of years. Furthermore many pharmacopoeial drugs that are purely synthetic in nature owe their origin to natural products that have acted as a template from which the synthetic chemists have designed their novel structures. Examples of these include numerous antibiotics, most local anaesthetics and antimalarials and the new skeletal muscle relaxant atracurium besylate (Fig. 37.3), a compound designed and synthesized using the natural muscle relaxant tubocurarine (Fig. 37.4) as a model but introducing features that gave a shorter duration of action, thus making it easier to control dosage (Stenlake *et al.* 1983).

Figure 37.3: *Atracurium besylate.*

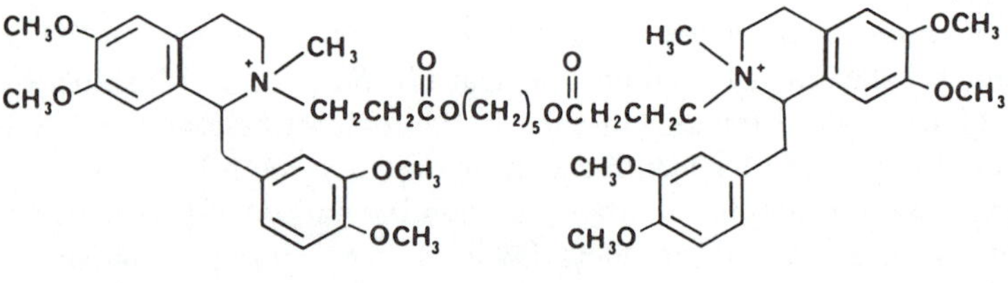

At present, there appears to be an increase in interest in medicines of natural origin, both within the pharmaceutical industry and by the general public. This is partly as a matter of fashion that currently suggests it is more desirable to use remedies from natural sources because they are 'natural', but partly through a genuine dissatisfaction with currently available synthetic drugs and progress in development of new products (Capasso 1985; Tyler 1986). Despite the decline in official recognition Phillipson (1981) was able to report the occurrence of over 5000 herbal products on the UK market, involving more than 500 different plants.

Phillipson (1986) estimates that in 1980, in addition to those grown *in situ*, western Europe imported about 400 different botanicals amounting to some 80 000 tonnes at a collective import cost for France, W. Germany, Holland and the UK of about $117 million. Tyler (1986) estimates that in 1981 'health-food' outlets in the USA sold $360 million of herbs. Yet despite the strong pharmaceutical research base that exists in the developed world very little is known about the true efficacy and potential dangers of many of these materials. Phillipson & Anderson (1984) suggest that the use of a number of traditional herbal remedies in the UK must today be regarded with some concern as chemical analyses has shown them to contain secondary metabolites known to be toxic without satisfactory verification of their efficacy. Efforts to legislate on herbal remedies using the criteria adopted for pure organic compounds (HMSO 1985) require that research into these materials be performed. Unfortunately the necessary funds for this are not forthcoming and the academic research community on which to draw for such a research programme is rapidly diminishing. This is most notable in schools of pharmacy where the amount of time allocated to the study of natural products has fallen drastically over the past 20 years.

The Developing World

In much of the developing world the population is generally still reliant on plant drugs for the treatment of disease and about 75 per cent of the world's population is faced with health care standards below an acceptable level. The criticisms levelled at our understanding of herbal drugs used in the west (see above) apply equally to almost all of the traditional medicines employed throughout many parts of the developing world. The World Health Organization has the ambitious aim of attaining minimal acceptable standards for all by the year 2000 and to achieve this it has recognized the need to integrate traditional systems of medicine with other health care programmes (Bannerman *et al.* 1983). This should involve the study of an estimated 20 000 species used in herbal remedies (Penso 1982), some of which still need to be adequately identified taxonomically let alone examined for pharmacological properties and chemical constituents. There are countries in which the development of relevant research programmes is well advanced, notably in the People's Republic of China where it is estimated that about 250 000 people are involved in the cultivation and administration of medicinal plants and many thousands more are involved in the scientific analysis of these materials. However, it must be recognized that the resources that are involved in doing this are enormous and the Chinese example cannot be matched in many developing countries, whatever the will to do so.

THE CASE FOR CONTINUING THE INVESTIGATION OF PLANTS FOR BIOACTIVE COMPOUNDS

Setting aside the moral question of the future of the 75 per cent of the world's population that continue to rely on drugs from natural sources, can a case be made for continuing or actually expanding our investigation of the natural world for new pharmaceutical compounds for the benefit of the developed world? I believe the answer is very much in the affirmative and although the following comments are addressed to drugs for the treatment of diseases in man they can be applied equally well with respect to veterinary medicines and pest control chemicals.

According to Baladrin *et al.* (1985) 5–15 per cent of higher plants have been surveyed for biologically active compounds. This figure will certainly be even lower for fungi. These statistics are of interest in that they indicate how little has been done, yet in reality represent a massive overestimate of our knowledge as, in all but a few cases, testing will have been restricted to only one or a few kinds of activity. However, it does represent an enormous body of data. Why then have so few new natural products been developed for use as pharmaceuticals over the past 20 years? Farnsworth (1984) has identified the major problem that has bedevilled work on natural products as 'an almost complete lack of interaction between chemists, botanists, biologists and physicians'. That this is the case is confirmed by an examination of the scientific literature. Reports of the isolation and identification of secondary metabolites are commonplace and the pharmacological literature on isolated compounds is also extensive, but the optimal procedure whereby a plant extract is obtained, the presence of activity confirmed and then the active principle(s) isolated and identified by following a bioassay guided procedure remains a relative rarity.

An example of the success of the 'research-team' approach is the recent isolation of the dimeric indole alkaloid yuehchekene (Fig. 37.5) from the roots of *Murraya paniculata* (Rutaceae) as part of the WHO Task Force Programme for the development of new birth control agents. The presence of anti-implantation activity in the root extracts of *M. paniculata* was established following an agreed WHO protocol (Kong *et al.* 1985a) and the responsible compound, yuehchukene, was then isolated by following the activity through column chromatography and high pressure liquid chromatography separation techniques. Yuehchukene proved to be highly active, being present only in trace amounts (about 30 ppm) but was relatively easy to identify using modern spectroscopic techniques (Kong *et al.* 1985b). This study is important for three reasons. First, it was achieved as part of a strongly goal-directed effort controlled by WHO. Second, it relied on the combined efforts of several different disciplines (botanists, ethnobotanists, biochemists, pharmacologists, chemists, spectroscopists) working together to establish the bioactivity of yuehchukene. Third, it benefited from wholehearted international collaboration.

Yuehchukene has since been synthesized (Cheng *et al.* 1985) and is now undergoing further testing.

Figure 37.4: Tubocurarine.   *Figure 37.5*: Yuehchekene.

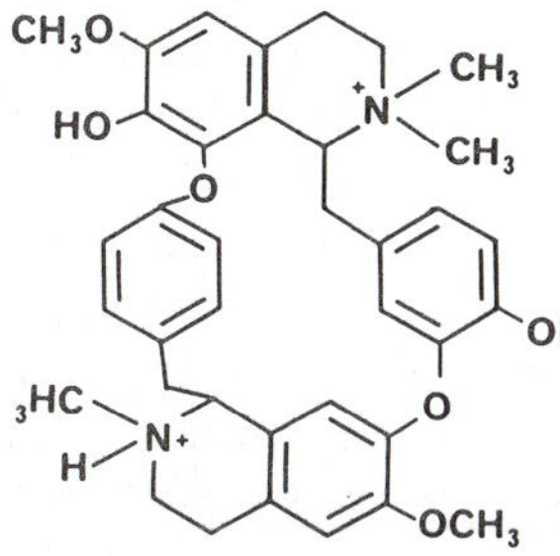

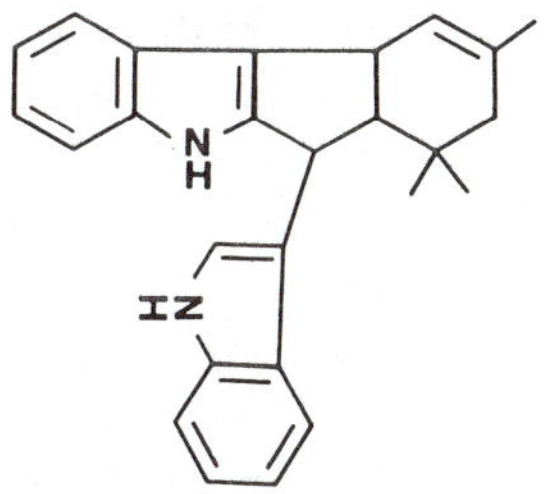

The yuehchukene story is also significant for two other reasons. First, whilst *Murraya paniculata* has many local uses (Kong *et al.* 1986) these did not include application as an arbortifacient so that there is no direct ethnobotanical link identifying it as a target species in a search for fertility control agents. Second, it is a plant species that has been subjected to numerous phytochemical investigations and is a very well known source of coumarins (Gray 1984), but had not previously been suspected to contain dimeric indole alkaloids as secondary metabolites.

The message that I take from this, and other similar investigations and which is supported by the extensive phytochemical and related pharmacological literature, is that the plant kingdom (and by extrapolation the fungal and parts of the animal kingdom) continues to be a potential major source of new bioactive compounds and that these can be found if a planned and rational methodology is adopted. Furthermore, the search should not be restricted to those species that are employed in local materia medica; while these offer an obvious place to start a search all plant species should be regarded as potential sources of a given type of bioactivity until it is proved otherwise. This view that there remains a vast potential for bioactive compounds in the plant kingdom is further sustained by our developing ideas of the role of many plant secondary metabolites as allelochemicals; that is, compounds with a function in defending the producer against attack by competing organisms (Harborne 1982). Such a function will have selected for maximum biological activity (Kubitzki & Gottlieb 1984) coupled with maximum structural diversity in the secondary metabolism of sympatric species (Waterman & McKey 1988).

THE FUTURE DEVELOPMENT OF RESEARCH ON MEDICINAL PLANTS

When considering the most appropriate methods for pursuing future research into bioactive compounds from natural sources a basic dichotomy must be taken into account. Research workers and their paymasters in the developed world are generally looking for advances on drugs that are in use and so their assessment of new compounds will clearly be judged critically against what is presently available. In the developing world the questions to be answered are more basic and more critical. Is the material being supplied to the local population efficacious, useless or dangerous? Can we produce substances from our own resources which will perform the same function as imported western medicines? Such substances need not meet the criteria that would be required for a new marketable drug in the developed world. This should not be interpreted as advocating lower standards for the developing world but takes the pragmatic view that production of a local replacement of equivalent efficacy to an imported drug represents a considerable advance for the producer.

It is research directed primarily toward the needs of the developing world that I wish to consider here. Except for rare and highly specific needs, drugs to combat AIDS being an obvious example, there is little evidence at present that major drug companies are going to invest heavily in research aimed at identifying new sources and so initial developments are likely to arise primarily from the success of programmes aimed at fulfilling the needs of the developing world.

Identifying Plants with Useful Activity

How do we find bioactivity in plants? Four different approaches can be considered:

(a) *The Ethnobotanical Approach.* Throughout the world considerable effort has been put into documenting local uses of plants. A major problem is that much of this work has been undertaken haphazardly by people without the necessary scientific training to assess what they are seeing or being told. These problems have been discussed in detail by Croom (1983) who lists a large number of points that need to be noted to obtain a useful ethnobotanical record. The two most critical points that need to be recorded without ambiguity are species identity and the exact method of preparation and use. Before using an existing ethnobotanical survey as a basis for a research programme it must be critically assessed, and all too often this leads to its rejection. Ideally, ethnobotanical surveys should be team efforts involving anthropologists collaborating with experts in the field of health care capable of interpreting methods of preparation and use.

(b) *Use of a Specific Screen.* Perhaps the most practical approach is to have research groups searching for particular types of bioactivity using bioassay techniques specifically designed for their assessment. The anti-fertility screen that detected yuehchukene (Kong *et al.* 1985a) is an example of this. In the last ten years there have been considerable new developments in screening techniques with greatly enhanced sensitivity and specificity. Through these it is now possible to set up laboratories equipped to test extracts against various parasite induced conditions (malaria, onchocerciasis, amoebic dysentery) and disease states (anti-inflammatory, cardiovascular and cytotoxicity screens) as well as for antimicrobial activity. Tests using cell cultures and monoclonal antibodies are rapidly being developed. Such screening procedures will remain relatively costly and should be set up on an international (geographical) rather than a national basis. The problem for the future should be primarily a political one; that of identifying priorities.

(c) *Chemotaxonomic Screening Programme.* Once a promising source of activity has been established then it is logical to examine allied species for similar compounds. For example after the identification of yuehchukene a search was made of other *Murraya* species in the hope of finding a better source. As a result the alkaloid was isolated from two other species although not in significantly greater quantities (Kong *et al.* 1986). More recently it has been isolated from the allied genus *Merrillia* (Kong *et al.* 1988). The Indian species *Commiphora mukul* (Burseraceae) is the source of a number of promising anti-inflammatory steroids, the guggulsteroids (e.g. Fig. 37.6) (Anand & Nityanand 1984). Many closely allied species of *Commiphora* occur in the arid area of the horn of Africa (Somalia, Ethiopia, Kenya). A recent phytochemical examination of one of these revealed a unique group of octanordammaranes such as mansumbinone (Fig. 37.7) (Provan & Waterman 1986). Clearly this section of *Commiphora* contains species capable of producing unusual triterpene derivatives and deserves further examination.

(d) *Random Screening.* In many countries areas of natural vegetation are being set aside in an attempt to preserve the natural flora. Such areas should not be regarded as museums and, as a long-term goal, efforts should be made to obtain an inventory of the major classes of constituents found in individual plant species in such areas. As well as providing data for future reference such programmes have the potential for throwing up useful compounds entirely by chance and they are also important in supplying material for the training of scientists in the necessary disciplines of isolation and structure elucidation of natural products and screening procedures to assess bioactivity.

Figure 37.6: *Guggulsteroid from*
Commiphora mukul.

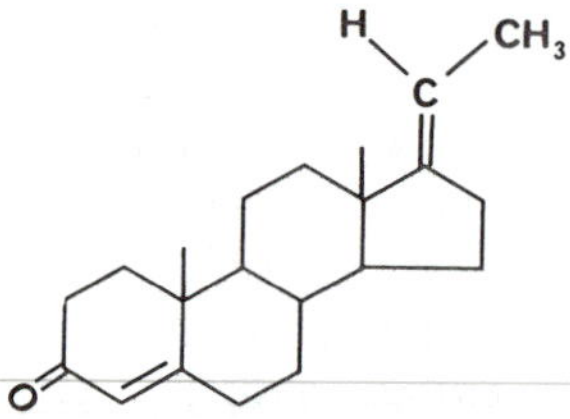

Figure 37.7: *Masubinone.*

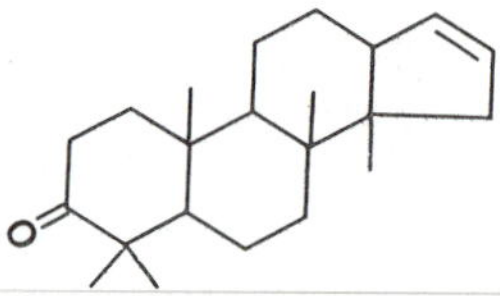

It should be stressed that such projects are long-term. As part of a survey of the Douala-Edea Forest Research and the Korup National Park in Cameroon several chemistry students working in my laboratory undertook research projects based entirely on material from those forests. The results of their studies, which encompass only four plant families and no more than about 5 per cent of the tree species, have recently been reviewed (Waterman 1986). Such forests clearly have the capacity to occupy a major phytochemical research facilities for decades!

Development and Exploitation of New Sources of Bioactive Compounds

In the happy event of a screening programme leading to the identification of material with interesting activity what then is the optimal route to proceed towards development and exploitation? The problems involved in this can be divided into five areas, outlined below. Parts (a) – (d) form a natural progression through which a new find could be developed by the discoverer(s). The final area (e) could stem directly from (a/b) in cases where bioactivity was interesting but for some reason the active compound was not directly usable.

(a) *Identification of Active Components.* Whether in the final analysis a new find is to be used as an extract or as a source of a fine chemical it will be necessary to isolate and characterize the active component(s). Today, this is usually a relatively simple task, given that access is available to sophisticated spectroscopic equipment, notably high-field nuclear magnetic resonance spectroscopy and mass spectrometry. For example, yuehchukene (Fig. 37.5) was characterized in three weeks from a sample that was never greater than 5 mg. The problem is that such equipment is very expensive to purchase and maintain and, with a few exceptions, is located in laboratories in the developed world.

(b) *Characterization of Mode of Action and Safety.* Pharmacologists will have already been involved in identifying the activity that has led to studies at this level. They now continue to play an important role by the elucidation of mechanism of activity and occurrence of side-effects and potential toxicity. This information will be central to final decisions about desirability of exploitation and the most appropriate methods to adopt.

(c) *Setting up Production Facilities.* An appropriate method of production needs to be established with a protocol of preparing the material and, if necessary, extracting it. Methods must be developed to allow the end-product of this process to be quality controlled.

(d) *Selection of Appropriate Form for Administration.* When a source of useful activity has been identified it becomes of prime importance to select the most appropriate dosage form for administration. To achieve this consideration has to be given to the conditions under which it will be used (it must retain activity), who will administer it (self-medication cuts down the possible routes) and an indication of potential problems and contraindications. To achieve this the initial research teams require the backing of a formulation group (pharmaceutical technologists, toxicologists, pharmacologists).

(e) *Further Exploitation Through Synthetic Procedures.* Included here are chemical modifications of the natural product in the hope of producing a more appropriate semisynthetic product and the development of purely synthetic processes designed to exploit a natural product lead compound. Given an interesting new find, this is an area that pharmaceutical companies are likely to step in and exploit. In an ideal world it would be expected that they would acknowledge the pioneering role of the discovering laboratory and arrive at some agreement on profit sharing, but this is probably a forlorn hope.

HOW SHOULD THE RESEARCH EFFORT BE ENCOURAGED?

If effort alone was enough then this section would not be needed. Very much is being attempted but with a few notable exceptions it is uncoordinated and, from an individual point of view, underfunded and frustrating. The impetus for such research must in future stem primarily from the developing world in view of the facts that (a) it is they who control most of the potential resources, and (b) it is they who have the most to gain from a successful exploitation of this area. I would identify the following developments as necessary prerequisites to success. Each of these is already being carried out, to some extent, to my knowledge but generally without sufficient funds and without the support and conviction that is needed to bring success.

(a) *Development of Research Teams.* Tyler (1986) stresses the need for teamwork in the attempt successfully to exploit local medicinal resources. At present this is rarely achieved. Most developing countries now have a nucleus of trained scientific personnel but rarely do they combine their efforts in the manner necessary to exploit natural resources. Far too often research workers trained in some aspect of value to such a project continue to pursue their own postgraduate research line on their return home without direction or guidance. I do not blame them for this as individuals, it is primarily the responsibility of those in authority to select areas into which efforts (and funds) will be directed. It is also incumbent upon those in authority to keep the best young scientists in research rather than burdening them with administration or other 'paper-pushing' functions.

(b) *A Systematic Approach.* It is imperative that research teams be given specific goals rather than an unspecified remit. What these goals are must be decided by the political and health authorities involved. Such decisions should often be taken on an international rather than a national basis; in many cases research facilities could profitably cover geographic regions rather than individual countries. At the very least countries with similar problems should coordinate their efforts.

(c) *Scientific Support.* The range of skills and equipment that are necessary to tackle the full range of problems that studies of this type require are considerable and it is not practical to try and cover all eventualities within one research group. The need for reliable support is an absolute prerequisite of any realistic programme. Such support can be considered in terms of interaction between research groups in developing countries (south-south cooperation) or interaction with scientific centres in the developed world (north-south collaboration).

North-south collaboration requires a development of contacts and understanding. It must be realized that the work of research groups in the developing world will be goal-directed, based at the practical level of getting the job done. Their output and requests for assistance should be judged with this in mind rather than in the terms used to assess academic research. Regular training programmes will be required to keep scientists engaged in these programmes up-to-date with the latest techniques in the various disciplines required. Access to sophisticated types of equipment, notably spectroscopic techniques, will be a major requirement for the foreseeable future. It can be argued that all these are at present available through personal contacts and various organizations such as the International Organization for Chemical Sciences in Development. However, these interactions generally work better at a more personal level and at present many scientists working under difficult conditions are frustrated by a lack of access to such techniques.

In the immediate future what seems to be required is a direct twinning between groups in the developed and developing world. In the long term, however, the importance of south-south collaboration cannot be overemphasized.

Regional planning leading to centres that collectively share the range of expertise and facilities required would represent important progress towards self-reliance. Progress in this direction has been greatest in South East Asia where the Regional Network for the Chemistry of Natural Products in South East Asia, sponsored by UNESCO and largely funded by Australia, has been responsible for the support and development of regional centres of excellence that are now able to perform many of the functions that were formerly restricted to Australian laboratories (Cannon 1988). In East Africa the recently formed NAPRECA (Natural Products Research Network for Eastern and Central Africa) offers a framework for the evolution of a similar network in that continent.

CONCLUDING COMMENTS

The search of the natural world for bioactive molecules capable of either direct use or of acting as lead compounds for the synthetic chemist has so far been relatively disappointing. Studies in this area have, to a large extent, been haphazard and ineffective. What is required is a new, cohesive approach to the search in which emphasis is placed on teamwork and collaboration between research groups encompassing a range of disciplines and scientists in both the developed and developing worlds. This would have most immediate positive results for health care in the developing world but in the long term will be to the benefit of mankind as a whole.

Finally it must be stressed that the organic chemistry performed by living organisms in the synthesis of secondary metabolites involves enzyme-mediated procedures that are generally beyond the scope of the organic chemist to repeat on an economic level, if at all; indeed the mechanisms involved are still not understood in many cases. For this reason, if no other, our present diversity of plant species must be conserved. Today we see the beginning of the development of techniques in biotechnology that should, in future, allow us to use these enzyme systems to drive synthetic organic reactions of a range and complexity that is beyond the scope of the chemist (Ellis 1986). The accelerating loss of plant species, particularly in the tropical rainforest zones, is in some respects analogous to the elimination of potential Nobel Prize winning organic chemists just as their careers are about to commence!

REFERENCES

Anand, N. and S. Nityanand (1984) Integrated approach to development of new drugs from plants and indigenous remedies. In: *Natural Products and Drug Development*, P. Krogsgard-Larsen, S. Brogger-Christensen and H. Koford (Eds), pp. 79–91. Munksgaard, Copenhagen.

Balandrin, M. F., J. A. Klocke, E. S. Wurtele and W. M. Bollinger (1985) Natural plant chemicals: sources of industrial and medicinal materials. *Science* 228: 1154–1160.

Bannerman, R. H., J. Burton and Wen-Chieh (Eds) (1983) *Traditional medicine and health care coverage*. World Health Organization, Geneva.

British Pharmacopoeia (1932) Pharmaceutical Press, London.

British Pharmacopoeia (1980) Pharmaceutical Press, London.

Cannon, J. R. (1988) The role of networks in the development of chemistry in South East Asia. *Fitoterapia* – (in press).

Capasso, F. (1985). Medicinal plants: an approach to the study of naturally occurring drugs. *J. Ethnopharm.* 13: 111–114.

Cheng, K-F., Y-C. Kong and T-Y. Chan (1985) Biometric synthesis of yuehchukene. *Chem. Comm.* 48.

Croom, E. M. (1983) Documenting and evaluating herbal remedies. *Econ. Bot.* 37: 13–27.

Ellis, B. D. (1986) Production of plant secondary metabolites without plants: a perspective. *Biotech. Advances.* 4: 279–288.

Farnsworth, N. R. (1984) The role of medicinal plants in drug developments. In: *Natural Products and Drug Development*. P. Krogsgaard-Larsen, S. Brogger-Christensen and H. Kofod (Eds), pp. 17–30. Munksgaard, Copenhagen.

Gray, A. I. (1984) Structural diversity and distribution of coumarins and chromones in the Rutales. In: *Chemistry and Chemical Taxonomy of the Rutales*. P. G. Waterman and M. F. Grundon (Eds) pp. 97–146. Academic Press, London.

Harborne, J. B. (1982) *Introduction to Ecological Biochemistry* (2nd Edn.) Academic Press, London.

HMSO (1985) Products containing herbal ingredients. *Medicines Act Leaflet (MAL 39)* DHSS Marketing Division, London

Kong, Y-C., K.-F. Cheng, R. C. Cambie and P. G. Waterman (1985b) Yuehchukene: a novel indole alkaloid with anti-implantation activity. *Chem. Comm.* 47–48.

Kong, Y-C., P. P-H. But, K-H. Ng., K-F. Cheng, K-L. Chang, K-M. Wong, A. I. Gray and P. G. Waterman (1988) The biochemical systematics of *Merrillia*; in relation to *Murraya*, the Clauseneae and the Aurantioideae. *Biochem. Syst. Ecol.* (in press).

Kong, Y-C., K-H. Ng, K-H. Wat, A. Wong, I.-F. Lau, K-F. Cheng, P. P-H. But and H-T. Chang (1985a) Yuehchukene – a novel anti-implantation indole alkaloid from *Murraya paniculata*. *Planta Medica* No. 4: 304–307.

Kong, Y-C., K-H., Ng, P. P-H. But, Q. Li, S-X, Yu, H-T. Zhang, K-F. Cheng, D. D. Soejarto, W-S. Kan and P. G. Waterman (1986) Sources of the anti-implantation alkaloid yuehchukene in the genus *Murraya*. *J. Ethnopharm.* 15: 195–200.

Kubitzki, K. and O. R. Gottlieb (1984) Phytochemical aspects of angiosperm origin and evolution. *Acta Bot. Neer.* 33: 457–468.

Penso, G. (1982) *Index Plantarum Medicinalium Totius Mundi Eorumque Synonymorum.* Editorale Farmaceutica, Milan.

Phillipson, J. D. (1981) The pros and cons of herbal remedies. *Pharm. J.* 227: 387–392.

Phillipson, J. D. (1986) Plant remedies in modern medicine. *Pharmacy Updata:* 378–380.

Phillipson, J. D. and L. A. Anderson (1984) Pharmacologically active compounds in herbal remedies. *Pharm. J.* 232: 41–44.

Provan, G. J. & P. G. Waterman (1986) The mansubinones: octanordammaranges from the resin of *Commiphora incisa*. *Phytochem.* 25: 917–922.

Sneader, W. E. (1985) *Drug Discovery: the Evolution of Modern Medicines.* John Wiley & Sons, Chichester.

Stenlake, J. B., R. D. Waigh, J. Urwin, G. Dewer and G. C. Coker (1983) Atracurium: conception and inception. *Brit. J. Anaesthesia* 55: 35–105.

Tyler, V. E. (1986) Plant drugs in the twenty-first century. *Econ. Bot.* 40: 279–288.

Waterman, P. G. (1986) A phytochemist in the African rain-forest. *Phytochem.* 25: 3–17.

Waterman, P. G. and D. B. McKey (1988) Herbivory and secondary compounds in rain-forest plants. In: *Ecosystems of the World*, H. Leith & M. J. A. Weger (Eds), (in press). Elsevier, The Hague.

38

Water Relations of Guayule and their Effect on Rubber Production

D. Mills, A. Benzioni and M. Forti

INTRODUCTION

Guayule (*Parthenium argentatum*) is a rubber-producing shrub native to the semi-arid zones of Mexico and Texas. Guayule produces significant quantities of a rubber which is chemically similar to the rubber produced by *Hevea*. Today the only commercial source of natural rubber is the hevea tree, which is confined to the humid tropics. Commercialization of guayule is not immediately feasible because of two major barriers: low rubber yields and the high cost of crop establishment.

The effect of irrigation on rubber production has been studied intensively since guayule cultivation was initiated by The International Rubber Co. at the beginning of this century. It was found that high irrigation encouraged biomass production but that rubber concentration remained low (Hammond & Polhamus 1965). High rubber yields could be achieved by alternating long periods of low and high water stress (Benedict *et al.* 1947). Although maximum production of rubber requires large quantities of water (Miyamoto & Bucks 1985), the efficiency of water use, that is, rubber yield per unit of irrigation water, is actually higher under high stress conditions (Miyamoto *et al.* 1984). Thus in places such as Israel, where a shortage of water prevails, growing guayule under low irrigation could be commercially advantageous.

We investigated the effects of irrigation on the canopy growth, rubber and resin production of three USDA guayule lines. The irrigation regimes chosen (250, 400 and 600 mm per annum) were similar to or lower than the amounts used for cotton and wheat in the Northern Negev of Israel.

MATERIALS AND METHODS

Field Procedures

Seedlings were transplanted in July 1985 at the Omer experimental field near Beer-Sheva. About 95–96 per cent transplant survival was obtained. A randomized block design with three irrigation treatments and four replicates was used. Each of the 12 subplots was 5·8 m × 10·0 m in area and consisted of six rows of three USDA varieties planted in random pairs. The spacing between plants in each row was 30 cm and the distance between rows 96 cm (32000 plants/ha). The USDA lines used were 11604, 12229 and 11591 (about 64 plants per line, treatment and replicate).

Plants were drip irrigated using one line per row and one dripper every 50 cm supplying 2 l/ha. Fertilizer was applied as follows: 150 kg/ha N, 20 kg/ha P_2O_5 and 40 kg/ha K (annual totals).

Soil water content was measured with a neutron moisture meter (Ronli Electronics Ltd., Model DMG 33) at depths of 20, 40, 60, 80, and 100 cm. The equipment was calibrated at the site. Soil was a loess characterized by a medium water-holding capacity. Field capacity was estimated at 24% by volume and wilting point at 6·5%, as determined by the lower limits of water depletion for guayule plants.

Leaves with petioles were sampled for determination of relative water content (RWC). Three leaves per plant were sampled from the upper part of the canopy; five plants were sampled at random from each irrigation treatment. Sampling took place between 09.30 and 10.00 hr local time. Leaves were immediately placed in preweighed, tightly stoppered plastic vials. Vials with leaves were weighed to 0·1 mg at the laboratory 30–60 min after picking. One ml of deionized water was added to each vial and the vials were restoppered. Hydration was carried out at 20 °C for 24 hr (this period was found to be adequate for full hydration of the leaves), and the leaves were then blotted dry on paper towel and weighed. Next the leaves were dried once again at 70°C for 48 hr and reweighed. Relative water content was calculated as described elsewhere (Ehler & Nakayama 1984)

Infrared thermometer (IRT) measurements were taken with TELATEMP Model AG-42 between 09.30 and 10.00 hr. The instrument was placed vertically at 10–20 cm from canopy top. Perpendicular readings were generally similar to angled ones due to high leaf coverage. Twelve individual plants were sampled at random from each irrigation treatment. Non-stressed plants were kept in a separate subplot as a base line for Crop Water Stress Index (CWSI) determination and irrigated once in ten days to maintain available soil water within the 85–95% range.

Growth was determined by measuring canopy height and diameter or spread, while biomass production was determined by measuring the dry matter of branches only. Note, however, that canopy dimensions may represent mainly branch elongation. For canopy measurement 20–30 plants per block per line per irrigation treatment were sampled (total of 240–360 per irrigation) and for biomass production determination two plants per block per line per water treatment were sampled (total of 24 per irrigation).

Determination of resin and rubber contents

Plant canopy was cut 2–3 cm above ground level (roots were not sampled), weighed and dried at 70 °C for 48 h. After defoliation the branches of single plants were cut into 2–3 cm pieces with pruning shears and the dry material was weighed. Next the sample was dipped into liquid nitrogen and ground in a hammer mill (Thomas ED5) equipped with a screen with 3 mm by 45 mm slits. One 2·00 g fraction was dried further for 24 hr at 70 °C for determination of residual moisture. Rubber and resin were then determined following the procedure described by Black *et al.* (1983). Two 2·00 g samples were weighed into polyethylene tubes and homogenized in 10 ml acetone with a Polytron mixer for 20 secs. The mixer was washed twice with 10 ml acetone, and the total homogenate was centrifuged at $3000 \times g$ for 10 min. The clear supernatant was filtered through filter paper into a preweighed disposable aluminium dish. The pellet was re-extracted with acetone and all filtered supernatants were combined. The acetone-soluble fraction was dried under a fan and over warm sand. The aluminium dish was weighed to determine resin content. Ten ml cyclohexane was added twice to the resin-free pellet to extract the rubber. The samples were handled as described above for determination of resins, and the rubber quantity was determined gravimetrically.

RESULTS

Water Management and Stress

The climatic conditions that prevail in the Northern Negev are presented in Fig. 38.1a. In contrast to guayule's native habitat (Chihuahua desert), none of the area's annual precipitation (200 mm on average) falls in summer, the season of maximum evapotranspirational demand. Plants were therefore irrigated from spring (March or April) to autumn (October). Water management is outlined in Table 38.1. During the first year available soil water before and after irrigation was 17–30%, 19–48% and 19–59% for low, medium and high treatments, respectively. In the second year available soil water fractions were higher (see Fig. 38.1a). It was found that by the end of each irrigation cycle the soil moisture profile was about the same as at the end of the previous cycle. Water application efficiency (soil water depletion divided by irrigation water) ranged between 86 and 96%. No rainfall occurred during irrigations and no drainage was detected. Soil water depletion was monitored during most irrigation cycles, but not in all. We assumed, therefore, that irrigation plus rainfall equals total evapotranspiration (ET). Values of evapotranspiration were about 15, 25 and 40% of class A pan evaporation (measured at the meteorological station in Beer-Sheva) for the dry, medium and wet treatments, respectively.

Figure 38.1: *Annual Distribution of Monthly Precipitation and Daily Class A Plan Evaporation (a) and Average Daily Minimum and Maximum Temperatures (b).*

Measurements were carried out for 30 years at the meteorological station at Beer-Sheva.

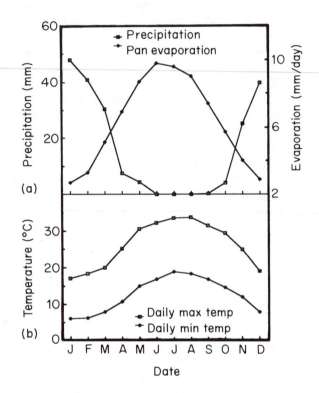

Table 38.1: *Outline of Water Management*

	Year 1985/86		Year 1986/87		
	Per irrigation	Per annum	Per irrigation	Per annum	Total for two years
Water (mm) by:					
Precipitation	–	145	–	201	346
Establishment	–	210	–	–	210
Irrigation					
Low	30·1	241	34	238	479
Medium	55·4	443	59·3	415	858
High	81·5	652	80·4	563	1215

Eight irrigations were given in 1985–86 and seven in 1986–87. Irrigation was applied every 3–4 weeks between April and October.

Figure 38.2*: Course of Soil Water Depletion (a), Plant Canopy Minus Air Temperature (b) and Relative Water Content of Leaves (c) during a Typical Irrigation Cycle (20/7–10/8) in the Second Year.*

Irrigation treatments are outlined in Table 38.1. The very wet treatment was kept under 85–95% of available soil water. Data are for line 11604.

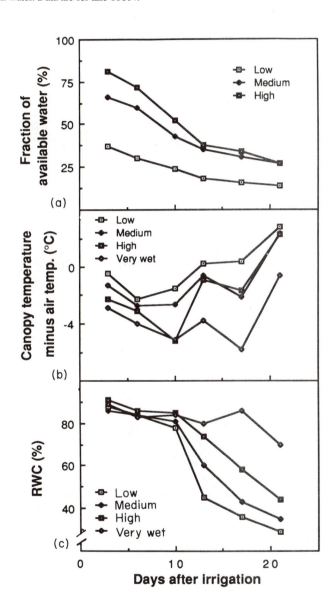

Figure 38.3: *Effect of Irrigation on the Pattern of Height Increase (a) and Biomass Production (b).*

Values are the average of three guayule lines. Irrigation treatments are outlined in Table 38.1. Time 0 = July 85.

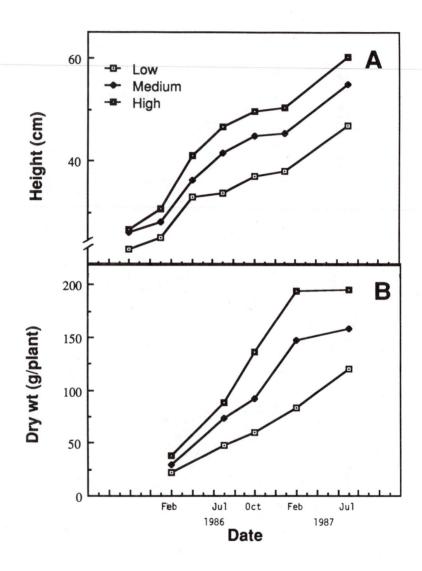

The depth of the root zone after the first year of growth evaluated from the profile of soil moisture depletion was about 50, 60 and 70 cm respectively for low, medium and high irrigation treatments. In the second year roots used water to a depth of 50, 70 and 90 cm for low, medium and high irrigation treatments, respectively.

The soil water depletion, relative water content and canopy-minus-air temperature of line 11604 during a typical irrigation cycle are presented in Fig. 38.2a-c. Up to six days after irrigation evapotranspiration rates (equivalent to soil water depletion rates) and canopy-minus-air temperatures were similar in all irrigation treatments. From the tenth day onwards, rates of evapotranspiration declined in low and medium irrigation treatments but remained steady in high treatment (Fig. 38.2a). At day 13 the canopy-minus-air temperatures of the high irrigation treatments and of the very wet treatment were similar (Fig. 38.2b). Relative water content was high for all treatments up to day 10 (80–90%) and began to decline from day 13, dropping most rapidly in the low irrigation treatment (Fig. 38.2c). Relative water content values of 30% and lower were registered on stressed plants. However these plants recovered after irrigation. These exceptionally low RWC values indicate that guayule is highly drought-resistant. During the course of each irrigation cycle leaf colour changed from relatively dark to light green and eventually to silver-grey, and turgidity gave way to a folded shape. The silver-grey, folded leaves had a wilted appearance; this phenomenon was observed at day 13, 17 and 21 after irrigation for the low, medium and high irrigation treatment, respectively, corresponding to a relative water content of about 44% (Fig. 38.2c).

The canopy-minus-air temperature measurements seem to indicate that plant stress sets in from about day 10 after irrigation for the medium and low treatments and from day 13 after irrigation for the high water treatment. Stress was more severe in the low irrigation treatment.

Growth and Biomass Production

The different degrees of stress experienced by guayule under the three irrigation regimes were reflected in growth. Higher water quantities resulted in higher growth (Fig. 38.3a). Statistically significant differences ($P < 0.05$) in canopy height were found in all measuring periods between plants irrigated with low or medium and plants irrigated with high amounts of water. Increase in plant height was rapid in spring and summer and slower in autumn and winter. Very similar growth patterns were observed in the southwestern USA (Bucks *et al.* 1985a).

Irrigation affected biomass production in very much the same way as it did height. Larger canopy and branch biomass was obtained in the high irrigation treatment (Fig. 38.3b).

Figure 38.4: *Effect of Irrigation on the Pattern of Rubber Concentration (a), Resin Concentration (b) and Rubber Yield (c).*

Values are the average of three guayule lines. Irrigation treatments are outlined in Table 38.1. Time 0 = July 85.

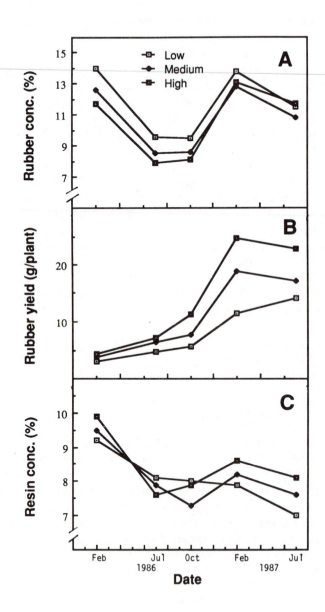

Rubber and Resin Production

High water quantities and the associated high growth rate resulted in decreasing rubber concentrations in the three lines tested (Fig. 38.4a). Rubber concentration also varied during the year, being highest at the end of winter after break of dormancy and lowest during spring and summer. In July 1987 there was no difference in rubber concentration between the low and high treatments, probably due to insufficient irrigation in spring 1987 (irrigation was resumed only at the end of April, five weeks later than the previous year). Lines differed in rubber concentration in all irrigation treatments. Concentration of rubber in line 11 604 was significantly higher than that in lines 11 591 and 12 229 (data not shown). Canopy rubber yield was higher when water quantity was increased (Fig. 38.4c), as might be expected from the fact that irrigation affects biomass production more than it does rubber concentration.

On the whole, plants of the high water treatment had a higher resin concentration (Fig. 38.4b) and yielded more resin (data not shown). In February 1986 and 1987 and in July 1987 there was a significant difference in resin yield between high and low water treatments, while no differences in resin concentration were observed between plants in the three water treatments in July and October. Line 11 604 had higher resin concentrations compared to the other lines (data not shown).

DISCUSSION

Our objective was to study the relation between plant water status and biomass and rubber production for three guayule lines under the conditions prevailing in the Northern Negev of Israel. All three irrigation regimes were designed to subject the plants to some degree of water stress, the assumption being that alternations of stress and irrigation enhance rubber production during stress and . biomass production during good water status (Benedict *et al.* 1947). The irrigation regimes selected also took into account the limited water supply in Israel and thus the quantities applied were similar to or lower than those used for other crops such as wheat and cotton. These water quantities are low compared with recent studies conducted in the southwestern USA (Miyamoto & Bucks 1985), in which irrigation of 500–2000 mm water per annum was applied to a population of 49 500 – 54 000 plants/ha.

Height increase, biomass production (branch dry matter) and canopy rubber production were found to be directly related to irrigation water supply (Figs. 38.3a&b). Water use efficiencies (dry matter or rubber yield divided by total evapotranspiration) ranged between 0·40 and 0·47 kg dry branches per m^3 water and between 46 and 55 g rubber per m^3 water for the low, medium and high irrigation treatments (Table 38.2). Water use efficiency was inversely correlated to water stress. Values reported in recent research carried out in the southwestern U.S.A. were lower: 16–30 (Bucks *et al.* 1985b), 31–37 (Bucks *et al.* 1985a),

Table 38.2: *Effect of Irrigation on Water Use Efficiency of Two-year-old Guayule Plants*

Treatment	Total ET	Branch biomass	Rubber yield kg/ha	Resin yield kg/ha	Biomass kg/m^3	Rubber g/m^3	Resin g/m^3
Low	825	3900	450	270	0·47	55	33
Medium	1204	5100	550	390	0·42	46	32
High	1561	6300	730	510	0·4	47	33

Total evapotranspiration (ET) equals irrigation + precipitation.

32–44 (Miyamoto *et al.* 1984), and 32–55 (Fangmeier *et al.* 1985) g rubber per whole plant per m^3 water. The higher water use efficiency in Israel may be due to the use of drip irrigation as well as to subjection of plants to alternating stress conditions. The main advantage of drip irrigation is the economy in water resulting from the fact that no water is lost by evaporation from droplets in the air and from wetted foliage as in sprinkler irrigation. Moreover, there is less evaporation from the soil surface than in furrow systems (Nir 1982).

Rubber concentration in branches ranged between 8 and 14% (higher at the end of winter and lower in summer). This is much higher than the concentrations of 5–8% reported in the literature for similar irrigation treatments or total evapotranspiration (Miyamoto & Bucks 1985). Rubber concentration generally increases with increasing water stress. This phenomenon was also found in our study (Fig. 38.4a) and in many other studies (Hammond & Polhamus 1965; Miyamoto & Bucks 1985 and references therein).

It is difficult to determine whether the plants in this study were under more severe stress than the plants in the studies conducted in the US. It seems that the two dry treatments (similar ET) in a study conducted in Arizona (Bucks *et al.* 1984) are comparable to our treatments (judging by the timing and extent of fall in canopy-minus-air temperature after irrigation). A proper comparison of the degree of stress will be possible once the base line for computation of the crop water stress index (CWS) is completed.

Resin concentration is usually not greatly affected by water stress (Miyamoto & Bucks 1985). In our study we found that in two-year-old guayule plants resin concentration increased somewhat with decrease in water stress (Fig. 38.4b). The same result was obtained in two previous analyses (statistical differences for $P < 0.05$).

ACKNOWLEDGEMENTS

We thank Ms. S. Avni and Mr. I. Knafo for their excellent technical work, and Mr. Dov Mills for supplying meteorological data. This study was conducted within the framework of the Cooperative Arid Lands Agriculture Research Program-Egypt-USA-Israel (CALAR), funded by the US Agency for International Development (Contract No. NEB-0170-A-00-2047-00) and administered by the San Diego State University Foundation.

REFERENCES

Benedict, H. M., W. L. McRary and M. C. Slattery (1947) Response of guayule to alternating period of low and high moisture stresses. *Bot. Gaz.* 108: 535–549.

Black, L. T., G. E. Hamerstrand, F. S. Nakayama and B. A. Rasnik (1983) Gravimetric analysis for determining the resin and rubber content of guayule. *Rubber Chem. Techn.* 56: 367.

Bucks, D. A., F. S. Nakayama and O. P. French (1984) Water management for guayule rubber production. *Trans. Amer. Soc. Agric. Engin.* 27: 1763–1770.

Bucks, D. A., F. S. Nakayama, O. P. French, B. A. Rasnick and W. L. Alexander (1985a) Irrigated guayule-plant growth and production. *Agric. Water Manag.* 10: 81–93.

Bucks, D. A., R. L. Roth, F. S. Nakayama, O. P. Gardner and B. R. Gardner (1985b) Irrigation water, nitrogen and bioregulation for guayule production. *Trans. Amer. Soc. Agric. Engin.* 28: 1196–1205.

Ehler, W. L. and F. Z. Nakayama (1984) Water stress status in guayule as measured by relative leaf water content. *Crop. Sci.* 24: 61–66.

Fangmeier, D. D., Z. Samani, D. Garrot, Jr. and D. T. Ray (1985) Water effects on guayule rubber production. *Trans. Amer. Soc. Agric. Engin.* 28: 1947–1950.

Hammond, B. C. and L. G. Polhamus (1965) *Research on guayule (Parthenium argentatum) 1942–1959.* USDA Technical Bull. 1327.

Miyamoto, S. and D. A. Bucks (1985) Water quantity and quality requirements of guayule: current assessment. *Agric. Water Manag.* 10: 205–219.

Miyamoto, S., J. Davis and K. Piela (1984) Water use, growth and rubber yields of four guayule selections as related to irrigation regimes. *Irrig. Sci.* 5: 95–103.

Nir, D. (1982) Drip irrigation. In: *Handbook of Irrigation Techniques* Vol. I, H. J. Finkel (Ed.), pp. 247–298. CRC Press Inc., Boca Raton, Florida.

39

The Possible Contribution of Ethnobotany to the Search for New Crops for Food and Industry

J. F. Barrau

INTRODUCTION

Technological and scientific advances in the field of plant industry have led, needless to say, to obvious and outstanding progress in agriculture and horticulture. However, they have also been responsible for a situation characterized today by:

(a) *A growing homogenization and specialization of the world's cultivated flora.* As an example, seven crops (wheat, rice, maize, potato, barley, cassava and sorghum) provide the bulk of the world's food supply today. In the course of the same process of agricultural homogenization and specialization, the number of cultivars (except in the case of ornamentals) has spectacularly decreased, particularly during the last thirty years. Moreover, some cultigens have fallen or are falling into disuse. This reduction of the biological diversity of the domesticated phytocoenoses has induced a growing vulnerability of modern agroecosystems, of which the productivity can only be maintained or increased through constantly augmented inputs of energy. Finally, some of the highly sophisticated 'miracle' cultivars created recently by plant breeders have proved to be rather fragile; the failure of a 'miracle' wheat in the summer of 1987 in France, due to poor weather conditions, illustrated that point rather spectacularly.

(b) *An indiscriminate imposition in many parts of the world of standardized agrotechnic and agroeconomic models* conceived in the ecological and historical contexts of advanced countries. This has been an active factor in the erosion of not only the diversity of the cultivated flora but also of local and often efficient systems of knowledge and practices concerning plant resources and their uses.

If, in the light of the above, one considers the very limited percentage, less than one per cent, of species used for our economic crops (Jones & Wolf 1961) since the 'revolution of domestication' (Barrau 1983), one can legitimately wonder if the remaining unused 99 per cent could not yield new crops and plant products useful to mankind. There lies an obvious potential for innovation in the utilization of plant resources. To that must be added the possibility of recovering that part of the domesticated flora that has been discarded during the course of agricultural 'progress' and also the possibility of discovering, or rather rediscovering plants, long used by man but have so far been ignored or neglected by the crop scientists

Of course, economic botany and agronomy can efficiently prospect this potential but they may also benefit from the assistance of ethnobotany, which takes care of the relations between human societies and the plant world in their anthropological, biogeographical and historical concepts.

The aims of ethnobotanical research are many and the concept of 'ethnobotany' has somewhat changed and been interpreted in different ways since it was first conceived by J. W. Harshberger in 1895 (see note at end of text; Barrau 1971). However, one of the main tasks of ethnobotany is to collect and analyse local, or traditional, systems of plant knowledge, their properties and uses. It can thus bring to scientific light useful data based on the practical experiences gained by human societies through ages of close association with their floristic environments and resources.

On what concerns the consciousness of biological diversity, ethnobotanical research has already shown how rich and detailed can be the folk-botany of some ethnic groups. A classical example of such a contribution is Conklin's (1954) study of the relations of the Philippines Hanuno to their plant world; in the tropical forest where they live, these shifting cultivators and gatherers distinguish more categories of plants than would a scientific botanist, and the uses they make of these plants are indeed very diverse.

To quote another and more recent example, Bahuchet (1985), who has made an in-depth ethnoecological study of Aka pygmies of the Central African forest, has shown the intimate knowledge that they have of their sylvan environment where they gather for their sustenance the tubers of nine wild species (including six *Dioscorea* spp.), the leaves of at least ten species, the kernels of fifteen species, the fruits of some eighteen species, and, last but not least, thirty species of mushrooms! To that must be added the many plant species used by these pygmies to build their shelters, or for making tools, utensils, weapons, hunting contraptions, arrow poisons, etc.

MEDICINAL PLANTS

Besides such comprehensive treatments, ethnobotanists have often undertaken more specialized investigations, such as studies of local pharmacopoeia and herbal medicine practices. Among many examples in that field, I would like to cite two recent French examples, that of Grenand *et al.* (1987) on traditional pharmacopoeia of French Guyana and, particularly, the outstanding research carried out by Lieutaghi (1986) on the folk-herbal medicine of Haute-Provence in France; for ethnobotany is no longer confined to the study of the relations between man and plants by 'primitive' or 'aboriginal' peoples living in past or strange exotic environments as stated by Harshberger in 1895 (see p. 403). Such ethnobotanical investigations on medicinal plants has sometimes contributed to increasing the drug arsenal of modern medicine. In that respect, the systematic ethnobotanical *cum* pharmacognosical surveys in the Soviet Union on folk-herbal medicine deserves consideration as they have resulted in bringing under cultivation and into official pharmaceutical use several plants studied during the course of the surveys.

What is happening in China today is also worth observing. On one hand there is a generalized reappraisal by modern scientific standards of the vegetable drugs used in traditional Chinese medicine; on the other hand, big pharmaceutical organizations (Merck, Rhone-Poulenc, Upjohn, etc.) have concluded scientific cooperation agreements (is this surprising?) with Chinese research institutions investigating their own pharmacopoeia.

The examples already quoted have shown how much can be learned from local systems of knowledge of plants and plant products through the mediation of ethnobotanical research.

FOOD PLANTS

The search for food crops is yet another example well suited to an ethnobotanical approach. Attention can thus be drawn to cultivated plants long ignored or underestimated by agricultural scientists. The relatively recent example of the 'rediscovery' of the winged bean, *Psophocarpus tetragonolobus* (Barrau 1956) or of the African niebe, *Vigna unguiculata* subsp. *unguiculata*, illustrate that point.

The example of the niebe perhaps deserves to be looked at in some detail as it may show the kind of contribution ethnobotany can make to the rehabilitation of neglected food crops. This pulse, now known to be of African origin (Steele 1976) and once baptized a little too hastily as *V. sinensis*, has a long and complex history. In the sixth-century Byzantine *Codex Aniciae Julianae*, based on the earlier classical works of protopharmacologists and protobotanists Crateuas (first century BC) and Pedanius Discorides (first century AD), there is an accurate drawing of *V. unguiculata* (cf. Arber 1953: 8–9, 186). In 1944, August Chevalier, the founder of our laboratory at the Paris Museum of Natural History, began to throw some light on the origin and diffusion of the niebe and on its status in the African food complex. As a matter of fact, this old African domesticate spread during antiquity to the Near East and India, then to the Mediterranean periphery and to South East Asia. In Southern Europe it was a food crop of some importance until replaced by the American *Phaseolus*. Through the African slave trade it was introduced into the New World. Its usefulness as a fodder, the cowpea (subsp. *cylindrica*) and as a green vegetable, the yard-long or asparagus bean (subsp. *sesquipedalis*) have long been recognized. Yet the niebe itself as a pulse crop did not receive all the agronomic attention it deserved. This has been particularly true in the continent of its origin, Africa. In his important study of the food plants of West Africa, Busson (1965), while drawing attention to the nutritional value of niebe, noted that its yield could easily be increased, and Steele (1976) thought that the protein content could also be improved. But for years, tropical agriculturalists working in tropical Africa overlooked the niebe. However, there has recently been a growing awareness of the potentialities of this food crop and it has been recommended that this valuable pulse be given priority in breeding programs designed to improve the food situation in Africa (cf. for example, Westphal *et al.* 1985). Needless to say, ethnobotanists could be of further assistance in the surveys of local cultivars of *V. unguiculata* subsp. *unguiculata* and the study of its traditional method of cultivation and uses. The same is true of North Africa. Our studies of the traditional food crops of Algeria indicate that, under the name of lumbia, *V. unguiculata* was there considered to be a pulse of significance in the local diet. The Algerian government has undertaken research and development programs to increase local pulse production. It may be of interest to include *V. unguiculata* in these programs and thus to rehabilitate a food plant which had been economically important in the past; it may even turn out to be better adapted to local conditions than other pulses now under experimentation.

In France, within the framework of investigations under the patronage of the recently created Bureau des Ressources Génétiques, the application of the methods of ethnobotanical research to the search for old local cultivars of fruit trees, vegetables, cereals, etc. which have fallen into disuse or which have survived only as rural isolates has proved to be very fruitful (Marchenay & Lagarde 1986). If I may quote a personal experience, in 1971 when I undertook ethnobotanical research in France after more than twenty years of such undertakings in the Malayo-Oceanic area, I decided to look at all the food plants in the mountains of Vaucluse, where my father's family had its origin. From childhood memories I knew that a kind of porridge was a popular peasant dish there in winter. I started looking for the cereal used for that purpose and found to my surprise that it was the neolithic einkorn, *Triticum monococcum*, which was still grown there, as well as in some localities of the Southern Alps, as a subsistence cereal of which the unground grain was used to prepare this porridge. This was unknown to my learned friends in French agricultural research. Today this relict prehistoric wheat begins to find an economic outlet in the 'biological-health-foods' and marketed at a price rather satisfying for the stubborn traditional growers who, through generations, had kept this einkorn in cultivation to satisfy their lasting taste for its porridge.

Non-cultivated edible plants obtained through gathering also deserve to be more thoroughly surveyed from an ethnobotanical and economic-botanical point of view. As a matter of fact, a lot can be learned about such plants by looking at the studies of anthropologists who have worked with food-gathering societies or described food-gathering activities of societies practising cultivation. Under the influence of ethnobotany, a growing number of anthropologists do now take care in providing precise ethnographic data about the uses of plants and even their exact botanical identification.

Although during the past fifty years, economic botanists and ethnobotanists have insisted on the potential agronomic importance of some non-cultivated food plants used through gathering (cf. on this subject Harlan 1975), much remains to be done, with, for example, the 'wild' cereals of dry Africa and Australia. The same could be said about many undomesticated or semi-domesticated and very locally exploited edible plant species of both the Old and New Worlds, even in the case of species used in the distant past. What do we know, for example, about the *Setaria* sp. which, according to palaeo-ethnobotanists (cf. Harlan 1975) was used as a cereal in Mesoamerica prior to the appearance of maize? As a matter of fact, the Old World's *Setaria* millet (*S. italica*), particularly its Asiatic and Malaysian cultivars, should also be worth further study; ethnobotanists have shown (for instance Conklin 1980) in the case of the Philippines Iguago both the advantages and the persistence of this millet, even in areas where rice cultivation is now dominant. This is of importance at a time when countries such as China, where *Setaria italica* had kept its varietal diversity and economic importance, have a tendency to disregard this productive crop and replace it with other cereals, and in the case of China, by maize which is considered as the symbol of western agronomic achievement.

Let us now look at a plant about which I have long attempted to draw attention (Barrau 1959 & 1962), the *Metroxylon* sago-palm of Malaysia and Oceania. It constitutes an enormous potential for starch production, yet this sago palm is virtually undomesticated and is still largely utilized through traditional methods of gathering and even processing, although there has been some development in Malaysia of small-scale 'industrial' extraction of the *Metroxylon* starch. Today the economic and agroecological potentialities of the sago-palm are scientifically recognized (cf. Yamada & Kainuma 1986). But if one wants to develop ecologically and technologically sound methods of improving, increasing and expanding sago starch production, one should perhaps know more about the phytotechnical 'savvies' of traditional sago gatherers. Here again ethnobotany can help.

However, and this should be an ethic rule to be observed in all applications of ethnobotanical research, an important comment has to be made at this point; it has been clearly worded recently by Graham Baines (1987), 'To gather and apply traditional knowledge for the benefit of mankind is a worthy objective...yet there is a moral imperative to ensure that this objective is achieved through mutually supportive relationships with traditional knowledge informants. Practical models for such partnerships are needed...to protect the rights and dues of those whose societies shaped and supplied traditional knowledge.'

But it also happens that ethnobotanical research can provide information on subjects of general interest, for instance on the WHAT, WHY and WHERE of the plant domestication process, this in the light of recent or contemporary instances. Let me take an example in which I was involved. In the Melanesian traditional subsistence gardens of New Caledonia, there was a yam grown apart from the noble old-time cultivated *Dioscorea* spp. Its vernacular name implied that it was a relative newcomer and it was not fitting into the botanical taxonomy of the Malayo-Oceanian area. Finally, it proved to be the Australian *D. transversa*, known to be a wild species gathered by the Australian aborigines (Bourret 1973). The story then unfolded that, at the time of the infamous 'black birding' (1847–1904), where Malaysian islanders were kidnapped to work as slave labour on the Queensland plantations, some of them in their new country of forced exile, deprived of one of their favourite tubers, found that the local aborigines were gathering a wild yam. They brought it under cultivation and, when a legal end was put to 'black birding', some of its survivors took the new domesticate back to the New Hebrides, now Vanuatu. At that time, Melanesians from these islands were brought to New Caledonia as 'indentured' labourers and they brought with them *D. transversa*, which was quickly adopted by the New Caledonian Melanesians, who were always ready to appreciate a new edible cultigen providing that it can fit within one of the traditional categories of food crops. They did the same recently with African and American cultivated yams, introduced from the West Indies. *Dioscorea cayenensis* subspp. *cayenensis* and *rotundata* (syn. *D. rotundata*), *D. trifida*, etc.

The above is only one example but it does show that ethnobotanical investigations can elucidate why a plant is domesticated and why a plant is accepted as a new crop. It may be useful to look in a similar way at other recent additions or readditions to the cultivated flora.

CONCLUSIONS

In what I have mentioned so far I have mostly quoted examples of medicinal and food plants. But the same can be said for all categories of economic plants. It can also apply to traditional processes of plant product utilization and preservation involving microorganisms (fermented foods and beverages, sour dough preservation, retting, etc.).

Ethnobotanical research is fundamentally a matter of field work combining the methods of social and natural sciences. I sometimes feel that it is nothing more than a somewhat new branch of the good old 'Natural History' of the eighteenth-century 'enlightenment'; a time when the boundaries between human and natural sciences and disciplines were not as rigid as they are today! Ethnobotany is basically an interdisciplinary field work enterprise, and has often to make use of available historical sources. This is the case, for example, on plants used by Australian aborigines about which accounts of early explorers, missionaries and the like can provide useful information.

A lesser known method of ethnobotanical investigation is the careful scrutinization of notes attached to herbarium specimens. Siri von Reis Altschul (1973) has made a brilliant demonstration of the efficiency of this method of treatment using the notes in the Harvard University Herbaria in the search for drugs and goods from little known plants. Personally I am finding it to be a very efficient tool in the work I am currently doing on the *Maranta* spp. cultivated by the Amerindians in South America and the Caribbean islands.

By studying from both the inside and outside, traditional botanical knowledge of plant resources, by describing processes of their utilization in precise ethnographic and naturalistic ways, by taking into account prehistorical and historical aspects of plant uses, ethnobotany can make available an array of facts and data which can help to improve and develop the mobilization of these resources to meet the needs of mankind. Yet, let us not forget that it must first devote its efforts to the betterment of life of those who are its informants.

Ethnobotanical research can also contribute to maintain or reinstate biological diversity in the World's economic and cultivated flora while providing ideas on methods of plant and environmental management based on those traditional systems which have been too often ignored or despised by modern agricultural science.

NOTE ON 'ETHNOBOTANY'

On Thursday, the 5th of December 1895, in the daily *The Evening Telegraph* published in Philadelphia, there was, on page 2, a short anonymous article entitled 'Some new ideas: the plants cultivated by aboriginal people and how used in primitive commerce'. It gave a brief account of a lecture given by J. W. Harshberger before the Archaeological Association of the University of Philadelphia. There, for the first time appeared the term 'ethno-botany' together with an abridged definition of what it meant for its inventor. The following year, 1896, both in the *American Antiquarian* 17, 2: 73–81 and the *Botanical Gazette* 21: 146–154, J. W. Harshberger gave a full account of what he considered to be 'the purpose of ethnobotany'. Restricted to the study of relations between men and plants in the case of 'primitive and aboriginal' peoples, Harshberger's ethnobotany would help:

–   to elucidate the cultural position of the 'tribes' having used or using some plants or plant products;
–   to clarify the past distribution of useful plants;
–   to define ancient trade routes followed in the exchange of products of vegetable origin, and
–   to discover useful plants which could be of interest for modern commerce and industry.

REFERENCES

Arber, A. (1953) *Herbals, their origins and evolution, a chapter in the history of botany.* Cambridge University Press, Cambridge.

Bahuchet, S. (1985) *Les Pygmées Aka et la forêt centrafricaine, ethnologie écologique.* SELAF, Paris.

Baines, G. (1987) Not another tool of exploitation! *Tradition, Conservation and Development,* Newsletter of IUCN/COE. Traditional Ecological Knowledge Working Group, No. 5, June 1987.

Barrau, J. (1956) Les légumineuses à tubercules alimentaires de la Mélànesie, *La Terre et la Vie,* Bulletin de la Société Nationale d'Acclimation de France, No. 1: ll-16.

Barrau, J. (1959) The sago-palms and other food plants of marsh dwellers in the South Pacific islands. *Econ. Bot.* 13(2): 151–163.

Barrau, J. (1962) *Les plantes alimentaires de l'Océanie, origines, distribution et usages.* Musée Colonial de la Faculté des Sciences, Université d'Aix-Marseille, Marseille.

Barrau, J. (1971) L'ethnobotanique au carrefour des Sciences Naturelles et des Sciences Humaines. *Bull. Soc. Bot. France* 118(3–4): 237–248.

Barrau, J. (1983) *Les hommes et leurs aliments, esquisse d'une histoire écologique et ethnologique de l'alimentation humaine.* Messidor-Temps Actuels, Paris.

Bourret, D. (1973) *Etude ethnobotanique des Dioscoreaceae alimentaires de Nouvelle-Calédonie.* Thèse de Doctorat du troisième cycle en Botanique Tropicale, Université Pierre et Marie Curie, Paris.

Busson, F. (1965) *Plantes alimentaires de l'Ouest africain, étude botanique et chimique,* Published by the Author, Marseille.

Chevalier, A. (1944) Le dolique de Chine en Afrique, son histoire, ses affinités, les formes sauvages et cultivées, son rôle dans l'alimentation indigène et en agriculture tropicale et sub-tropicale. *Rev. Bot. Appl. Agric. Trop.* No. 272–273–274: 128–152.

Conklin, H. C. (1954) *The relation of Hanunóo culture to the plant world.* Ph.D. Dissertation. Yale University, New Haven (Available from University Microfilms, Ann Arbor).

Conklin, H. C. (1980) *Ethnographic atlas of Ifugao, a study of environment, culture and society in Northern Luzon.* Yale University Press, New Haven and London.

Grenand, P., C. Moretti and H. Jacquemin (1987) *Pharmacopées traditionelles en Guyane: Créoles, Palikur, Wayâpi.* Memoire No. 108, ORSTOM, Paris.

Harlan, J. R. (1975) *Crops and Man.* American Society of Agronomy, Foundations of Modern Crop Science Series, Madison.

Jones, Q. and I. Wolf (1961) Using germ plasm for new products. In: *Germ Plasm Resources,* R. Hodgson (Ed.), pp. 265–278. Publication No. 66, American Association for the Advancement of Science, Washington.

Lieutaghi, P. (1986) *L'herbe qui renouvelle, un aspect de la médecine traditionelle en Haute-Provence.* Editions de la Maison des Sciences de l'Homme, Paris.

Marchenay, P., and M. F. Lagarde (1986) *A la recherche des variétés locales de plantes cultivées, guide méthodologique.* Groupe de Recherche et de Developpement sur le Patrimoine Génétique Animal et Végétal de la Région Provence-Côte d'Azur, Conservatoire Botanique, Porquerolles.

Reis Altschul, S. von (1973) *Drugs and Food from Little-known Plants, Notes in Harvard University Herbaria.* Harvard University Press, Cambridge.

Steele, W. M. (1976) Cowpeas, *Vigna unguiculata,* Leguminosae-Papilionoideae, In: *Evolution of Crop Plants,* N. W. Simmonds (Ed.), pp. 183–185. Longman, London and New York.

Westphal, E., J. Embrechts, J. D. Fewerda, H. A. E. van Gils-Meeus, H. J. W. Mastaers and J. M. C. Westphal-Stevels (1985) *Cultures vivrières tropicales avec référence spéciale au Cameroun*, PUDOC, Wageningen.

Yamada, N. and K. Kainuma (Eds) (1986) *Sago '85: Proceedings of the Third International Sago Symposium*. The Sago Palm Research Fund, The Mitsui Trust and Banking Co. Ltd., Tokyo.

40

Economic Botany and Kew in the Search for New Plants

G. E. Wickens

INTRODUCTION

The earliest evidence of the use of plants by Man is believed to be the association of the hackberry, *Celtis australis,* with Pekin Man in the Middle Pleistocene, ca. 500 000 BP (Renfrew 1973). Although there are still enormous gaps in our knowledge, important palaeoethnobotanical contributions on the use of plants by Man include the evolution of cultivated crops in Europe and the Near East by Renfrew (1973), Egypt by Hadidi (1985) and surveys of New World palaeoethnobotany by Yarnell (1969) and Callen (1969). The written evidence from the earlier civilizations include the Sumerian ideograms from ca. 4000 BC, the Ehrya from China from ca. 3000 BC, the sacred Vedas of the Aryan peoples of northern India dating from ca. 1500 BC and the Ebers Papyrus from Pharaonic Egypt, also from ca. 1500 BC (Schultes 1960).

The development of agriculture in prehistoric Britain is described by Fowler (1983), and near to the site of this Conference, some 30 km to the east of Southampton, Iron Age Butser, by Reynolds (1980). Man has undoubtedly been dependent on plants to provide the necessities for his survival on earth since the dawn of prehistory. Plants have been used for food, fuel, shelter, fibre, clothing, medicines, domestic products, biochemicals, etc., as well as providing fodder and shade for his livestock. Plants also protect and maintain the environment against erosion, maintain and improve fertility as well as preventing atmospheric imbalance, etc.

ECONOMIC BOTANY

The study of the use of plants is known as economic botany. Economic plants can be defined as those plants which are utilized either directly or indirectly for the benefit of Man. Indirect usage includes the needs of Man's livestock and the maintenance of the environment; the benefits may be domestic, commercial, environmental or aesthetic. This is a much broader concept than has previously been accepted; the narrow view previously held in the United Kingdom that economic plants are largely synonymous with tropical plantation crops is no longer tenable (see Wickens, in prep. for further information).

411

Coles (1970) states that 'Every food plant of major importance to mankind was grown in the Neolithic stage of culture'. I go further, for it is my belief that there are very few plants whose uses are not already known to Man, apart from certain new biochemicals required for previously inconceivable usages. Some uses in prehistory may never have survived to be recorded. Even today ethnobotanists, working among primitive tribes with no written language, are discovering 'new economic plants'. In other examples the uses have been superseded by changing technology, political considerations, economics or social structure and await new demands and applications. Thus changing technology has led to a replacement of many of our natural fibres by synthetic fibres; political considerations have, for example, resulted in the suppression by the Spanish Conquistadors of the grain amaranths as a major crop in South America (Cole 1979; National Research Council 1984). The high cost of harvesting wild *Landolphia* species for their rubber content has led to its replacement by para rubber in West Africa. Likewise changes in the social structure, especially in the developing countries, has often resulted in the replacement of traditional foods by more 'socially acceptable' foods, such as wheat being preferred to millets and sorghum. The examples given are an over-simplification of what has happened, no single factor is entirely responsible. Similarly new demands for edible oils has resulted in a renewed interest in *Linum usitatissimum*, a crop that has been cultivated for more than 6000 years (Renfrew 1973; van Zeist & Bakker-Heeres 1975). The liquid wax content of jojoba seeds, *Simmondsia chinensis*, was known as far back as 1895 (Diguet 1895; Baker 1964). Although used by the natives of Mexico as a hair restorer (Saunders 1934), it is only comparatively recently that it has found an application in the western cosmetics industry and, when produced in sufficient quantities and at a competitive price, as a substitute for sperm whale oil.

Written information may be available but untapped, inaccessible because of poor library facilities, written in an unfamiliar language or obscure journal, or just lost under the enormous volume of ever accumulating literature. For example, poor library facilities combined with rather obscure journals has meant that much of the information of *Leptochloa fusca* (syn. *Diplachne fusca*), a grass widespread through Africa and southern Asia, is unavailable; the use of this salt-tolerant grass in land reclamation and for fodder (Booth 1983) is virtually unknown outside India and Pakistan. Similarly, the recent interest in medicinal plants has, for example, brought together a great deal of information into English that was formerly only available in Arabic (Boulos 1983) and Chinese (Duke & Ayensu 1985). How many economic botanists today have the time to search anthropological and similar non-botanical literature? It has been suggested that Flemming's discovery of the antibiotic properties of *Penicillium notatum* in 1928 would have been much earlier if due attention had been given to the scrolls of the Pharaohs or the use of mouldy bread for the dressing of wounds in medieval Europe (Böttcher 1963); similarly the biblical references to Moses (Leviticus 11 *passim*) and the drawing up of a code for food hygiene can now be shown as a safeguard against the presence of mycotoxins (Schoental 1980).

The general economic depression has encouraged many nations, especially those of the developing countries, to place an increasing emphasis on a greater utilization of plant resources and their conservation for the future. Much of this is, for reasons already stated, largely one of rediscovery and reappraisal of earlier uses and their reintroduction into present-day societies. There is an increasing need for the identification of such potential economic plants and for the gathering and ordering of all available information necessary for their development (Lucas & Wickens 1988). There is a need for the development of databases and other forms of plant information services, both nationally and internationally for the efficient handling of the information and to make it readily available to all who need to know. The plant information services currently available for arid lands is discussed by Bisby (1985).

THE ROYAL BOTANIC GARDENS, KEW

For two centuries Kew has been involved in the study and distribution of economic plants around the world, beginning in 1787 with the ill-fated attempt by Captain Bligh in His Majesty's Armed Vessel Bounty to convey breadfruit, *Artocarpus altilis,* from Tahiti to the West Indies (Wickens 1986). The history of Kew's economic bŏtanists and their work is discussed by Wickens (in press) and Simmonds (in press) respectively.

Following its inception in 1759 as a Royal Botanic Garden by Augusta, Princess of Wales, Kew became public property in 1840 and the responsibility of the Commissioners for the Department of Woods and Forests; it was opened to the public the following year. Over the years the inclusion of neighbouring gardens has resulted in the Royal Botanic Gardens now occupying ca. 120 ha. The addition in 1965, at a peppercorn rent from the National Trust, of Wakehurst Place, some 45 km to the south of Kew, added a further 200 ha.

To undertake research and supervise the gardens Kew employs 110 scientific staff, 35 curatorial and supervisory horticultural staff and 19 technicians. Their varied activities are organized into three Divisions, Herbarium, Jodrell Laboratory and Living Collections Divisions. Although these will be considered separately, in practice they perform a closely integrated role in the study of economic plants. Two further Divisions, Administration and Information & Exhibitions Division, although essential for certain of Kew's functions, will not be considered here.

Herbarium Division

The main function of the Herbarium is the identification and correct naming of plants in both living and preserved specimens at Kew and the preparation of floras and monographs to further future identification, correct naming and classification of plants worldwide. Accurate identification and uniformity of nomenclature are clearly basic requirements for all research. Far too much potentially valuable research is rendered totally worthless because of inadequate safeguards by means of voucher specimens to ensure future confirmation of identity, especially where researchers obtain conflicting results. The application of the International Code of Botanical Nomenclature (Stafleu 1983) ensures, with a few exceptions, international agreement as to the correct acceptable name for a taxon, thereby avoiding the confusion that can be caused when different researchers apply different names to the same taxon.

The value of floras for the naming of plants, both in the field and laboratory, requires little justification and Kew is rightly proud of its contribution to the floras of the world. The Flora of Iraq edited by Guest *et al.* (1966-cont.) is of particular interest because of its meticulous observations on economic uses.

The monographic treatment of families and genera leads to a better understanding of floristic delimitations and plant relationships, essential basic information for the plant breeder, especially when seeking wild relatives of cultivated plants for his breeding programmes. The recent review of the wild species of *Allium* at Kew for IBPGR is an example of such a study (Nielsen, in press). Thus although the work of plant taxonomists at Kew may appear esoteric to a superficial observer, closer examination will reveal a far more practical application, especially now that research is being concentrated on the more economically important families, such as the Leguminosae, Gramineae and Palmae.

In the Leguminosae the classification of the important economic genus *Indigofera* has been reviewed for tropical Africa by Gillet (1958). The small genus *Psophocarpus*, revised by Verdcourt & Halliday (1978) includes the economically important winged bean, *P. tetragonolobus*, a cultivated species not known in the wild and consequently its possible wild relatives are of particular interest to the breeder. The revision has led to improved identification and knowledge of species distribution, which, together with label information from herbarium specimens as to optimum fruiting period for seed collecting, has enabled Kew collectors to collect seeds of wild relatives of the winged bean for the Seed Bank.

The genus *Macrotyloma*, revised by Verdcourt (1982), contains a number of important pasture legumes, including the horsegram, *M. uniflorum* and the Archer Dolichos, *M. axillare*. The former is cultivated for green manure and pulses as well as being an important fodder crop, with the cultivar Leichhardt now being widely grown for fodder and pasture improvement in both Australia and Texas. The Kersting groundnut, *M. geocarpum* (syn. *Kerstingiella geocarpum*) from West Africa is an undervalued potential food crop (NAS 1979).

The carob, *Ceratonia siliqua,* is an economically important Mediterranean tree which had been considered monotypic until Kew botanists recognized a second species, *C. oreothauma*, from Oman and Somalia, thereby opening up interesting possibilities for the plant breeder (Hillcoat *et al.* 1980). To meet the needs of the forester the taxonomy and distribution of a number of important browse and gum producing African species of *Acacia* have been revised by Brenan (1984); the publication has already proved useful to the collector of germplasm.

Mention must be made of the contributions made to legume science and systematics arising from the Kew International Legume Conference 1978, sponsored by Kew, Missouri Botanical Garden and the University of Reading (Summerfield & Bunting 1980; Polhill & Raven 1981), with a sequel to the latter from invited papers edited by Stirton (1987).

Research in the Gramineae has resulted in the recent definitive publication by Clayton & Renvoize (1986) of keys and descriptions to all sub-families, tribes and genera of the grasses, together with their full synonymy, C3/C4 characteristics, etc., thereby leading to a better understanding of their taxonomic relationships. It was during the research for this publication that *Diplachne* was found to be congeneric with *Leptochloa, L. fusca* being a species mentioned earlier in this paper. Accounts for the grasses have been prepared for a number of regional floras, including the Floras of West Tropical Africa, Tropical East Africa and Zambesiaca, as well as for Iraq, Iran and Pakistan, with others in preparation. These studies have proved invaluable for germplasm collectors in their search for wild relatives of rice, sorghum, etc. The genus *Hyparrhenia*, which contains a number of valuable African range grasses, has been revised by Clayton (1969).

The third major economic family, the Palmae, has been very much neglected by taxonomists in the past, mainly because of the inherent difficulties in collecting good herbarium specimens for study. In recent years research at Kew has concentrated on the taxonomy, biology and conservation of Asiatic rattans. Forest destruction and an increasing demand for cane furniture has greatly reduced many rattan populations and the cultivation of elite species and the conservation of existing wild populations is now urgently required (Dransfield 1981, 1985 & in press). Local manuals on rattans (Dransfield 1979 & 1984) as well as an outline classification of the palms have also been prepared (Dransfield & Uhl 1986).

Work in other families includes a revision of Cucurbitaceae by Jeffrey (1979 & 1980), with special reference to their economic importance. In the Compositae, the taxonomically difficult Senecionae in East Africa, important for their toxins, has also been revised by Jeffrey (1986), and the *Vernonia galamensis* complex in Africa revised by Gilbert (1985), a species that is currently being considered as a potential oil crop (see Perdue in these proceedings). Other important contributions to economic botany include *Amaranthus* of Southern Africa by Brenan (1981), *Acokanthera* (Apocyanaceae) by Kupicha (1982), species of which are noted for their use as arrow poisons and consequently of interest as a potential source of toxins, and a revision of *Moringa* by Verdcourt (1985), currently being used by Dr. Samia Al Azharia Jahn in her study of medicinal plants for the GTZ Water Purification Project.

PALYNOLOGY UNIT

Taxonomic research in the Leguminosae, Palmae, etc. is supported by the Herbarium's Palynology Unit. For example, the very distinctive and highly specialized pollen of *Psophocarpus tetragonolobus* is considered indicative of its recent arrival in the evolutionary scale, a finding considered in keeping with its presumed status as a recently evolved cultivated species (Poole 1979).

MYCOLOGY SECTION

Work on the taxonomy of the macro-fungi at Kew, especially for tropical Africa, has produced, as a spin-off, an enumeration of the edible fungi of Zambia (Pegler & Piearce 1980), some of which may be worthy of cultivation.

ECONOMIC AND CONSERVATION SECTION

The history of Kew's involvement in economic botany and of the herbarium assuming responsibility in 1985 for the economic plant collections and all research in economic botany and of the development of the survey of Economic Plants for Arid and Semi-arid Lands (SEPASAL) is related elsewhere (Wickens 1986, in press & in prep.). The SEPASAL database has now been transferred from a Wang to Kew's Prime computer and is currently in the process of having a number of data fields upgraded in order to cope with an increased range of queries and in anticipation of its future role as a global database for economic plants.

The full potential of plants and their uses cannot be appreciated until all relevant information on habit, habitat, distribution, uses, etc., has been brought together. For example it is not unusual for certain uses to be restricted to localized communities and remain unknown to their neighbours. Thus, the role of the SEPASAL unit is to gather, collate and distribute information on economic plants, in particular the lesser known minor crops, semi-domesticated and wild species. The computer program is designed to answer the various permutations of the question 'WHICH plant will grow WHERE and produce WHAT?' and is available to all who are interested in the use of plants. It is our experience that there are very few accounts of economic plants, particularly of the lesser known, where additional information could not be found at Kew.

In-depth accounts of plants considered to be of particular potential value include the previously mentioned *Leptochloa fusca* (Booth 1983) and *Cordeauxia edulis* (Wickens & Storey 1984). The latter species is regarded as having a potential as a dessert nut in addition to its present value as an important food plant for the arid tropics. More general studies for arid and semi-arid Africa include a survey of forage and browse plants (IBPGR/Kew 1984) and a study on the alternative uses of trees and shrubs for FAO, which is nearing completion. Work, however, is not confined to the arid lands; all economic and potentially economic plants, both tropical and temperate are the responsibility of the Section. The Useful Plants of West Tropical Africa (Burkill 1985) is an example of the Section's wider interest.

Mention should also be made of the Kew International Conference on Economic Plants for Arid Lands (Wickens *et al.* 1985) which initiated a number of similar international conferences on what had previously been a much neglected subject.

Jodrell Laboratory Division

Founded in 1876, the Jodrell Laboratory owes its name to T. J. Phillips Jodrell, who donated the building to Kew (Metcalfe 1976; Metcalfe & Jones 1986); Jodrell Bank is a more recent construction and owes its name to another branch of the family (Field *et al.* 1987). The present building, officially opened in 1965, now houses the anatomy, cytogenetics and biochemistry sections, the physiology section being housed at Wakehurst Place.

Anatomy Section

The section's major contribution is to systematic anatomy and the identification of plants by their anatomical structure. The associated large reference collection of anatomical slides is of immense value for the identification of both recent and archaeological plant material and artifacts. For example, the legal and economic problems arising from damage to buildings and drains by the roots of trees and shrubs and the need for their identification stimulated research at Kew by Cutler & Richardson (1981) and Cutler *et al.* (1987), thereby providing town planners and landscape architects with vital information on the suitability or not of trees and shrubs for planting near buildings, etc.

Information on wood anatomy and usages is currently being incorporated into a computer databank linked with the ECOS databank on economic plants. When developed it will then be possible to relate anatomical characteristics to uses and properties and identify suitable species plus information on their distribution, etc., thereby aiding the better use of local timber resources or indicating alternative sources to endangered species.

Less obvious contributions to economic plants include the proceedings of the Kew/Linnean Society Symposium on the plant cuticle (Cutler *et al.* 1982) which may appear to some to be of little relevance in the search for new economic plants; the role of the cuticle as a mechanical protection against insect predators or as a permeable or nonpermeable medium for agrochemicals is, however, of importance to the plant breeder. The evolution of the lactifers in the economically important Euphorbiaceae is discussed by Rudall (1987), is also of particular interest to the breeder of *Hevea*, etc.

Cytogenetic Section

Since its inception the Cytogenetic Section has provided basic information on chromosome complements for the taxonomist, often essential to the taxonomist for understanding plant relationships as in, for example, the economically important mints, *Mentha* spp., by Harley & Brighton (1977), or for the breeder of the horticulturally important daffodils, *Narcissus* spp., by Brandham (1986 & 1987), Brandham & Kirton (1987) and Brandham & Stocks (in press).

Biochemistry Section

A systematic survey of *Aloe* leaf-exudate compounds has been carried out in recent years (Reynolds 1985a&b), improving our understanding of taxonomic relationships in the distribution of compounds of pharmacological interest, and of *A. vera* in particular by Grindlay & Reynolds (1986).

Other research has concentrated on the search for wild and semi-cultivated legumes with biochemical resistance to bruchids of interest to the plant breeder (Birch *et al.* 1985). Further progress in this search is presented in a paper by my colleagues, Drs. Monique Simmonds and Linda Fellows at this Conference.

The Bicentenary Joint Meeting of the Linnean Society with the Phytochemical Society at Kew last year on the Euphorbiales is in keeping with the belief that 'Economic uses of a plant are almost always reflected in its chemistry' (Reynolds & Cutler 1987).

Physiology Section

The Section, based at Wakehurst Place, Ardingly, Sussex, has two roles, carrying out research into problems of seed storage and seed germination, and the running of the Seed Bank.

Seed storage research has concentrated on tropical timber trees, especially recalcitrant species with particular emphasis on the Dipterocarpaceae, the dominant trees of the South East Asian rainforests. Studies into the mechanisms controlling seed dormancy clearly have important implications regarding the introduction of new species into cultivation, especially in the tropics. The interactive effects of light and temperature and in particular the role of the photoreversible pigment phytochrome are under investigation.

The Seed Bank contains ca. 6000 collections of ca. 2750 species from more than fifty families. The three main roles of the Seed Bank are to improve technology by making use of the associated physiology research, to aid researchers by making high quality wild seed readily available, and finally to conserve germplasm under threat of extinction. The Seed Bank also acts as an IBPGR recognized staging post for the germplasm collections in transit from the country of collection to an International Centre in another (Linington 1982).

In recent years seed collection has concentrated on economically important wild species in the arid and semi-arid tropics, such as Somalia, Mali and, later this year, in Botswana, working in conjunction with ECOS.

Living Collections Division

Kew's living plant collections contain some 80000 accessions, representing a broad range of the world's vegetation (Field *et al.* 1987), a wealth of living plant material for the researcher. One aspect of the work of the Division which is that of particular interest to the economic botanist is the Micropropagation Unit.

Micropropagation Unit

The three objectives of the Unit are the propagation of plants which are difficult to increase by conventional means, the 'rescue' of plants which are dying or diseased and their return to cultivation as healthy plants, and finally the propagation and subsequent distribution of rare and endangered species (Fay & Muir, in press). The work of the Unit in the multiplication of a number of potential arid land plants is discussed by Woods (1985). Of particular interest is the rare *Ceratonia oreothauma* subsp. *oreothauma*, a browse species from Oman and potentially important as a source of genetic material for crossing the carob, *C. siliqua*; sufficient material of which has now been multiplied for distribution to researchers. Attempts to multiply from rather poor seed of *Cordeauxia edulis* has so far proved difficult; its potential is discussed by Wickens & Storey (1984).

CONCLUSION

In the past Kew has been responsible for introducing numerous economic crops around the world. With the development of more specialist, crop-oriented research institutions Kew no longer needs to fulfil its former role. It remains however as a prime source of basic botanical knowledge for all new crops as well as for the wild relatives of existing crops.

REFERENCES

Baker, H. G. (1964) *Plants and Civilization*. Macmillan, London.

Birch, N., B. J. Southgate and L. E. Fellows (1985) Wild and semi-cultivated legumes as potential sources of resistance to bruchid beetles for crop breeder: a study of *Vigna/Phaseolus*. In: *Plants for Arid Lands*, G. E. Wickens, J. R. Goodin & D. V. Field (Eds), pp. 303–320. Allen & Unwin, London.

Bisby, F. A. (1985) Plant information services for economic plants of arid lands. In: *Plants for Arid Lands*, G. E. Wickens, J. R. Goodin & D. V. Field (Eds), pp. 413–425. Allen & Unwin, London.

Böttcher, H. M. (1963) *Miracle Drugs. A History of Antibiotics*. (English translation from the German edition 1959) Heinemann, London.

Booth, F. E. M. (1983) *SEPASAT Dossier No. 1: Leptochloa fusca (L.) Kunth*. Economic & Conservation Section, Royal Botanic Gardens, Kew. (mimeo).

Boulos, L. (1983) *Medicinal Plants of North Africa*. Reference Publications. Algonac, Mich.

Brandham, P. E. (1986) Evolution of polyploidy in cultivated *Narcissus* subgenus *Narcissus*. *Genetics* 68: 161–167.

Brandham, P. E. (1987) Bigger and better; the evolution of polyploid *Narcissus* in cultivation. *Daffodils* 1986–7: 53–59.

Brandham, P. E. and P. R. Kirton (1987) The chromosomes of species, hybrids and cultivars of *Narcissus* L. (Amaryllidaceae). *Kew Bull.* 42: 65–103.

Brandham, P. E. and K. Stocks (in press) A solution to current problems in Jonquil breeding. *Daffodils* 1987–88.

Brenan, J. P. M. (1981) The genus *Amaranthus* in Southern Africa. *J. S. Afr. Bot.* 47: 451–492.

Brenan, J. P. M. (1984) *Taxonomy of Acacia species*, FAO, Forestry Division, Rome.

Burkill, H. M. (1985) *The Useful Plants of West Tropical Africa. Vol. 1, Families A-D.* Royal Botanic Gardens, Kew.

Callen, E. O. (1969) Diet as revealed by coprolites. In: *Science in Archaeology. A Survey of Progress and Research*, D. Brothwell and E. Higgs (Eds), pp. 235–243. Thames & Hudson, London.

Clayton, W. D. (1969) *A revision of the genus Hyparrhenia. Kew Bull. Add. Ser. II.*

Clayton, W. D. and S. A. Renvoize (1986) *Genera Graminum. Grasses of the World.* Royal Botanic Gardens, Kew.

Cole, J. N. (1979) *Amaranths from the Past for the Future*. Rodale Press, Emmaus, Penn.

Coles, S. (1970) *The Neolithic Revolution*. British Museum (Natural History), London.

Cutler, D. F. and I. B. K. Richardson (1982) *Tree Roots and Buildings*. Construction Press, Harlow, Essex.

Cutler, D. F., K. L. Alvin and C. E. Price (1981) *The Plant Cuticle*. Academic Press, London.

Cutler, D. F., P. F. Rudall, P. E. Gasson and R. Gale (1987) *Root Identification Manual of Trees and Shrubs*. Chapman & Hall, London.

Diguet, L. (1895) Le jojoba (*Simmondsia californi* Nutt) C. A. *Rev. Sci. Nat. Appl.* 1895: 685-687 (1889–95).

Dransfield, J. (1979) *A Manual of Rattans of the Malay Peninsula*. Forest Department: Malaysian Forest Records No. 29, Kuala Lumpur.

Dransfield, J. (1981) The biology of Asiatic rattans in relation to the rattan trade and conservation. In: *The Biological Aspects of Rare Plant Conservation*, A. Synge (Ed.), pp. 179–186. Wiley, Chichester.

Dransfield, J. (1984) *The Rattans of Sabah*. Sabah Forestry Record No. 13.

Dransfield, J. (1985) Prospects for better known canes. In: *Proceedings of Rattan Seminar, Kuala Lumpur, 2–4 October, 1984*, K. M. Wong and N. Manokaram (Eds), Forest Research Institute, Kepong.

Dransfield, J. (in press) Prospects for rattan cultivation. In: *Proceedings of Symposium on Economic Botany of Palms; Biology Utilization and Conservation 13–16 June 1986 New York Botanical Garden*. Advances in Economic Botany, New York Botanical Garden.

Dransfield, J. and N. W. Uhl (1986) An outline of a classification of palms. *Principes* 30: 3–11.

Duke, J. A. and E. S. Ayensu (1985) *Medicinal Plants of China,* 2 Vols. Reference Publications, Algonac, Mich.

Fay, M. F. and H. J. Muir (in press) The role of micropropagation in the conservation of plants. Paper presented at Conference on Techniques for the Conservation of Threatened Plant Species in Botanic Gardens of Mediterranean Area. Cordoba Botanic Garden, Spain, 10–14 May 1987.

Field, D. V., L. E. Fellows and M. Stanniforth (1987) Beauty and utility – two faces of Kew Gardens. *J. Roy. Agric. Soc. England* 148: 101–112.

Fowler, P. J. (1983) *The Farming of Prehistoric Britain.* Cambridge University Press, Cambridge.

Gilbert, M. G. (1985) Notes on East African *Vernonieae (Compositae.* A revision of the *Vernonia galamensis* complex. *Kew Bull.* 41: 19–35.

Gillet, J. B. (1958) *Indigofera (Microcharis) in Tropical Africa. Kew Bull. Add. Ser. 1.*

Grindlay, D. and T. Reynolds (1986) The *Aloe vera* phenomenon: A review of the properties and modern uses of the leaf parenchyma gel. *J. Ethno-pharm.* 16: 117–151.

Guest, E., A. Al-Rawi and C. C. Townsend (Eds) (1966-continued) *Flora of Iraq.* Ministry of Agriculture, Baghdad.

Hadidi, M. N. El (1985) Food plants of prehistoric and predynastic Egypt. In: *Plants for Arid Lands,* G. E. Wickens, J. R. Goodin and D. V. Field, (Eds) pp. 87–92. Allen & Unwin, London.

Harley, R. M. and C. A. Brighton (1977) Chromosome numbers in the genus *Mentha* L. *Bot. J. Linn. Soc.* 74: 71–96.

Hillcoat, D., G. Lewis and B. Verdcourt (1980) A new species of *Ceratonia* (Leguminosae-Caesalpinioideae) from Arabia and the Somali Republic. *Kew Bull.* 35: 261–271.

IBPGR/Kew (1984) *Forage and Browse Plants for Arid and Semi-Arid Africa.* International Board of Plant Genetic Resources, Rome.

Jeffrey, C. (1979) The economic potential of some Cucurbitaceae and Compositae of Tropical Africa. In: *Taxonomic Aspects of African Economic Botany,* G. Kunkel (Ed), pp. 35-38. Publ. Excmo. Ayuntamiento de Las Palmas, Las Palmas.

Jeffrey, C. (1980) A review of the Cucurbitaceae. *Bot. J. Linn. Soc.* 81: 233–247.

Jeffrey, C. (1986) The *Senecioneae* in East Tropical Africa. Notes on Compositae IV. *Kew Bull.* 41: 873–943.

Kupicha, F. K. (1982) Studies on African *Apocyanaceae*: the genus *Acokanthera. Kew Bull.* 37: 41–67.

Linington, S. (1982) Theory in practice: the work of the Kew Seed Bank. *Brit. Assoc. Seed Analysts Bull.* No. 18.

Lucas, G. L. and G. E. Wickens (1988) Arid land plants – the data crisis. In: *Arid Lands: Today and Tomorrow.* E. E. Whitehead, C. F. Hutchinson, B. N. Timmermann and R. G. Varady (Eds), pp. 113–126. Westview Press, Boulder, Colorado.

Metcalfe, C. R. (1976) History of the Jodrell Laboratory as a centre for systematic botany. *Leiden Bot. Ser.* No. 3: 1-19.

Metcalfe, C. R. and K. Jones (1976) *Jodrell Laboratory Centenary 1876–1976*. Royal Botanic Gardens, Kew.

NAS (1979) *Tropical Legumes: Resources for the Future*. National Academy of Sciences, Washington D. C.

National Research Council (1984) *Amaranth: Modern Prospects for an Ancient Crop*. National Academy Press, Washington, D. C.

Nielsen, H. (in press) *Economically important species of the genus Allium*. IBPGR, Rome.

Pegler, D. N. and G. D. Piearce (1980) The edible fungi of Zambia. *Kew Bull*. 35: 475–491.

Polhill, R. M. and P. H. Raven (Eds) (1980) *Advances in Legume Systematics*, Parts 1 and 2. Royal Botanic Gardens, Kew.

Poole, M. M. (1979) Pollen morphology of *Psophocarpus* (Leguminosae) in relation to its taxonomy. *Kew Bull*. 34: 211–220.

Renfrew, J. M. (1973) *Palaeoethnobotany. The prehistoric food plants of the Near East and Europe*. Columbia University Press, New York.

Reynolds, P. J. (1980) *Butser Ancient Farm Impressions*. Archaeological Research, Petersfield, UK.

Reynolds, T. (1985a) The compounds in *Aloe* leaf exudates: a review. *Bot. J. Linn. Soc*. 90: 157-177.

Reynolds, T. (1985b) Observations on the phytochemistry of the *Aloe* leaf-exudate compounds. *Bot. J. Linn. Soc*. 90: 179–199.

Reynolds, T. and D. F. Cutler (Eds) (1987) Proceedings of the Linnean Society/ Phytochemical Society Joint Meeting on the Euphorbiales. *Bot. J. Linn. Soc*. 94: 1-326.

Rudall, P. J. (1987) Laticifers in Euphorbiaceae – a conspectus. *Bot. J. Linn. Soc*. 94: 143–163.

Saunders, C. F. (1934) *Useful Wild Plants of the United States and Canada*. McBride, New York (reprinted 1976 as Edible and Useful Wild Plants of the United States and Canada. New York, Dover Publications).

Schoental, R. (1980) A corner of history. Moses and mycotoxins. *Preventative Medicine* 9: 159-161.

Schultes, R. E. (1960) Tapping our heritage of ethnobotanical lore. *Econ. Bot*. 144: 257–262.

Simmonds, N. W. (in press) Kew's past, present and future role in tropical agriculture. *Trop. Agric. Assoc. Newsletter*.

Stafleu, F. A. (Ed.) (1983) *International Code of Botanical Nomenclature*. Bohn, Scheltema & Holkena, Utrecht/Antwerp, and Junk, The Hague.

Stirton, C. H. (Ed.) (1987) *Advances in Legume Systematics*, part 3. Royal Botanic Gardens, Kew.

Summerfield, R. G. and A. H. Bunting (Eds) (1980) *Advances in Legume Science*. Royal Botanic Gardens, Kew.

van Zeist, W. and J. A. H. Bakker-Heeres (1975) Evidence for linseed cultivation before 6000 BC. *J. Arch. Sci*. 2: 215–219.

Verdcourt, B. (1982) A revision of *Macrotyloma* (Leguminosae). *Hooker's Icones Plantarum*. 38 4. Bentham-Moxon Trustees, Kew.

Verdcourt, B. (1985) A synopsis of the Moringaceae. *Kew Bull*. 40: 1-23.

Verdcourt, B. and P. Halliday (1978) A revision of *Psophocarpus* (Leguminosae-Papilionoideae). *Kew Bull*. 33: 191–227.

Wickens, G. E. (1986) Breadfruit to computers: economic botany at Kew. *Span* 29, 2: 62–64.

Wickens, G. E. (in preparation) What is economic botany?

Wickens, G. E. (in press) Two centuries of Economic Botanists at Kew. *Trop. Agric. Assoc. Newsletter.*

Wickens, G. E. and I. N. J. Storey (1984) *SEPASAT Dossier No. 5: Cordeauxia edulis Hemsley.* Economic & Conservation Section, Royal Botanic Gardens, Kew (mimeo).

Wickens, G. E., J. R. Goodin and D. V. Field (Eds) (1985) *Plants for Arid Lands.* Allen & Unwin, London.

Woods, A. (1985) The potential for the *in vitro* propagation of a number of economically important plants for arid areas. In: *Plants for Arid Lands,* G. E. Wickens, J. R. Goodin and D. V. Field (Eds), pp. 333–342. Allen & Unwin, London.

Yarnell, R. A. (1969) Palaeo-ethnobotany in America. In: *Science in Archaeology. A Survey of Progress and Research.* D. Brothwell and E. Higgs (Eds), pp. 215–228. Thames & Hudson, London.

General Index

acai palm 317, *See also: Euterpe oleracea*
adhesives 76, 130
aerial yam *See:* yam, aerial
aflatoxin 24, 26
Africa 47, 66, 73, 79, 94, 123–134, 198, 205, 239, 246–255 *passim*, 266, 272, 277, 405–406, 412, 415, 417
  Central 33, 93, 124–125, 247, 266, 266–267, 389
  East 125, 201, 272, 389, 415
  Horn of 385
  North 247, 405
  South 131, 253
  Southern 94, 197, 416
  tropical 123–134, 250, 266, 272, 405, 414–417
  Tropical East 415
  West 72–73, 124–125, 130–132, 200, 265, 405, 412, 415, *See also:* Gulf of Guinea
African oil palm 61, 150, 151, 155, 323, 326, *See also: Elaeis guineensis*
African plum 265–271, *See also: Dacryodes edulis*
agalactia 88–89, 91
agarwood 89, *See also: Aquilaria malacensis*
agroforestry 54, 68, 72, 80, 81, 85, 87, 89, 92, 143, 146, 150, 154, 161, 168, 319
AIDS 150, 368, 384
ajowan 90, *See also: Trachyspermum ammi*
alcohol 126, 127, 132, 145, 294
alcohol production 38, 144, 154, *See also:* gasohol
alcoholic beverage 130, 142, 235, 272, 304, *See also:* fermented beverage
alfalfa 283, 284, 297, *See also: Medicago sativa*
algae 333–339
algarrobo 280–287, *See also: Prosopis chilensis*
Algeria 405
alkalinity 99
alkaloids 84, 87–88, 97–98, 103, 106, 131, 147, 189, 311, 366, 368, 371, 378–379, 382–383, 385
allelochemicals 358, 383

aloe *See: Aloe* spp.
amaranth 139, 141, 147, 191, 412, *See also: Amaranthus* spp.
Amazon basin *See:* Amazonia
Amazonia 150–163, 304–305, 317, 326, *See also:* America, South
ambrette 90, *See also: Abelmoschus moschatus*
America 73, 125, 166, 170, 266, 288–289, 301, 304
  Central 126, 160
  Latin 124, 133, 159, 304, 318, 330
  North 223, 233
  South 61, 78, 124, 205, 239, 304, 327, 408, 412, *See also:* Amazonia
amino acids 98, 103, 105, 143–144, 167, 169–170, 211, 222, 235, 268, 296, 311–312, 336
anaesthetics 380
Andes 159, 222–233
Angola 125, 267
animal feedstuffs 24, 31, 36, 101–107 *passim*, 110, 110–113, 124–134 *passim*, 170, 174, 176–180, 182, 186, 191, 225, 246, 249, 251–253, 275, 288–292, 297, 301, 314, 338, 345, 368, 370, *See also:* fodder
anise 89, 90, *See also: Pimpinella anisum*
anti-fertility 277, 385
anti-inflammatory 329, 385
antiamoebic 91
antibacterials 89, 91
antibiotic 378–380, 412
antidiarrhoeals 88
antifeedant 354, 370–372
antimalarials 380
aphrodisiacs 91
apple 24
aquaculture 333, 337
aquatic plants 105, 177
Arabia 272
Archer Dolichos 414, *See also: Macrotyloma axillare*
Argentina 166, 280, 288
Arizona 400
aroma chemicals 76, *See also:* cosmetics
arrow poison 131, 403, 416
asafoetida 89, *See also: Ferula* spp.

Asia 42, 49, 60, 73, 89, 124, 126, 138,
170, 174, 246–250 *passim*, 266, 288,
303, 412
  South East 47, 125, 133, 346, 389,
  404, 418, *See also:* Malaysia
asparagus bean *See:* bean, yard-long
Australia 36, 38, 43–47, 49–50, 80, 139,
143, 147, 202, 208, 211, 250, 366,
389, 406, 414, *See also:* Queensland,
South Australia, Victoria, Western
Australia
azadirachtin 354–357, 362

B vitamin *See:* vitamins, B
babassu 154–155, 157, 158, *See also:*
Orbignya spp.
bagasse 179–180, 182, 295
bambara groundnut 139, 143, *See also:*
Vigna subterranea
bamboo 25, 139, 144, 147, 318, *See also:*
Bambusa spp.
banana 54, 62, 181, 346
Bangladesh 252
barley 219, 222, 223, 239, 348, 402
basil 90, *See also: Ocimum* spp.
baskets 327
bay oil *See:* oil, bay
bean 167, 217, 242, 311
  faba 237
  lima 157, *See also: Phaseolus lunatas*
  mung 24
  navy 187, 225, *See also: Phaseolus*
  vulgaris
  Phaseolus 222
  rice 139, 140, 147, *See also: Vigna*
  umbellata
  soya 23, 25–26, 36, 45–47, 97, 112,
  187, 191, 221, 223, *See also: Glycine*
  max
  velvet 38
  winged 25–26, 108–114, 139–140,
  147, 253, 404, 414, *See also:*
  Psophocarpus tetragonolobus
  yam 123, 124, 127, *See also:*
  Sphenostylis stenocarpa
  yard-long 404, *See also: Vigna*
  unguiculata subsp. *sesquipedalis*
Belgium 188
Belize 327
benzoin 89, *See also: Styrax* spp.
bergamont *See: Citrus aurantinum* subsp.
  bergamia

beverage 69, 126, 127, 250, 300, 326,
407
beverage plants 137, *See also:* alcoholic
  beverages
bioactive compounds *See:* biologically
  active compounds
biochemicals 358, 411–412, 418
biofertiliser *See:* green manure
biogas 102–103, 182, *See also:* fuel
biological activity 100, 328, 354, 356,
  361, 365, 368, 383
biologically active compounds 338, 378,
  382–384, 386
biomass 53–54, 101–105, 107, 133, 143,
  150, 171–172, 175–176, 183, 301,
  338, 391–392, 397–399
bird-seed *See:* pet food
birth control 382, *See also:* contraceptive
bitter-orange 81
black pepper 77, *See also: Piper nigrum*
Blue Mallee 80, *See also: Eucalyptus*
  polybrachtea
blueberry 190, *See also: Vaccinium*
  corymbosum
Bolivia 78, 227, 280, 328
borage 189, 219, *See also: Borago*
  officinalis
botanicals *See:* medicinal plants
Botswana 419
bow strings 327
brackish water 333, *See also:* sea water
  and tidal swamps
Brazil 78, 133, 147, 150, 155–156,
  160–162, 307, 309, 317–319,
  324–328, 330
Brazil nut 155, 157, 160, 211, *See also:*
  Bertholletia excelsa
bread wheat *See:* wheat
breadfruit 413, *See also: Artocarpus*
  altilis
Britain *See:* UK
brooms 159, 327
browse 68, 99, 415, 417, 419
bruchid 367, 369, 418
brushes 327
buckwheat 139, 142, 147, 191, 366, *See*
  also: Fagopyrum spp.
bulbil-bearing yam *See:* yam,
  bulbil-bearing
Burundi 125
bush-butter tree *See:* African plum

C3 species 106

C4 species 106
cabbage 104, *See also: Brassica oleracea*
cabinet work 270
caiaué 316, *See also: Elaeis oleifera*
calamus 89, *See also: Acorus calamus*
Cameroon 125, 265–267, 386
camphor 89, *See also: Cinnamomum camphora*
Canada 84
cananga 90, *See also: Cananga odorata*
candles 146
cane furniture 303, 415
canihua 223, *See also: Chenopodium pallidicaule*
caraway 90, *See also: Carum carvi*
cardamom 81, *See also: Elettaria cardamomum*
cardiovascular 17, 89, 91, 385
Caribbean 47, 166, 176, 408, *See also:* West Indies
carob 415, 419, *See also: Ceratonia siliqua*
carotenoids 103, 106, 153, 155, 163, 190, 311, 313, *See also:* vitamins, A
carrot 123–125, *See also: Daucus carota*
cascara sagrada 91, *See also: Rhamnus purschiana*
cashew 313, *See also: Anacardium occidentale*
cassava 34, 44, 98, 123–133 *passim*, 144, 150, 153–154, 157, 161, 163, 261, 304, 327, 402, *See also: Manihot esculenta*
cassava flour 128, 129, 156–157, 315
cassie 89, *See also: Accacia farnesiara*
cedar wood 89
celery 90
Central Africa *See:* Africa, Central
Central America *See:* America, Central
Centre for Arid Zone Studies 244
cereal surplus 151
cereals 24, 36, 95, 98, 104, 126, 128, 142, 158, 174, 176, 179, 183, 186–188, 191, 222, 225, 235, 235–236, 239–244, 246–255 *passim*, 295, 314–316, 405–406
charcoal 182, 278, 301, 327, *See also:* firewood *and* fuel
chemical coatings 206
chenopods 139, 142, 147, 223, *See also: Chenopodium*
chicken feed 314–315

chickpea 38, 191, 358–361, *See also: Cicer arientinum*
Chile 223, 226, 280–287, 291, 302, 337
chillies 82
China 14–15, 78, 80, 109, 141, 147, 247–248, 250, 338, 381, 404, 406, 411
chokeberry 190, *See also: Aronia melanocarpa*
cinchona 346
cinnamon 90, *See also: Cinnamomum camphora*
citronella 90, *See also: Cymbopogon* spp.
citrus 72, *See also: Citrus* spp.
clothing 188, 411
cloudberry 190, *See also: Rubus chamaemorus*
clove 90, *See also:* spices
coal slurries 252
cocoa 25, 54–56, 58–62, 127, 150, 156, 160, *See also: Theobroma grandiflorum*
cocona 257–263, *See also: Solanum sessiliflorum*
coconut 54, 58–61, 110, 129, 133, 156, 157, 265
coconut oil *See:* oil, coconut
coconut palm 323, *See also: Cocos nucifera*
cocoyam 123–125, 127, 132, *See also: Xanthosoma* spp.
coffee 54, 62, 110, 127, 171, 218, 346
colocynth 147, *See also: Citrullus colocynthis*
Colorado 233
Columbia 313
common millet *See:* millet, common
Commonwealth Potato Collection 346
condiments 137, 161, *See also:* spices
Congo 125, 267
conservation 53, 71, 74, 135, 148, 181, 265, 324–325, 330, 346–351, 413, 415–416, *See also:* erosion
Consultative Group on International Agricultural Research 346
contraceptive 91, 131, *See also:* birth control
pill 126, 127, 219
copra 26, 58, 60
coriander 90
corn *See:* maize
corticosteroid drugs 126, 127, 131

cosmetics 23, 37, 85, 89, 130, 252, 265, 412, *See also:* aroma chemicals
Costa Rica 163, 309–320 *passim*, 324
cotton 43, 188, 278, 391, 399
cotton seed cake 278
cotton seed oil *See:* oil, cotton seed
cover crop *See:* crops, cover
cowpea 157, 404, *See also: Vigna unguiculata* subsp. *cylindrica*
crambe 37, *See also: Crambe hispanica*
crates 169
crop staples 125, 174
crop subsidies 217
crops
    cover 54, 93, 96–97, 110, *See also:* green manure
    food 34, 54, 82, 162, 169, 203, 236–237, 246, 346–347, 404–405, 407, 415
    fruit 96, 136, 190, 210
    nut 136
    oil 155–162
    root 123–134, 190
    seed 168, 187
    staple food 346
    tuber 123–134, 136, 144
    vegetable 136, *See also:* green vegetable
Cuba 179, 183
cubeb 90, *See also: Pier cubeba*
cumin 90
cuphea 139, 146, 147, *See also: Cuphea* spp.
curative baths 328
curcuma 90, *See also: Curcuma* spp.
Czechoslovakia 112

daffodil 418, *See also: Narcissus* spp.
date palm 72, 323, *See also: Phoenix dactylifera*
Denmark 101, 224, 233
desert locust 354, *See also: Schistocerca gregaria*
detergents 138, 284–285, 299
dhaincha 246, 250, *See also: Sesbania bispinosa*
dill 90, *See also: Anethum graveolens*
diosgenin 127, 131, 219, 274, 277
diseases 14, 17, 33, 38, 41, 43, 47, 51, 61–63, 72, 79, 110, 141–143, 147, 150, 154, 156, 159, 163, 191, 205, 216, 225, 236–237. 239–240, 242, 248–250, 258, 261, 265, 304,

309–310, 316, 319, 348, 352, 358, 362, 372, 378, 381–382, 385
domestic products 411
Dominican Republic 179, 182
durum 240–242
dye 138

East Africa *See:* Africa, East
economic plants 135, 350, 407, 411–413, 416–418
Economic Plants for Arid and Semi-Arid Lands 416
Ecuador 227
edible fungi *See:* fungi, edible
edible oil *See:* oil, edible
EEC 14–15, 23, 26, 56, 84, 185–187, 191, 217
eel-grass 99, *See also: Zostera marina*
effluent flocculents 252
eggplant 257
Egypt 79, 411
einkorn 405, *See also: Triticum monococcum*
elephant grass *See:* Napier grass
elephant yam *See:* yam, elephant
erosion 54, 81, 93, 98, 166, 402, 411, *See also:* conservation
erucic acid 37
essential oil *See:* oil, essential
ethanol 24, 103, 144, 186–187, 275, 295
Ethiopia 197–205 *passim*, 233, 235–238, 250, 272, 385
ethnobotanical survey 384
eucalyptus 80–81, 90, 253, *See also: Eucalyptus* spp.
eucalyptus oil *See:* oil, eucalyptus
Europe 34, 54, 79, 85, 89, 125, 133, 151, 160, 185–189, 191, 239, 247–249, 381, 404, 411, 412
European chestnut 313, *See also: Castanea sativa*
evening primrose 150, 189, 216–221, *See also: Oenothera* spp.
evening primrose oil *See:* oil, evening primrose
explosives 15, 252

false yam *See:* yam, false
fatty acids 156–157, 160, 191, 211–214, 216, 265, 268, 312–313, 316, 325, 338
feed *See:* animal feedstuffs
fennel 90
fenugreek 90, 219

fermented beverage 288, *See also:* alcoholic beverage

fertilizers 29, 33, 53, 58, 60, 63, 71, 78, 104, 142, 203, 225, 236, 237, 239, 258, 357, 392

fever 270

fibre 38, 49, 58, 102, 104, 107, 127–128, 130–133, 137, 144, 158–159, 167, 178, 182, 188–189, 222, 251–255, 274, 284, 285, 291, 292, 298–300, 308, 313, 323–324, 327, 370, 411–412

ficin 90

fig 90

finger millet *See:* millet, finger

firewood 68, 71, 73, 137, 143–144, 166–167, 171, 172, 251, 280, 297, 301, 328, 357, *See also:* charcoal

fish poison 272

fishing lines 327

fishing spears 327

flavours 62, 76, 133, 152–162 *passim*, 189, 257, 268, 313, 316, 378

flax 188–189, *See also: Linum usitatissimum*

Florida 166

flour 112, 127–133 *passim*, 141, 153–154, 191, 248, 293–295, 298–300, 304–319 *passim*

fodder 69, 97, 137, 139, 143, 146, 148, 157, 187, 235, 246, 248, 251, 255, 280, 283–284, 404, 411–412, 414, *See also:* animal feedstuffs

folk medicine *See:* traditional medicine

food additive 69, 251, 252, 338

food crops *See:* crops, food

food industry 13–22, 89, 98, 242, 297–298, 338

forage 38, 68–69, 73, 99, 110, 112, 143, 166–170, 172, 176, 182, 223, 235, 248, 255, 280, 284, 286, 301, 318, 350, 417, *See also:* animal feedstuffs *and* fodder

foxtail millet *See:* millet, foxtail

France 79, 84, 101, 146, 188, 381, 402–405

frankincense 90, *See also: Boswellia* spp.

fruit 14, 20–21, 53, 56–62, 67–69, 71–73, 87, 113, 136–137, 146, 152–163 *passim*, 168, 170, 185, 190, 208–211, 257, 265–268, 270, 272–278, 280–287, 303, 304, 306–311, 313–319, 324, 326–330, 403, 405

fruit crops *See:* crops, fruit

fuel 24, 71, 81, 102, 130, 134, 138, 142–146, 150, 169, 170, 175–183, 186, 248, 278, 323, 327–329, 411, *See also:* biogas, charcoal *and* firewood

fuelwood *See:* firewood

fungi 72, 86, 129, 362, 379, 382

   edible 416, *See also:* mushrooms

fungicides 33, 208, 239, 269

furniture 56, 169

Gabon 125, 266–267

galactagogues 89

galactomannan 38, 295

galactomannan gum 295–296

galangal 90, *See also: Alpinia* spp.

gamma linolenic acid 216

gamma linolenic oil *See:* oil, gamma linolenic

garlic 90, *See also: Allium* spp.

gasohol 126, 130, *See also:* alcohol production

gastric disorders *See:* stomach ailments

genetic resources 135–139, 147, 265, 346–351

geranium 25, 90, *See also: Pelargonium* spp.

geranium oil *See:* oil, geranium

Germany 84, 278, 362, 381

   West 233

germplasm 71, 139–163 *passim*, 197, 198, 203, 205, 218, 225, 233, 247, 310, 314, 317–318, 330, 415, 419

germplasm collection 41, 50, 139, 141–144, 148, 197, 225, 227, 250–251, 255, 305, 307, 306, 419, *See also:* seed bank

Ghana 112, 125, 139, 147, 203

ginger 82, 89, 123–124, *See also: Zingiber officinale*

ginger grass 90, *See also: Cymbopogon* spp.

ginseng 90, *See also: Panax ginseng*

global surplus 14–16

glucosidase 366–368, 371

glycosides 84, 87–88, 98, 379

gold of pleasure 191, *See also: Camelina sativa*

grain 26, 36, 54, 95–96, 99, 102, 130, 137, 139–142, 167, 176–177, 179, 183, 187, 191, 246–251 *passim*, 291, 358, 405, 412
grain legume 126, 191, 358, *See also:* pulse
grass 25, 85, 90, 94, 99–100, 181, 235, 286, 412
Greece 224
green manure 94, 96, 166, 251, 414, *See also:* crops, cover
green vegetable 47, 110–112, 127, 404, *See also:* crops, vegetable
guar galactomannan 295
guar gum 252, 296, 297
guarana 162, *See also: Paulloninia cupana*
Guatemala 81
guava 160, *See also: Psidium guajava*
guayule 37–38, 139, 145, 147, 391–400, *See also: Parthenium argentatum*
guggulsteroids 385
Gulf of Guinea 267, *See also:* Africa, West
gum 72, 76, 79–80, 138, 162, 250–253, 295–296, 301, 415
gum naval stores 79
gum rosin 76, 79

hackberry 411, *See also: Celtis australis*
hammocks 327
handicrafts 327
harvest index 95–96
harvesting 21, 26, 38, 47–49, 57, 59, 62, 79, 85, 97, 110, 152, 154, 159, 170, 173, 187, 190, 258, 289, 292, 297, 301, 412
Hausa potato 123–124, 127, *See also: Solenostemon rotundifoius*
headache 270
health food 102, 225, 233, 255, 333
heglig 272, *See also: Balanites aegyptiaca*
herbal
    drugs 84–92, 381
    medicine 84, 403
herbicides 21, 33, 189, 225, 230, 239
Himalayas 223, 233
hog food 250
hog millet *See:* millet, common
Holland *See:* Netherlands
Honduras 81, 157, 304

Hong Kong 84
hops 90
Horn of Africa *See:* Africa, Horn of
horsegram 414, *See also: Macrotyloma uniflorum*
huauzontle 223, *See also: Chenopodium berlandieri* subsp. *nuttalliae*
hunting contraptions 403

ICDUP *See:* International Council for the Development of Underutilized Plants
ice plant 190, *See also: Mesembryanthenum crystallimum*
illipe nut 56
illuminant 146
India 14–15, 38, 78–79, 81, 84, 90, 125, 135–148, 197, 208, 247–252, 358, 362, 404, 411–412
Indonesia 49, 78, 84, 139, 147
industrial oil *See:* oil, industrial
industrial starch 154
insect attractants 352
insect protein 323
insecticidal 91
    rotenoids 366
insects 38, 41, 43, 47, 48, 51, 72–73, 103, 205, 236, 248, 268, 304, 319, 352–362, 365, 368–372, 418
International Council for Research in Agroforestry 66
International Council for the Development of Underutilized Plants 108, 109, 111, 113
International Legume Database 350
Iran 415
Iraq 414–415
Irish potato *See:* potato
irrigation 14, 45, 49, 71, 99, 250, 260, 391–400
Israel 147, 272, 391, 399–400
Italian milllet *See:* millet, foxtail
Italy 101, 147, 190, 224, 250

Jamaica 202
jantar *See:* dhaincha
Japan 14, 56, 84, 93, 101, 247, 248
Japanese mint 78, *See also: Mentha arvensis*
jasmine 79, 90
jasmine concrete 79, *See also:* oil, essential
Jerusalem artichoke 104, 123, 125, *See also: Heliathus tuberosus*

jojoba 37, 38, 99, 139, 145, 147, 347,
  412, *See also: Simmondsia chinensis*
juniper berry 90

kairomone 352–362
kale 104, *See also: Brassica oleracea*
kenaf 38, 45, 49–50, *See also: Hibiscus
  cannabinus*
Kenya 125, 200, 202–203, 385
Kersting groundnut 414, *See also:
  Macrotyloma geocarpum*
Kew Gardens *See:* Royal Botanic
  Gardens, Kew
koracan *See:* millet, finger
kudzu 93, *See also: Pueraria lobata*

labour 25, 32, 40, 56, 59, 68, 71, 77,
  79–80, 89, 111, 133, 152, 175, 179,
  407
lactogenic preparation 88
land reclamation 93–100, 412
larvicidal 91
latex 145–146, 155, 161, 169
Latin America *See:* America, Latin
lauric acid 325
lauric oil *See:* oil, lauric
lavender 90
laxative 91, *See also:* purgative
leaf nutrient concentrate 101, 103
leaf protein concentrate 101–106, 129
legumes 36, 38, 42, 45, 47, 95, 97, 99,
  143, 166, 170, 173, 177, 180–182,
  242, 246, 250, 255, 414, 415, 418
lemon 90, *See also: Citrus* spp.
lemon grass 90, *See also: Cymbopogon
  citratus*
lentil 167, 191, *See also: Lens culinaris*
licorice 90, *See also: Glycyrrhiza* spp.
lima bean *See:* bean, lima
linoleic acid 103, 212, 265
living fence 168
LNC *See:* leaf nutrient content
locoweed 369–370, *See also: Astragalus*
  spp.
LPC *See:* leaf protein concentrate
lubricant 37, 145, *See also:* oil, industrial
lumber *See:* timber
lumbia 405, *See also: Vigna unguiculata*
lupin 24, 36, 97–98, 191, 217, 221, 225,
  See also: Lupinus spp.
lupin pearl *See:* pearl lupin
lysine 105, 143, 167, 169, 191, 212, 222,
  235, 253, 268, 311

maize 24–25, 31, 61–62, 78, 126, 133,
  154, 158, 167, 169, 188, 222, 294,
  297, 312–315, 318, 348, 402, 406
maize flour 315
malaria 385
Malaysia 61, 156, 303, 406, *See also:*
  Asia, South East
Mali 419
mandarin 90, *See also: Citrus* spp.
market 13, 15, 17–20, 23–28, 29–32, 34,
  39–50 *passim*, 56–60, 71, 77–81,
  84–85, 89–91, 110–112, 114, 142,
  150–162 *passim*, 185–192, 197, 206,
  217–221, 225, 228, 233, 236, 246,
  255, 265, 277, 288, 313–320, 381
marketing 18–19, 21, 25, 29–31, 34,
  36–50 *passim*, 57, 62–63, 72, 82, 85,
  88, 110, 114, 148, 152, 162, 186, 220,
  257, 290, 313
Mauritania 272
Mauritius 180
meadowfoam 37, *See also: Limnanthes
  alba*
mealybug 368, *See also: Pseudococcus
  longispinus*
medicinal oil *See:* oil, medicinal
medicinal plants 85–86, 142, 189, 347,
  381, 404, 412, 416
medicinal products *See:* pharmaceuticals
medicinal properties 257
medicine 76, 85, 87–88, 106, 137,
  169–170, 189, 323, 328, 338, 346,
  380–382, 384, 403, 404, 411
Mediterranean 125, 249, 404, 415
menthol 78, 90
mesquite 290, *See also: Prosopis* spp.
methanol 78, 354
methyl benzoate 210–211
Metroxylon starch 406
Mexico 99, 131, 145, 147, 166, 168–169,
  174, 223, 239, 244, 288, 291, 300,
  302, 391, 412
Middle East 240, 247, 277, 289, *See
  also:* Near East
milk 32–33, 110, 112–113, 128–130,
  177, 181, 192
millet 26, 246–248, 255, 406, 412, *See
  also: Setaria italica*
  common 247, 249–250, *See also:
  Panicum miliceum*
  finger 94, 247–249, *See also: Eleusine
  coracana*

foxtail 247, 249, *See also: Setaria italica*
Italian *See:* millet, foxtail
pearl 348
Siberian *See:* millet, foxtail
miner's lettuce 190, *See also: Montia perfoliata*
mint 78, 90, 123, 418, *See also: Mentha* spp.
molasses 24, 177–183
mulberry 366, *See also: Morus nigra*
mung bean *See:* bean, mung
muscle relaxant 380
mushrooms 102, 403, *See also:* fungi, edible
mustard 104
myrrh 90, *See also: Commiphora* spp.

Napier grass 104, *See also: Pennisetum purpurem*
natural products 20, 23, 197, 354, 378–382, 385–387, 389
natural rubber *See:* rubber
navy bean *See:* bean, navy
Near East 404, 411, *See also:* Middle East
neem 73, 354, 356–357, *See also: Azadirachta indica*
nematocide 357
Nepal 141, 147
Netherlands 84, 381
nets 33, 71, 176, 327, 334, 360
New Caledonia 407
New Guinea *See:* Papua New Guinea
New World 327–330, 404, 406, 411
New Zealand 97, 101
niébé *See: Vigna unguiculata* subsp. *unguiculata*
Nigeria 124, 125–126, 130, 139, 143, 147, 267, 272
Nikolai Vavilov *See:* Vavilov, Nikolai
nitrogen 54, 58, 103–104, 133, 166, 170, 175, 250, 251, 254, 366, 393
North Africa *See:* Africa, North
North America *See:* America, North
nut 59, 69, 72, 147, 154, 157, 160, 162, 166, 210–211, 214, 275, 417
nut crops *See:* crops, nut
nutmeg 90, *See also:* spices

oat 141, 297, 300
Oceania 406

oil 37, 78–79, 89, 127, 166–168, 190–191, 308–317 *passim*, 323–329, 357, 379
balanites 275
bay 89, *See also: Pimenta* spp.
borage 219
coconut 26, 60, 130, 210, 323, 325
cotton seed 278
edible 110, 146, 168, 275, 412
essential 25, 76–82, 84, 89, 189–190
eucalyptus 80
evening primrose 216, 217
gamma linolenic 189, *See also:* gamma linolenic acid
geranium 79
industrial 37, 38, 168, *See also:* lubricant
jojoba 145
lauric 155–156, 312, *See also:* lauric acid
lupin 221
medicinal 80–82, *See also:* pharmaceuticals
palm 55–56, 60, 128, 131, 156, 265, 323, 325, *See also: Elaeis* spp.
pine 90
sandalwood 56, *See also: Santalum album*
santalum 211–213
semi-drying 146
soya bean 45
sperm whale 38, 145, 191, 412
sunflower 25
tangerine 90
vegetable 26, 56–57, 189, 191
vernonia 197–200, 202, 415
winged bean 112, 139
oil palm 25, 53–62 *passim*, 72, 110, 265, 312, 316, 325, 330, 346
oilseed rape 24, 151, 188, 191, *See also:* rape *and Brassica napus*
Old World 222, 323
oleic acid 212, 265, 312, 325
oleoresin 76, 90
olibanum *See:* frankincense
Oman 415, 419
onchocerciasis 385
onion 123, *See also: Allium cepa*
oral complaints 270
orange 90, 131, 158, 208, 209, 304, 326, *See also: Citrus* spp.
otitis 270

ovulatory preparations 89
oxalate crystals 311, 313

Pakistan 202–203, 254, 412, 415
palm 53–63, 66, 71, 74, 90, 154–158, 161, 303, 304, 308, 312, 323–331, 415
palm heart 60, 155, 161, 162, 308, 317–319, 324, 330
palm kernel 155–157
palm oil *See:* oil, palm
palmito *See:* palm heart
paper 38, 49, 76, 102, 107, 127, 130, 144, 251, 253
Papua New Guinea 56, 58, 60, 82, 139, 147
Paraguay 78, 81, 155
particle board 278
passion fruit 160
patchouli 90, *See also: Pogostemon cablin*
pea 20, 24, 47, 95, 188, 217, 242, 360–361, *See also: Pisium* spp.
pea aphid 370, *See also: Acyrthosiphon pisum*
peach palm *See:* pejibaye
peach tomato *See:* cocona
peanut 24, 157, 274, 275, 278, 300, *See also: Arachis hypogea*
pearl lupin *See: Lupinus mutabilis*
pearl millet *See:* millet, pearl
pejibaye 58, 60, 61, 154, 304–320, *See also: Bactris gasipaes*
pejibaye flour 314–315
pepper 90, 129, 257
peppermint 190, *See also: Mentha piperita*
Peru 78, 90, 143, 227–228, 257, 266, 280, 317, 320, 323
pest control 352, 362, 365, 382
pesticides 21, 33, 61, 76, 106, 352, 362, 365, 367
pet food 295, *See also:* bird-seed
petitgrain 81
pharmaceuticals 37–38, 84–87, 107, 131, 169, 189, 219, 252, 277, 295, 301, 357, 380–382, 387, 404, *See also:* medicinal plants and sythetic drugs
Phaseolus bean *See:* bean, *Phaseolus*
pheromones 358, 361, 362
Philippines 26, 109, 112, 139, 143, 144, 147, 403, 406
phytohaemagglutinins 103

pigeonpea 36, 44, 47–49, 358–360, *See also: Cajanus cajan*
pine 76–77, 79–80, 82, 98, *See also: Pinus* spp.
pine oil *See:* oil, pine
pineapple 90, 160
pollination 60, 259, 268, 304, 348
porridges 235, 248, 405
Portugal 80
potato 24–26, 34, 104, 123–125, 127, 131, 154, 169, 187, 222, 257, 258, 313, 402, *See also: Solanum tuberosum*
potato blight 26, *See also: Phytophthora* spp.
poultices 328
printing ink 76
printing paper 253
printing paste 252
processing 14, 18, 21, 34, 39, 49, 59, 62, 66, 71, 77–78, 87–90, 98, 103, 110, 112, 124–134, 148, 152, 175, 176, 187, 202, 217, 223–226, 255, 257, 275–276, 288–290, 295, 297–298, 300, 317–318, 368, 406
propagation 62, 71, 73, 82, 152, 208, 262, 268, 269, 419
proso millet *See:* millet, common
Prosopis flour 298, 300
Prosopis galactomannan 295, 296
prostaglandin hormone 216
protein 36, 47, 69, 97–99, 101–107, 110, 126–128, 131–132, 139–141, 156–158, 161, 163, 166–172, 174, 179–182, 186–188, 191, 210–211, 217, 221, 222, 233, 235, 240–242, 246–253, 274, 285, 290–300, 311–312, 314, 318, 336, 357, 378, 405
pruning 54, 61, 150, 290, 393
psyllium 91, *See also: Plantago arenaria*
pulp 38, 49–50, 80, 130, 147, 156, 159–160, 253, 258, 267–270, 310, 326
pulping 49, 79
pulse 141, 404–405, 414, *See also:* grain legume
pupunhadine 311
purgative 147, 272, *See also:* laxative

quandong 208, 209, *See also: Santalum acuminatum*
Queensland 38, 45, 48–49, 125, 407
Queensland arrowroot 124, *See also: Canna edulis*

quinoa 191, 222–233, *See also: Chenopodium quinoa*

raddish *See: Rhaphanus sativas*
ragi *See:* millet, finger
rape 104, 187, 221, 225, *See also:* oilseed rape *and Brassica napus*
rapeseed 187, *See also:* oilseed rape
raspberry 190, *See also: Rubus ideaus*
rattan 56, 62, 303, 327, 415, *See also: Calamus* spp.
reproductive disorders 91
resin 138, 145, 254, 329, 379, 391–393, 399
respiratory 89, 91
rhubarb 91, *See also: Rheum* spp.
rice 24, 25, 54, 60, 112, 127, 141, 174, 180, 225, 250–251, 300, 402, 406, 415
rice bean *See:* bean, rice
risga 123, 124, 127, *See also: Plectranthus esculentus*
root crops *See:* crops, root
rope 327
rose 90, 223, 268
rosemary 90
Royal Botanic Gardens, Kew 94, 109, 346, 350, 413
rubber 37–38, 54, 58–59, 61–62, 110, 145, 150, 187, 303, 346, 391, 393, 398–400, 412
rubber products 76
rumba 139, *See also: Citrullus colocynthis*
Russell lupin 97, *See also: Lupinus* spp.
Rwanda 125
rye 242

Sabah 60
saffron 77, 79
sago 55, 58–59, 61, 406, *See also: Metroxylon* spp.
sago starch 56, 406
salsify 190, *See also: Tragopogon porrifolius*
salt marsh 99, *See also:* brackish water *and* tidal swamps
sandalwood oil *See:* oil, sandalwood
santalbic acid 212, 213
saponin 103, 106, 147, 223–228, 232, 274–277, 379
sassafras 89
sea-buckthorn 190, *See also: Hippophea rhamnoides*

sea water 243, 333–337, *See also:* brackish water *and* tidal swamps
secondary metabolites 353–354, 358, 381–383, 389
secondary products 347
seed bank 96, 414, 418–419, *See also:* germplasm collection
seed crops *See:* crops, seed
semi-drying oil *See:* oil, semi-drying
semiochemicals 353, 358, 361
SEPASAL 350, 416
sesbania 250–254, *See also: Sesbania bispinosa*
sesbania gum 252–253
sex stimulants *See:* aphrodisiacs
shade 68, 73, 150, 161, 170, 171, 411
shelter 323, 403, 411
Siberian millet *See:* millet, foxtail
Sikkim 141
Singapore 84
soap 127, 146, 275
soil conditioner 252
soil conservation 93–100, *See also:* erosion
Solomons 58, 82
Somalia 200, 272, 385, 415, 419
sorghum 26, 38, 169, 177, 223, 237, 402, 412, 415
soup 128–129, 131–132, 136, 235, 318
South Africa *See:* Africa, South
South America *See:* America, South
South Australia 209, 211
South East Asia *See:* Asia, South East
Southern Africa *See:* Africa, Southern
Soviet Union *See:* USSR
soya bean *See:* bean, soya
soya bean oil *See:* oil, soya bean
Spain 79, 101, 127, 233
spearmint *See also: Mentha* spp.
sperm whale oil *See:* oil, sperm whale
spermatogenics 91
spices 56, 69, 76–82, 131, 137, *See also:* clove, condiments *and* nutmeg
Sri Lanka 88, 102, 112, 139, 147
staple food crops *See:* crops, staple food
staples *See:* crop staples
star anis *See: Illicium verum*
starchy staple 162
stearolic acid 212
steroids 87, 91, 275, 277, 353, 379, 385
stinging nettle 104, *See also: Urtica dioca*

Stokes aster 37, *See also: Stokesia* spp.
stomach ailments 272
strawberries 25
Sudan 200, 272
sugar 24–25, 50, 55, 102, 129, 130–132, 176–182, 187–188, 211, 250, 290, 292–295, 359, 366, 369, 371–372, *See also: Saccharum* spp.
sugar beet 187, 230, *See also: Beta vulgaris*
sugarcane 49, 139, 144, 176–183, 295, 346
sunflower 24, 36, 104, 187–188, 191, 225, *See also: Helianthus annus*
surface coatings 76
swainsonine 367–373
Sweden 107, 233
sweet potato 123–126, 127, 131–133, 139, 144, *See also: Ipomoea batatas*
sweet potato starch 132
Switzerland 84
synthetic drugs 86–87, 379, 380, *See also:* pharmaceuticals

Tahiti 413
Taiwan 141, 147, 202
tangerine oil *See:* oil, tangerine
tannia 124–125, 132, *See also: Xanthosoma* spp.
tannin 103, 105, 138, 180, 223, 357, 370
Tanzania 125, 147, 200, 202
taro 124, 132–133, *See also: Colocasia esculenta*
tea 29, 32–33, 54, 59, 62, 69, 328
t'ef 235–238, *See also: Ergrostis tef*
textile sizing 252
textiles 126, 130, 132, 188
texturised vegetable protein (TVP) 251
Thailand 24, 42, 44, 49–50, 58, 84, 139, 147
thatch 323
therapeutic drops 328
three-leaved yam *See:* yam, three-leaved
tidal swamps 55, *See also:* brackish water *and* sea water
tigernut 123, *See also: Cyperus esculentus*
timber 62, 80, 138, 143, 166, 270, 301, 357, 417–418
tobacco 32, 104, 113, 257
tolu balsam 90, *See also: Myroxylon balsamum* vars. *balsamum* and *pereirae*

tomato 161, 257
tool handles 169
tools 22, 87, 371, 403, 408
toxins 97, 415–416
traditional medicine 84–85, 131, 272, 381
triglycerides 156, 197, 213, 313
triticale 242, 244
tropical Africa *See:* Africa, tropical
Tropical East Africa *See:* Africa, Tropical East
trypsin inhibitor 311, 313, 314, 318
tuber crops *See:* crops, tuber
tubers 82, 110–112, 123–134, 136–137, 144, 403, 407
turmeric 82, 124–125, *See also: Curcuma longa*
turnip 187
turpentine 76–77, 79, 82, 89
turpentine oil *See:* oil, turpentine

Uganda 125, 197, 201, 247–248, 267
UK 14, 26–27, 84, 147, 185–188, 190, 192, 217, 218, 223, 225, 226, 233, 242, 244, 346, 350, 381, 411
USA 14–15, 26, 37, 49, 84, 93, 101, 114, 132, 145–147, 185–187, 190, 197, 205, 225, 249–250, 291, 301, 328, 333, 334, 350, 379–381, 389, 397, 399–400
USSR 101, 146, 147, 404
utensils 403

valerian 90, *See also: Valeriana* spp.
vanilla 60, 90, 346, *See also: Vanilla planiflora*
Vavilov, Nickolai 346
vegetable crops *See:* crops, vegetable
vegetable oil *See:* oil, vegetable
vermifuge 272
vetiver 90, *See also: Vetiveria zizaniodes*
Victoria 209
Vietnam 112
vitamins 69, 76, 87, 132, 161, 190, 257, 379
 A 111, 153, 315, 326, *See also:* carotenoids
 B 311
 C 131, 159, 209, 274, 311

water yam *See:* yam, water
wax 138, 146, 324
 liquid 145, 412
weapons 403

West Africa *See:* Africa, West
West Germany *See:* Germany, West
West Indies 159, 163, 250, 407, 413, *See also:* Caribbean
Western Australia 49, 209
wheat 25, 36, 38, 141, 167, 174, 185, 186, 192, 217, 219, 222–223, 239–244, 248–250, 296, 297, 300, 315–316, 318, 348, 391, 399, 402, 405, 412, *See also: Triticum* spp.
wheat flour 132, 192, 294, 294–295, 297, 299, 315
white Guinea yam *See:* yam, white Guinea
white lupin 191, *See also: Lupinus albus*
white pepper 77, *See also: Piper nigrum*
whole plant drugs 379
wild species 240–241, 247, 347–348, 403, 407, 414, 416, 419
winged bean *See:* bean, winged

yam 123–133, 154, 407, *See also: Dioscorea* spp.
  aerial 123–124, *See also: Dioscorea bulbifera*
  bean *See:* bean, yam

bulbil-bearing *See: Dioscorea bulbifera*
elephant 124, *See also: Amorphophallus* spp.
false 124, *See also: Icacina senegalensis*
Guinea 124
three-leaved 123–124, *See also: Dioscorea dumetorum*
water 124–125, *See also: Dioscorea alata*
white Guinea 123–124, *See also: Dioscorea cayenensis* subsp. *rotunda*
yellow 123, *See also: Dioscorea cayenensis*
yamogenin 274
yard-long bean *See:* bean, yard-long
yellow yam *See:* yam, yellow
yuehchukene 382–383, 385–386

Zaire 125, 266–267
Zambesiaca 415
Zambia 29–35, 147, 202, 205, 247, 248, 416
Zimbabwe 197, 200, 202–205

Taxonomic Index

Accepted names in roman, synonyms in *italics*

HIGHER PLANTS

Abelmoschus moschatus Medic. 90
Abutilon indicum (L.) Sweet 137
Acacia
    albida Del. 146
    farnesiana (L.) Willd. 89
    jaquemontii Benth. 138
    leucophloea (Roxb.) Willd. 138
    nilotica (L.) Willd. ex Del. 138
    pennatula (L.) Willd. 166, 170
    senegal (L.) Willd. 138
    spp. 99, 415
Acokanthera spp. 416
Aconitum heterophyllum Wall. 137
Acorus calamus L. 89
Acrocomia
    aculeata (Jacq.) Mart. 162
    sclerocarpa Mart. 328
    totai Mart. 326, 330
    spp. 154–155, 325
Aegle marmelos (L.) Correa 91, 136
Agave spp. 91
Agropyron junceum (L.) P. Beauv. 242
Aiphanes caryotifolia (Kunth) H. Wendl. 326
Albizia
    amara (Roxb.) Boivin 139, 143
    lebbek (L.) Benth. 166, 170–172
    procera (Roxb.). Benth. 139, 143
Allium
    cepa L. 123
    leptophyllum Schur 137
    tuberosum Rottl. ex Spreng. 136
    spp. 137, 414
Alocasia cucullata (Lour.) G. Don 136
Aloe
    barbadensis Mill. 137
    vera (L.) Burm. f. 418
    spp. 91, 418
Alpinia spp. 90, 91
Alternanthera sessilis (L.) R. Br. ex Roth 137
AMARANTHACEAE 104, 106
Amaranthus
    caudatus L. 141
    cruentus L. 141
    edulis Speg. ex Miq. 141
    gracilis Desf. ex Poir. a 137
    hybridus L. 137
    hypochondriacus L. 141
    leucocarpus S. Watson 191
    lividus L. 136
    polygonoides L. 136
    spinosus L. 137
    viridis L. 137
    spp. 104–105, 137, 416
Amomum aromaticum Roxb. 137
Amomum xanthioides Wall. 137
Amorphophallus spp. 124
Anacardium occidentale L. 73, 313
Anethum sowa Roxb. ex Flem. 137
Angylocalyx spp. 366
Annona muricata L. 159, 162
Annona squamosa L. 91
Anogeisus pendula Edgew. 138
APOCYANACEAE 378, 416
APONOGETONACEAE 123
Aquilaria malaccensis Lam. 89
ARACEAE 123
Arachis hypogea L. 265
Areca triandra Roxb. 137
Aronia melanocarpa (Michx.) Elliott 190
Artemisia spp. 89
Artocarpus
    altilis (Parkinson) Fosberg 413
    lakoocha Roxb. 136
Arundo donax L. 105
Asparagus
    racemosus Willd. 88, 91, 137
    sarmentosus L. 136
Astragalus
    lentiginosus Dougl. ex. Hook. 370
    oxyphysus A. Gray 369
    spp. 366
Astrocaryum
    aculeatum G. F. W. Meyer 159, 162
    ayri Mart. 328
    jauari Mart. 161
    murumuru Mart. 155, 324, 325
    tucuma Mart. 324–327
    vulgare Mart. 155, 156, 327
    spp. 155, 324–327
Atriplex hortensis L. 105
Atriplex spp. 104–105
Atropa belladona L. 90
Attalea funifera Spreng. 324

Attalea spp. 327
Aucoumea klaineana Pierre 266
Azadirachta indica Adr. Juss. 73, 91,
  138, 354, 357
Bactris
  gasipaes Kunth 58, 154–162 *passim*,
    304–320, 325–326, 330
  insignis (Mart.) Baillon 328
  minor Jacq. 328
  spp. 325
BALANITACEAE 272
Balanites
  aegyptiaca (L.) Del. 272–278
  pedicellaris Mildbr. & Schlecht. 272
  rotundifolia (Van Tiegh.) Blatter 272
  wilsoniana Dawe & Sprague 272
Bambusa
  arundinacea Retz. 144, 253, 254
  balcooa Roxb. 144
  multiplex (Lour.) Raeusch. 144
  nutans Wall. 144
  pallida Munro 144
  spinosa Roxb. 136
  tulda Roxb. 136, 138, 144
  vulgaris Wendl. ex. Nees 136
BASSELLACEAE 123
Berberis aristata DC. 91
Bertholletia excelsa Humb. & Bonpl.
  155, 157, 160
Beta vulgaris L. 125
Beta spp. 104
Bixa orellana L. 155, 162
Boerhavia diffusa L. 91, 137
Borago officinalis L. 189
Borassus flabellifer L. 136, 329
Boswellia serrata Roxb. ex Colebr. 138
Boswellia spp. 90
Brassica
  hirta Moench 105
  napus L. 105
  oleracea L. 191
  spp. 104–105
Breynia patens (Roxb.) Benth. 88
Breynia retusa (Dennst.) Alston 88
Brosimum alicastrum Swartz 166,
  168–170
Bryonopsis laciniosa sensu auct. non (L.)
  Naud 137
BURSERACEAE 266, 385
Butea monosperma (Lam.) Kuntze 137,
  138
Butia spp. 324–325

Cajanus cajan (L.) Millsp. 44, 358, 359
Calamus
  caesius Bl. 303
  floribundus Griff. 329
  rotang L. 329
  trachycoleus Becc. 303
Calathea allouia (Aubl.) Lindl. 154
Calligonum polygonoides L. 138
Calotropis gigantea Dryand 138
Calotropis procera (Ait.) Ait. f. 138
Camelina sativa (L.) Cranz 95, 191
Cananga odorata (Lam.) Hook. f. &
  Thomson 90
Canarium
  edule Hook. 266
  mubafo Ficalho 266
  saphu Engl. 266
  schweinfurthii Engl. 266
Canavalia
  polystachya (Forssk.) Schweinf. 136
  virosa (Roxb.) Wight & Arn. 136
  spp. 104
Canna edulis Ker-Gawl. 124–125
CANNACEAE 123
Capparis decidua (Forssk.) Edgew. 138
Cardiospermum halicacabum L. 91
Carica papaya L. 90
Carissa congesta Wight 136
Carum carvi L. 90
Caryocar
  cuneatum Wittmark 156
  nuciferum L. 160
  villosum (Aubl.) Pers. 156
  spp. 162
Caryodendron orinocense Karst. 160
Caryota urens L. 137
Cassia auriculata L. 138
Cassia sturtii R. Br. 139, 143
Castanea sativa Miller 313
Castanospermum australe A. Cunn. &
  Fraser 366
Casuarina
  cristata Miq. 144
  cunninghamiana Miq. 144
  equisetifolia L. 144
  spp. 147
Catharanthus roseus (L.) G. Don 90, 378,
  379
Celosia spp. 104
Celtis australis L. 411
CENTROSPERMAE 104

Cephaelis impecacuanha (Brot.) Tusscac 90
Ceratonia oreothauma Hillcoat Lewis & Verdc. 415, 419
Ceratonia siliqua L. 72, 415, 419
Ceropegia bulbosa Roxb. 137
CHENOPODIACEAE 104, 106–107
Chenopodium
    album L. 137–138, 142, 223–225
    berlandieri Moq. subsp. nuttalliae (Saff.) H. Wilson & Heiser 223
    pallidicaule Aellen 223
    quinoa Willd. 105, 142, 191, 222–233
    spp. 104, 105
Cicer arietinum L. 38, 191, 358, 361
Cicer songarium Stephen ex DC. 137
Cinchona spp. 90
Cinnamomum camphora (L.) T. Nees & Eberm. 89
Citrullus colocynthis (L.) Schrader 137–139
Citrullus lanatus (Thunb.) Mansf. 137
Citrus aurantium (Christm.) Swingle subsp. bergamia (Risso & Poit). Wight & Arn. 90
Citrus spp. 73, 136
Claytonia perfoliata Donn ex Willd. 190
Clitoria spp. 104
Coccinia grandis (L.) Voigt 137
Cocculus hirsutus (L.) Diels 137
Cocos nucifera L. 265, 316, 323, 329
Coleus
    dysentericus Bak. 123
    esculentus (N. E. Br.) G. Taylor 123
    forskohlii (Poir.) Briquet 136
Colocasia esculenta (L.) Schott 125, 127, 132, 136
Colocasia spp. 123, 132
Colophospermum mopane (J. Kirk ex Benth.) J. Kirk ex Léonard 139, 143
Commiphora
    mukul (Hook. ex Stocks) Engl. 385
    wightii (Arn.) Bhandari 138
    spp. 385
COMPOSITAE 123, 415
    subtribe SENECIONAE 415
Copernicia cerifera (Arruda) Mart. 324, 328
Corchorus capsularis L. 136
Cordeauxia edulis Hemsley 417, 419
Cordia
    dichotoma Forst. f. 138

gharaf (Forssk.) Ehrenb. ex Aschers. 137–138
    sinensis Lam. 137, 138
Couepia
    bracteosa Benth. 159
    longipendula Pilger 160, 162
    spp. 160
Couma utilis (Mart.) Muell. Arg. 159, 161–162
Couma spp. 155, 161
Crambe
    abyssinica R. E. Fries 37
    cordifolia Steven 136
    hispanica L. 37
Crateva nurvale Ham. 91
Crotolaria burhia Ham. ex Benth. 138
CRUCIFERAE 104, 106
Cucumis callosus (Rottl.) Cogn. 137
CUCURBITACEAE 104, 415
Cuphea spp. 147, 156
Curcuma
    angustifolia Roxb. 136
    longa L. 124, 125
    spp. 90
Cymbopogon
    citratus (DC.) Stapf 90
    martinii (Roxb.) Wats. 90
    spp. 90
CYPERACEAE 123
Cyperus
    bulbosus Vahl 137
    esculentus L. 123, 125
    rotundus L. 90
    scariosus R. Br. 90
    spp. 91, 127
Dacryodes
    Sect. Archidacryodes 266
    Sect. Curtisina 266
    Sect. Pachylobus 266
    buettneri (Engl.) H. J. Lam 266
    edulis (G. Don) H. J. Lam 265–271
        var. edulis 266
        var. parvicarpa Okafor 266
    igaganga Aubrév. & Pellegr 266
    macrophylla (Oliv.) H. J. Lam 266
    spp. 266, 270
Daemonorops draco (Willd.) Mart. 329
Dalbergia sissoo Roxb. 138
Datura spp. 90
Daucus carota L. 123, 125
Dendrocalamus
    asper (Schultes) Backer ex Heyne. 136

giganteous Munro 144
sikkimensis Gamble ex Oliv. 144
Derris spp. 366
Desmodium spp. 104
Desmoncus rudentum Mart. 328
Desmoncus spp. 327
Dichrostachys
glomerata (Forssk.) Chiov. 139, 143
nutans (Pers.) Benth. 139
Digitalis lanata Ehrh. 379
Digitalis spp. 90
Digitaria cruciata (Nees & Steud.) A.
Camus var. esculenta Bor 136
Dioscorea
alata L. 124–125, 144
bulbifera L. 123–124
cayenensis Lam. 123–124
subsp. cayenensis 407
subsp. rotundata (Poir.) Miège 123,
144, 407
dumetorum (Kunth) Pax 123–124
elephantipes Spreng. 131
esculenta (Lour.) Burkill 124–125,
144
opposita Thunb. 131
praehensilis Benth. 124
rotundata Poir. 124, 407
sylvatica Kunth 131
transversa R. Br. 407
trifida L. 407
spp. 90, 123–124, 127, 144, 154, 277,
403, 407
DIOSCOREACEAE 123
Diplachne fusca (L.) P. Beauv. ex
Roemer & Schultes 412
DIPTEROCARPACEAE 418
Dolichos uniflorus Lam. 136
Duboisia spp. 90
Echinochloa
colona (L.) Link 136, 137
crusgalli (L.) P. Beauv. 137
Eichhornia crassipes Solms 105
Elaeis
guineensis Jacq. 55, 72, 155, 156, 265,
312, 316, 323, 329
oleifera (Kunth) Cortés 156, 316, 325
spp. 325
Elaeocarpus floribundus Blume 136
Eleocharis tuberosa Schutt. 136
Elettaria cardamomum Mill. 81, 90
Eleusine
africana Kennedy-O'Byrne 247

compressa (Forssk.) Asch. &
Schweinf. 137
coracana (L.) Gaertn. 94, 246,
247–248
indica (L.) Gaertn. 94, 100, 247
subsp. africana (Kennedy-
O'Byrne) S. M. Phillips 247
Elymus farctus (Viv.) Runemark ex
Meldris 242, 243
Emblica officinalis (L.) Gaertn. 136
Emila coccinea (Sims) G. Don 136
Emila sagittata DC. 136
Enterolobium cyclocarpum (Jacq.)
Griseb. 171–172
Eragrostis tef (Zucc.) Trotter 235–238
Erythrina
americana 171, 172
glauca Willd. 180
poeppigiana (Walp.) C. F. Cook 181
Eucalyptus
globulus Labill 80
polybrachtea R. T. Baker 80
spp. 80
Eugenia
stipitata McVaugh 159–160, 162
uniflora L. 160, 162
Euphorbia
antisyphilitica Zucc. 146
caducifolia Haines 137, 146
lathyrus L. 146
neriifolia L. 146
nivulia Ham. 146
thompsoniana Holmboe 138
tirucalli L. 146
trigona Haw. 146
spp. 139, 146, 147
EUPHORBIACEAE 123, 166, 367, 418
Euryale ferox Salisb. 136
Euterpe
edulis Mart. 137, 161
oleracea Mart. 155–156, 158, 161,
317, 324–330
spp. 324, 326
FABACEA *See:* LEGUMINOSAE
Fagopyrum
esculentum Moench 137, 142, 191
tataricum (L.) Gaertn. 137, 138, 142
Faidherbia albida (Del.) A. Chev. 146
Ferula spp. 89
Flemingia procumbens Roxb. 136, 138
Galanga spp. 91

Taxonomic Index

Garcinia
   pendunculata Roxb. 136
   tinctoria (DC) W. F. Wright 136
Gliricidia sepium (Jacq.) Walp. 166–171,
   172, 180, 181
Glycyrrhiza spp. 90
GRAMINEAE or POACEAE 414, 415
   subfamily CHLORIDOIDEAE
      tribe ERAGROSTIDEAE, subtribe
      ELEUSININAE 235
Grewia
   asiatica L. 136
   tenax (Forssk.) Fiori 137
Haloxylon salicornicum (Moq.) Bunge
   137
Hancornia speciosa Gomes 160, 162
Hardwickia binata Roxb. 139, 143
Helianthus annuus L. 104–105, 191
Helianthus tuberosus L. 104, 123, 125
Hevea brasiliensis (Willd. ex A. Juss)
   Meull. Arg. 187
Hevea spp. 391, 418
Hibiscus cannabinus L. 38
Hibiscus sabdariffa L. 58
Hippophae rhamnoides L. 190
Holarrhenia antidysenterica (Roth) A.
   DC. 91
Houttuynia cordata Thunb. 136
Hygrophyla
   auriculata (Schumach.) Heine 91
   spinosa T. Anders. 91
Hygroryza aristata (Retz.) Nees ex Wight
   & Arn. 137
Hyoscyamus niger L. 91
Hyoscyamus spp. 90
Hyparrhenia spp. 415
Hyphaene thebaica (L.) Mart. 329
Icacina senegalensis A. Juss. 124
Illicium verum Hook. f. 89
Indigofera
   cordifolia Heyne ex Roth 138
   spp. 414
Inga
   edulis Mart. 159
   paterno Harms 157
   spp. 157
Ipomoea
   aquatica Forssk. 105
   batatas (L.) Lam. 123, 125, 127
Ixiolaena brevicompta F. Muell. 213
Jatropha curcas L. 139, 166–168

Jessenia
   bataua (Mart.) Burret 156, 325, 328
   spp. 158, 325
Kaempferia galanga L. 137
Kerstingiella geocarpum Harms 414
Kochia indica Wight 138
Kochia spp. 104
LABIATAE or LAMIACEAE 123, 127
Lactuca indica L. 136
LAMIACEAE *See:* LABIATAE
Landolphia spp. 412
Lecythis usitata Miers 160
LEGUMINOSAE 106, 250, 280, 366,
   414, 416
   subfam. PAPILIONOIDEAE 123
Lens culinaris Medic. 191
Leopoldinia piassaba Wallace 327
Leptadenia
   pyrotechnica (Forssk.) Decne 138
   reticulata (Retz.) Wight & Arn. 88,
      137
Leptochloa fusca (L.) Kunth. 412, 415,
   417
Leucaena
   leucocephala (Lam.) de Wit 139,
      170–172, 181
   pulverulenta (Schlechtend.) Benth.
      143
Limnanthes alba Hartweg ex Benth. 37
Limonia acidissima L. 136
Linum usitatissimum L. 188, 412
Lonchocarpus spp. 366, 369
Lupinus
   albus L. 191
   mutabilis Sweet 191
   spp. 104
Lychnis indica Roxb. 138
Macrotyloma
   axillare (E. Mey.) Verdc. 414
   geocarpum (Harms) Maréchal &
      Baudet 414
   uniflorum (Lam.) Verdc. 414
   spp. 414
Madhuca
   indica Gmel. 137
   longifolia (Koenig) MacBride 137
Malpighia
   coccigera L. 136
Mangifera indica L. 70, 73
Manicaria
   saccifera Gaertn. 56, 325, 328
   spp. 325

440

Manihot esculenta Crantz 123–125, 127, 304
Maranta spp. 408
MARANTACEAE 123
Mauritia flexuosa L. f. 56, 154, 156, 323–324, 326–327
Mauritia spp. 325
Maximiliana maripa (Correa de Serra) Drude 328
Medicago sativa L. 104–107, 283
MELIACEAE 357
Mentha
  arvensis L. 78–79
  piperita L. 190
  spp. 90, 418
Merrillia spp. 385
Mesembryanthemum crystallinum L. 190
Metroxylon spp. 406
Millettia spp. 366
Miscanthus sinensis Anderss. 107
Moghania vestita (Benth. ex Baker) Kuntze 136
Monochoria hastata (L.) Solms-Lamb. 137
Montia perfoliata (Donn ex Willd.) Howell 190
MORACEAE 166, 366
Morinda citrifolia L. 138
Moringa oleifera L. 136–138
Moringa spp. 416
Morus alba L. 136
Morus nigra L. 366
Mucuna
  capitata Wight & Arn. 136
  cochin-chinensis (Lour.) A. Chev. 136
  pruriens (L.) DC. 91
    var. utilis (Wall. ex Wight) Bak. ex Burckll. 136
  utilis Wall. ex Wight 136
Murraya paniculata (L.) Jack 382, 383
Murraya spp. 385
Myrciaria dubia (Kunth) McVaugh 159
Myroxylon balsamum (L.) Harms
  var. pereirae (Royle) Baill. 90
Narcissus spp. 418
Nasturtium
  indicum (L.) DC. var. *apetala* 136
  officinale R. Br. 105
Nelumbo nucifera Gaertn. 136
Nymphea spp. 105
Nypa fruticans Warmb. 55
Ocimum spp. 91

Oenocarpus bacaba Mart. 156, 326–327
Oenocarpus spp. 325
Oenothera biennis L. 189
Oenothera spp. 150, 216–221
OLACACEAE 211
Orbignya phalerata Mart. 324–327, 330
Orbignya spp. 154–155, 325, 327
Osmanthus fragrans Lour. 137
OXALIDACEAE 123
Pachira aquatica Aubl. 157, 160
Pachylobus
  edulis G. Don 266
    var. *mubafo* (Ficalho) Engl. 266
  saphu Engl. 266
Pachyrhizus tuberosus (Lam.) Spreng. 154
PALMAE 414, 415, 416
Panax ginseng C. A. Meyer 90
Panicum
  maximum Jacq. 172
  miliaceum L. 246–249
  sumatrense Roth ex. Roem. & Schult. 136
  spp. 255
Parkia
  javanica (Lam.) Merr. 136
  platycephala Benth. 157
Parthenium
  argentatum A. Gray 37, 391–400
  incanum Kunth 145
  stramonium Greene 145
  tomentosum DC.
    var. stramonium 145
    var. tomentosum 145
Paspalum scrobiculatum L. 136
Paullinia cupana Kunth 162
Pelargonium spp. 79
Pennisetum purpureum Schumach. 104
Pentaclethra macroloba (Willd.) Kuntze 157
Pereskia grandifolia Haw. 136, 161
Phaseolus
  acutifolius 99
  lunatus L. 157
  vulgaris L. 99
  spp. 104, 222, 404
Phoenix dactylifera L. 70, 323, 329
Phoenix sylvestris Roxb. 136, 329
Phragmites australis (Cav.) Trin. ex Steud. 105
Phragmites communis Trin. 105

Phyllanthus
  distichus (L.) Muell. Arg. 136
  emblica L. 91, 136
Pimenta spp. 89
Pimpinella anisum L. 90
Pinus radiata D. Don. 70
Piper
  cubeba L. 90
  longum L. 137
  nigrum L. 77
Piscidia
  communis (Blake) Harms 166
  piscipula (L.) Sarg. 166, 170–172
Pistia stratiotes L. 105
Pisum spp. 104
Plantago arenaria L. 91
Platonia esculenta (Arruda da Camara) Rickett & Stafleu 159
Plectranthus esculentus N. E. Br. 123, 127
Pogostemon cablin (Blanco) Benth. 90
Polygonum viviparum L. 137
Polygonum spp. 105
Poraqueiba sericea Tulasne 154
Portulaca oleracea L. 137
Portulaca tuberosa Roxb. 137
Pourouma cecropiifolia Mart. 159
Pouteria
  caimito (Ruiz & Pav.) Radlik. 159
  spp. 154
Prosopis
  chilensis (Molina) Stuntz 280–287, 291, 295, 296
  cineraria (L.) Druce 138
  glandulosa Torrey 291, 296, 298–300
  juliflora (Sw.) DC. 138, 171, 291
  pallida Kunth 291
  pubescens Benth. 285
  tamarugo F. Philippi. 285, 291, 296
  velutina Wooten 285, 291, 293–297
  spp. 285, 288–302
Psidium angulatum DC. 159, 162
Psidium guajava L. 73
Psophocarpus tetragonolobus (L.) DC. 108, 404, 414, 416
Psophocarpus spp. 104, 414
Pueraria lobata (Willd.) Owi 93
Quararibea cordata (Humb. & Bonpl.) Vischer 159
Raphanus sativus L. 124, 125
Raphanus spp. 104
Rauvolfia spp. 90, 91, 378

Rhamnus purshiana DC. 91
Rheum spp. 91
Rhodomyrtus tomentosa (Ait.) Hassk. 136
Rorripa indica (L.) Hiern 136
Rottboellia cochinchinensis (Lour.) W. D. Clayton 94
Rottboellia exaltata L. f. 94
Rubus
  chamaemorus L. 190
  ellipticus Smith 136
  fruticosus L. 190
  idaeus L. 190
  lasiocarpus Smith 136
RUTACEAE 382
Sabal serrulata (Michaux) Nutt. ex Schulter 329
Saccharum
  bengalensis Retz. 138
  munja Roxb. 138
Salicornia brachiata Roxb. 137
Salix spp. 107
Salvadora
  oleoides Decne 137–138
  persica L. 137
  spp. 138
SANTALACEAE 208, 211, 212
Santalum
  acuminatum (R. Br.) DC. 208–214
  album L. 208, 211
  lanceolatum R. Br. 210
Santiria trimera (Oliv.) Aubrév. 266
Saxifraga callosa Sm. subsp. callosa 91
Saxifraga ligulata Bellardi 91
Scheelea
  martiana Burret 155, 161
  princeps Karst 328
  spp. 325, 327
Scorzonera mollis M. Bieb. 137
SCROPHULARIACEAE 379
Sedum tibeticum Hook. f. & Thoms. 137
Senna alexandrina P. Mill 91
Serenoa repens (Bartram) Small 329
Serenoa serrulata 329
Sesbania
  aculeata (Willd.) Poir. 250
  bispinosa (Jacq.) W. Wight 246, 250, 254
  cannabina (Retz.) Pers. 250
  spp. 38, 252–255
Setaria
  italica (L.) P. Beauv. 246–249, 406

verticillata (L.) P. Beauv. 137
spp. 406
Sida
carpinifolia L. f. 138
cordifolia L. 91, 138
rhombifolia L. 138
Silene griffithii Boiss. 138
Silene indica Roxb. ex. Otth 138
Simarouba glauca DC. 139
Simmondsia chinensis (Link) C.
Schneider 37, 412
SOLANACEAE 104, 106, 123, 257
Solanum
khasianum 91
myricanthum Dunal 91
sessiliflorum Dunal 257–263
topiro Humb. & Bonpl. 159, 161
tuberosum L. 105, 123–125
spp. 91, 104
Solenostemon rotundifolius (Poir.)
Morton 123, 124, 127
Soreindeia deliciosa A. Chev. ex Hutch.
& Dalz. 266
Sphenostylis stenocarpa (Hochst. ex A.
Rich.) Harms 123, 124, 127
Spinacia oleracea L. 106
Spinacia spp. 104
Spondias lutea L. 159, 162
Stokesia spp. 37
Stylochiton
lancifolius Kotschy & Peyr. 124
warneckei Engl. 124
Styrax spp. 89
Swainsona spp. 366
Swartzia spp. 157
Syagrus
coronata (Mart.) Becc. 324
picrophylla Barbosa Rodrigues 328
spp. 324–325
Syzygium cumini (L.) Skeels 136
Talinium triangulare (Jacq.) Willd. 161
Talisia esculenta (St. Hil.) Radlk. 159
Tecomella undulata (Sm.) Seem. 138
Theobroma grandiflorum (Willd. ex
Spreng.) Schum. 156, 159, 162
Tithonia
rotundifolia (Miller) S. F. Blake 104
tagetiflora Desv. 104
Trachyspermum ammi (L.) Sprague 90
Tragopogon porrifolius L. 190
Trianthema portulacastrum L. 137
Trifolium pratense L. 105

Trifolium spp. 104, 106
Triticum
dicoccoides (Koern. ex Asch. &
Graebn.) Aaronsohn 240–241, 244
monococcum L. 405
Typha latifolia L. 105
UMBELLIFERAE 123
Urtica dioica L. 104, 107
Vaccinium corymbosum L. 190
Valeriana spp. 91
Vanilla fragrans (Salisb.) Ames 168
Vanilla planiflora Andr. 168
Vernonia
anthelmintica (L.) Willd. 197–198,
205
filisquama M. Gilbert 200
galamensis (Cass.) Less. 197–206, 415
subsp. afromontana (R. E. Fries)
M. Gilbert 201
subsp. galamensis 200
var. australis M. Gilbert 200
var. ethiopica M. Gilbert 199,
203
var. galamensis 200
var. petitiana (A. Rich.) M.
Gilbert 200
subsp. gibbosa M. Gilbert 201
subsp. lushotoensis M. Gilbert 201
subsp. mutomoensis M. Gilbert
201
lasiopus O. Hloffm. 197
pauciflora (Willd.) Less 198
Vetiveria zizanioides (L.) Nash 90
Vicia spp. 104
Vigna
aconitifolia (Jacq.) Maréchal 136
angularis (Willd.) Ohwi & Ohashi 136
capensis (E. Mey.) Walp. 141
pilosa (Willd.) Bak. 141
radiata (L.) Wilczek var. sublobata
(Roxb.) Verdc. 141
sinensis (L.) Savi ex Hassk. 404
trilobata (L.) Verdc. 136
umbellata (Thunb.) Ohwi & Ohashi
136, 141
unguiculata (L.) Walp. 157, 404–405
subsp. cylindrica (L.) van Eseltine
404
subsp. sesquipedalis (L.) Verdc.
404
subsp. unguiculata (L.) Wilczek
404, 405

vexillata (L.) Benth. 141
spp. 104
Virola spp. 157
Vitex negundo L. 91
Withania somnifera (L.) Dunal 91
Wolffia arrhiza (L.) Wimm. 136
Wrightia tinctoria R. Br. 138
Xanthocercis spp. 366
Xanthosoma
  brasiliense (Desf.) Engl. 161
  sagittifolium (L.) Schott. 124
  spp. 123–125, 127, 132
Yucca spp. 91
Zingiber officinale Roscoe 90, 123
ZINGIBERACEAE 123
Ziziphus mauritiana Lam. 136
Ziziphus nummularia (Burm. f.) Wight &
  Arn. 137, 138
Zostera marina L. 99
Anethum graveolens 137

LOWER PLANTS

Ferns

Arachnoides standishii (Moore) Ching
  366
ASPIDIACEAE 367

Fungi

Botrytis cinerea Pers. ex Fr. 191
Fusarium spp. 61
Helminthosporium miyiakei Nisikado
  236
Metarhizium spp. 366
Phytophora drechsleri Tucker 154
Piricularia spp. 248
Saccharomyces cerevisiae Meyen ex
  Hausen 295
Uromyces eragrostidis Tracey 236

Algae

Chlorella spp. 333
Spirulina
  maxima 334
  platensis (Gom.) Geitler 334, 335
  subsalsa Oerst. var. crassior 334–339
  spp. 99, 333–334, 338
Synechococcus elongatus Naegeli var.
  vestitus Copeland 334–338

Bacteria

Lactobacillus spp. 133
Streptomyces spp. 366

ANIMALS

Fish

Tilapia spp. 337–338

Insects

Acanthoscelides aureolus Horn 369
Acyrthosiphon pisum (Harris) 370
Antherigonia hyalipennis Emden 236
Apis mellifera L. 268
Callosobruchus maculatus F. 367, 369,
  370
COLEOPTERA 370
Ctenocolum tuberculatum Motschoulsky
  369
DIPTERA 369
Epilachna varivestis Muls. 354, 356
Heliothis
  armigera (Hubn.) 358–360, 362, 371
  virescens (F.) 370, 371
  zea (Boddie) 362, 369
  spp. 359–362
HOMOPTERA 368–369
LEPIDOPTERA 369–370
Locusta migratoria L. 356, 369, 371
Manduca sexta 369
Melanagromyza obtusa (Malloch) 358
Myzus persicae Sulz. 370
Pseudococcus longispinus (Targioni) 368
Rhodnius prolixus 357
Schistocerca gregaria (Forsskal) 236,
  354, 369, 371
Schizagraphis graminum Rondani 370
Spodoptera
  exempta (Walker) 236, 369–371
  littoralis (Boisduval) 369–372
  spp. 369
Tribolium castaneum (Herbst.) 299
Tribolium confusum J. du Val 369, 370